The Institute of British Geographers
Special Publications Series

29   Geography and National Identity

𝕁𝔹

The Institute of British Geographers
Special Publications Series

EDITORS: Felix Driver, Royal Holloway, London University, and Neil Roberts, Loughborough University

★ available in paperback edition

For a complete list see pp. 390–391

Geography and National Identity

Edited by David Hooson

BLACKWELL
Oxford UK & Cambridge USA

Copyright © The Institute of British Geographers 1994

First published 1994

Blackwell Publishers
108 Cowley Road
Oxford OX4 1JF
UK

238 Main Street
Cambridge, Massachusetts 02142
USA

British Library Cataloguing in Publication Data

A CIP catalogue record for this book is available from the British Library.

Library of Congress Cataloging-in-Publication Data

Geography and national identity / edited by David Hooson.
    p.   cm. – (The Institute of British Geographers special publications series: 29)
   Includes index.
   ISBN 0–631–18935–1 (cloth: acid-free paper) – ISBN 0–631–18936–X
(pbk.: acid-free paper)
   1. Political geography. 2. Nationalism. I. Hooson, David J. M. II. Series: Special publications series (Institute of British Geographers) : 29.
JC319.G42   1994
320.5 ' 4–dc20   93–17857
                                                                    CIP

Typeset in 10 on 12pt Plantin
by Apex Products, Singapore
Printed in Great Britain by Hartnolls Ltd, Bodmin, Cornwall

This book is printed on acid-free paper

Contents

List of Figures

Preface

David Hooson

As our world has been turned inside out in the last five years or so, two key elements which had been curiously overlooked or down-played in the ideological simplicities of the long Cold War have burst upon the scene – nationalism and a geographical awareness. Both are primordial in their meaning for people, evoking love and loyalty but also, in some cases, hate and rejection of others.

The assertive upsurge of national, regional and ethnic *identities* has upset the received wisdom of classical-liberal and Marxist thought, which held that, in the context of the shrinking, interdependent and homogenizing world of economics and communications, such bourgeois anachronisms as nationalism would inevitably melt away. Similarly, the effects of ideological reductionism tended to pre-empt the *geographical* dimension of world affairs with its vivid particularities of place so close to the heart of actual communities, thereby ironing out the wrinkles of the real world. Compounding the impact of these two resurgent forces (really the return of geography and the return – anything but 'the end' – of history) is the fact that they have always been inseparable. In several cases, most notably in the regions of the former Soviet Union, it can be shown that nationalist movements have been initially ignited by the perceived despoliation of the natural environment of a beloved homeland.

I grew up in Wales, a nation with an ancient language, culture and love of the land, but one without any autonomous power over its affairs, and which formed a tiny part of a far-flung conglomerate known as the British Empire, ruled from London. I have now lived in America for over thirty years, where I have been largely occupied with trying to understand the other erstwhile 'super-power', the USSR. With their philosophies and goals of 'the melting pot' and 'Soviet man', respectively, national identity seemed diffuse, elusive and ideologically contrived. Today the melting pot has perceptibly hardened in America, after a quarter-century of a transformed immigrant ethnic make-up, and multi-culturalism is on the rise, while the ex-USSR, which was ostensibly constituted on nationality principles seventy years ago, has now reaped the whirlwind of actually suppressed national feelings, exacerbated by devastated natural environments. In the context of the European Community, Wales seems to have rejuvenated its national identity, as have other small nations of Europe. The dark side of nationalism, in Europe and elsewhere, has been only too terribly displayed in recent years but there is the other side, probably more typical, which is not only benign and legitimate but also vital and energizing. The celebration of Catalonian national identity at the 1992 Olympic Games, with the whole world watching, was the most spectacular recent example of this.

Several valuable books on nationalism and national identity have appeared recently but none has explored the geographical connection. With 26 authors from 14 different countries, this is an attempt to begin to fill that gap. Its inception goes back to what now seems the ante-diluvian world of 1988, when a symposium on the present theme was held in Australia under the auspices of the Commission on the History of Geographical Thought, as part of the International Geographical Congress there. Since then world-wide awareness of both national identity and of the importance of the geographical angle of vision has been greatly enhanced by the cataclysmic events of recent years. We have tried to rise to the new challenges and opportunities and I am very grateful for the patience and cooperation of my fellow contributors in the face of frustrating delays. My gratitude also goes out to Lisa Husmann and Charles Hadenfeldt of the Geography Department at Berkeley for valuable assistance in the whole project, and finally to John Davey of Blackwell Publishers, without whose wise and witty counsel, gentle prodding and persistent encouragement, the end of the tunnel might never have been reached.

Introduction

David Hooson

Whatever the shape and meaning of the still hazy New World Order of the early 1990s, there can be no doubt that it is bound to be fundamentally different from that of the frozen and ruinously expensive bipolar superpower hegemony which prevailed throughout most of the 1980s and for the previous generation. In particular, the amazingly rapid and complete break-up of the supposedly impregnable and menacing Soviet Union and its East European 'bloc' has not only transformed the perceived balance of power in the world but reaffirmed the extraordinary resilience of sentiments of national and regional identity across this vast area, as in the rest of the world.

Identification with the shared experiences of like-minded groups, or cultures, within a familiar and stable territorial setting has, of course, been a natural process inherent in the history of human settlement. Communities have come to inhabit particular places and, over the centuries of occupation, have gradually come to identify with their regional environments, perceived as archetypal, endowed with love and celebrated in song and poetry, as well as understood in terms of appropriate land use and economic development. This sense of national identity in its recognizable modern form, took root and crystallized mainly in nineteenth-century Europe, in the Age of Empires and virulently competitive expansionist or rejuvenated 'nation-states'. Within these states and empires, 'minorities', whose cultures were submerged or denied, preserved and nurtured their identities in whatever ways they could, towards the day when they would be able to flower freely again.

The simultaneous collapse of the empires of Russia, Austria and Germany in 1918, together with a new clarion-call for 'self-determination', led to the birth, resurrection or amalgamation of many long-standing cultural groups into new nation-states in Europe, such as Czechoslovakia, Yugoslavia, Poland and Latvia, following the formation of the 'new' German and Italian states half a century earlier. Most were later trampled over by the expansion of Germany and Russia, and many stayed effectively stifled for a further half-century, though those who, by the luck of the draw, emerged to the west of the so-called Iron Curtain, have moved haltingly but ultimately with inexorable logic, towards integration and voluntary surrender of sovereignty.

However, the plans for this integration of the European Community (of twelve states), towards the magic date of the end of 1992, were fully in place by late 1989, when the demolition of the Berlin Wall occurred. Following this dramatic, symbolic event, the collapse of the Eastern European communist regimes and the unification of Germany came quickly and 1990 and 1991 were euphoric indeed in the region, culminating in the incredible withering away of the Soviet Union itself.

Thus it was that when 1992 rolled around, the plans of the EC, which had been comfortably bounded by the Atlantic and Iron Curtain, suddenly had to reckon with the insistent pleas for admission by another dozen states, and perhaps millions of impoverished refugees, pushing at the Golden Doors. The euphoria has given way to 'morning-after' depression and suspicion in Western Europe, which has led to a revival of nationalism and a new wariness towards integration, at the same time as nationalism is rearing its (sometimes) ugly head from Central Europe to Central Asia.

It seemed reasonable to expect that the inevitable globalization of culture and economy after the Second World War and the rapid growth of communications, would lead to a progressive dilution or even disappearance of national and regional identity feelings. It might also be expected that such an outcome would have been accelerated by the spread of 'multinational' capitalism and the accompanying tidal wave of English as a new international *lingua franca*. Moreover, on the 'other shore', where a third of the world's population has been ruled by Marxist ideology for most of the last half-century at least, concepts such as 'proletarian internationalism' and 'Soviet man' were assumed eventually to supersede outdated 'bourgeois' notions of nationalism and regionalism.

Such prognoses have been confounded by the further proliferation of 'new nations', with a need for each to reassert or create its identity, following the collapse of the European empires. Since 1989, with ever-increasing intensity, the long-delayed participation in this process by the peoples of the former Soviet realm has taken place. So far from heralding 'the end of history', what we have been witnessing is actually the *return* not only of history but also of geography, with an unmistakable upsurge of what may be called identity-consciousness in many guises, by turns exhilarating and destructive, but always clamouring irresistibly to be reckoned with, countervailing theories and beliefs to the contrary notwithstanding.

The last half of the twentieth century will go down in history as a new age of rampant and proliferating nationalisms of a more durable nature than the dreadful but now banished tyrannies which have also characterized our century. One has only to register that the United Nations now counts four times as many members – all considered 'sovereign states' – as it did in 1950. Moreover, if it were to live up to its name by admitting other legitimate 'nations', that is, of people who strongly assert their cultural identity and who regard themselves as being imprisoned against their will within an alien state, the number of members would swell still more. As it is, the recent admission in principle of the successor republics of the former Soviet Union, within boundaries routinely gerrymandered by Stalin, as well as those of former Yugoslavia, has raised the question of national identity in relation to sovereign statehood to a new plane. *Realpolitik* factors such as propinquity and relative size, as well as strategic importance to powerful states, will clearly continue to prevail. The relative ease with which Bangladesh broke away from Pakistan to become a recognized state contrasts with the long-stalled and vexed case of Kashmir, for instance, while the obvious desire of people like the Tibetans, East Timorese or Ossetians to be independent entities continues to be suppressed by force. Final success came to Eritrea's long struggle in 1993.

Times are changing rapidly, and it may be that the way is more open than it has been for some recasting to be done – preferably peacefully – of the arbitrary patchwork of units bequeathed by the crumbling empires to the United Nations line-up of states. The continuing sacrosanctity of the already accredited sovereign states has been reaffirmed in the Persian Gulf War. The distinction in this respect between the position accorded to the Kuwaitis compared with that of the much more numerous Kurds or Palestinians is worth pondering. The genie of national identity, with its infinite variety of scales, intensities and predicaments, is out of the bottle everywhere, and through its interaction with issues of sovereignty, whether in a relatively sedate democratic context such as Quebec, or even Scotland, or in bloodier confrontations in many parts of the world, will be hard to put back. The urge to express one's identity, and to have it recognized tangibly by others, is increasingly contagious and has to be recognized as an elemental force even in the shrunken, apparently homogenizing, high-tech world of the end of the twentieth century.

THE RELEVANCE OF GEOGRAPHY AND ITS HISTORY

Geography as a focused curiosity or, more formally, academic discipline, by its nature, history and very etymology has been nothing if not global in conception, and the inexorable advent of a unified world system should presumably breathe new life into its traditional way of thinking. On the other hand, geography became institutionalized as a departmental discipline in universities in the three or four decades before the First World War, initially in Germany and then in the other countries of Europe, Russia and the United States, at a time of competitive nationalisms, the final scramble for colonies, mass emigration, free trade, urbanization and universal education. All this encouraged the assertion and mobilization of national cultures and the 'state idea', and even the most idealistic of the new scholarly leaders, committed in theory to an impartial, objective global view of geography, were inescapably caught up in the web of the particular preoccupations, priorities and perceptions of their home country as they elaborated the thought-structure of the 'new' discipline.

Thus it was that *national identity* reared its partly chauvinistic, partly benign head, guided by the new cultural elite of urbanizing countries, continuing a patriotic love and knowledge of a motherland and familiar landscapes with delusions of nationalistic grandeur, sometimes threatening the existence of neighbouring nations and states. A man of the world – in theory geography's ideal – without a strong national loyalty such as, perhaps, Elisée Reclus, could become an object of suspicion, while geographers in the new German Reich were open to be used as agents, unwitting or not, of a resurgent national-cultural expansionism. The older-established imperial powers, Britain and France, were of course no strangers to this way of thinking, if less strident with it; and a major aim of their new geographies was to reassert and emphasize their own colonial legacies and the commercial networks they embodied. But French geographers, in particular, evoked the diversity, distinctiveness and harmony of their long-established regions (*pays*), with history – natural and human – richly explored and reaching the hearts and minds of the public in, most famously, Vidal's *Tableau de la géographie de la France*. In a somewhat different style, the national identity of Britain, necessarily with imperialistic overtones and within a world view, was expressed geographically by Mackinder and Herbertson. Also in that archetype of a land-based empire, Russia, geography was remarkably prominent, and a national identity developed, based upon profound examination of the natural zones of its vast spaces, combined with a concern for exploring the impact of man on nature and for ameliorating the lot of the peasants who were the overwhelming majority of the population. The Russian Geographical Society, which was at least as popular as any other learned society there, played a major part in fostering a spirit of national and imperial identity.

At the same time the young United States of America was experiencing a surge of new-found nationalism after the Civil War, illustrated by the Homestead Act, the threading of railways across the continent and the boundless naive optimism of the 'endless frontier'. The implanting of environmentalist ways of thought, together with an introspective system of landform evolution (the Davisian cycle), as the major trends of American geography in its institutionalizing phase in the early twentieth century may perhaps be attributed to a subliminal reaction to the shocking revelation of the 'Great American Deserts' following the former unquestioned assumption about man's omnipotent attitude to nature. A more sober nationalism eventually supervened, but the history of American geography still may not have fully recovered from the dead-end theories and the lack of an inspiring durable tradition in the field compared, say, with the case of the French or Russian schools.

The newly independent states of our century have found in geography a necessary tool for clarifying and fostering their national identity. Following the First World War, under the banner of national self-determination countries such as Poland, Estonia and Yugoslavia illustrate this, and the elaboration of a new or rejuvenated national identity exercised the minds of some of the most notable geographers of these countries. Since the Second World War the much more numerous ex-colonial states of Asia and Africa suddenly have had to come to terms with their geography in defining their identity. Since successor states to units arbitrarily set up by foreign powers disregarded the real national or tribal units, this has often been very difficult, as illustrated in the Biafra war and in Sudan, Burma, Indonesia and many other countries. Complete revision has sometimes become necessary, as in the case of Pakistan–Bangladesh after 1971, and conflicts between different nations over the same land, of which the

Israeli–Palestinian example is the most excruciating, necessitate quite different versions of their geography and therefore its history in those regions.

Within sovereign states, authentic nations, such as Quebec, Catalonia or Wales, continue to assert their own geographical as well as cultural identity, bolstered by a solid body of scholarly investigation. Other, more submerged, nations, without such buttressing scholarship, may be equally worthy of study in terms of the identity they claim for themselves, regardless of the over-arching identity proclaimed for them by their governing states. Often they are the majority occupier of a particular territory, which makes it much easier to determine their identity in full conformity with an environment modified by and adapted to them. No less important is the attempt to determine the identity of peoples who have now been overrun by others, but whose whole being derives from their ideas about places – homelands – in the past. For example, in Australia there is a lot of interest in this with respect to the aborigines, whose land it was exclusively for thousands of years before 1788 and whose identity was – and still is in a fundamental way – inseparable from it.

A special place must be made for China, the history of whose geographical scholarship goes back at least as far as that of the Western world. The fact that the separation between them was almost complete for most of this time and that China's institutions and science have been stagnating during the period of Western predominance over the last 500 years, makes it all the more necessary for the Chinese to resurrect and reinterpret the history of their geography, together with ideas about their national identity. The first university geography department in China was founded only 65 years ago and most of the inter-vening time has been consumed in devastating wars and 'cultural' and other revolutions. Only in the last decade have the Chinese geographers, along with their society as a whole, been able to rediscover and restore their national heritage and attempt to get out from under an obsolete Soviet model which was being changed at its point of origin just when the break with Moscow took place around 1960. The spirit of Taoism, with its long-standing philosophy of the wholeness and harmony of man and nature which has pervaded the age-old practices of the peasantry as well as scholarship in general, now seems clearly superior to Western (and Soviet) ideas about the conquest or trans-formation of nature. The authoritarianism, which raised its ugly head publicly again in 1989, as well as the contemptuous and repressive attitude to non-Han nationalities, are long-standing. However, it may be that the liberal tendencies which followed the death of Mao Zedong, picking up the threads of some of the kinder and gentler Chinese traditions, will yet reassert themselves in the next few years. Whatever happens, the landscapes occupied by nearly a quarter of the world's population are saturated with the results of this philosophy, and their identities cry out for interpretation.

Although modern geography crystallized in Western Europe around the turn of the last century, it seems that the last four decades or so, when most of the Congresses of the International Geographical Union have in fact been held outside Europe, have witnessed the spread of a welcome polycentrism in world geography. In particular the rejuvenation of the subject in Japan, China, Russia and many other countries is casting increasing doubt on the continued wisdom of relating exclusively to a Western, or Anglo-American, model in the future. At its best, the creative tension between the pervading globalization of our lives and the resurgence of interest in nationalism and national identity presents a great opportunity for geography to bring back some of the colour, style and philosophy native to particular peoples and landscapes and to move the spotlight more freely across the world. The threat of nuclear annihilation should have made everyone appreciate more their own precious environment and heritage and the fragility of our only world. The study of national identity, in the broad sense, can help stimulate a rejuvenation of historically based, problem-orientated, living regional geography.

The geographical subject matter has always been remarkably complex and, like history, has therefore found the discovery and formulation of universally viable theories peculiarly inaccessible. Although 'this naughty world' seems to be becoming even more resistant to simple classification and interpretation, it is never dull, and the continuing search for national identity – at many different scales but within a living geographical context – full of meaning and often charged with emotion, shows no sign of calming down, let alone disappearing. The full return of geography, from both its scientific and aesthetic angles, to provide among other things a more or less rational underpinning and understanding of insistent but volatile identity questions across the world, seems needed more than ever.

GEOGRAPHY AND IDENTITY: A SYMBIOSIS

From the time of the ancient Greeks, geography has been concerned with the interpretation of lands and peoples at many levels, from the global to the local, with perhaps particular emphasis on the intermediate or regional scales, where factors of significance can come to focus and reveal recognizable woods beyond the horizon of the local trees, towards the piecing together of a meaningful mosaic of worlds on the surface of our planet. These units, which can all claim to possess some distinctive identity, have not usually been able or needed to establish hard-and-fast boundaries. Typically, whether they were primarily natural or human regions, or a blend of both, they exhibited a distinctive core, which was separated from the core of the next region by a transitional zone. Only in the nineteenth century, when the newly galvanized self-important and often inflamed 'nation-states' proclaimed their identity and felt the need to separate themselves from their neighbours, did political frontiers become formalized and fortified, with the full panoply of passports, visas and customs.

Coincidentally with the crystallizing of this tendency, mainly after 1870, departments of geography were launched within the new and modernized universities of Europe. It was not surprising, therefore, that the new geographers felt the desire, or duty, to help define and give flesh to the emerging national identity of their country and its place in the world. The interplay between geography, or the sense of place, and ideas of national identity provide the themes for the varied chapters of this book.

Europe looms large in this collection, simply because it is where modern geography and modern nationalism exhibited their most intense coincidental development. Germany, as the leader in geographical thought and practice before the First World War and as the most prominent new nation after 1870, is obviously of first importance. Sandner shows how geography helped to lay the educational foundations for the new German national identity, in a sense legitimizing it. Concepts such as the 'Fatherland', 'blood and soil', *'volk'* (including people of German culture outside the borders of the German state), *'lebensraum'* (living space) and keen attention to its newly acquired colonies, were emphasized at all educational levels, with dire consequences all too well known in the twentieth century. Berlin was the capital of the Empire, as it is today, bearing the marks of victory and defeat, of destruction and reconstruction, of the division and reunification of the German nation. Rössler picks up the story here with a discussion of old and new ideas about the German identities, through the prism of the Bonn–Berlin capital debate.

By contrast, Claval traces the evolution of the identity or 'personality' of France, as seen by scholars from Michelet to Braudel, an essentially stable entity after 1815 but not conceived primarily as an ethnic unit. A high point in this evolution was the emergence of the coherent French school of geography, with Vidal de la Blache as its leader. Robic here analyses a particularly well-known work of Vidal's from the point of view of national identity, emphasizing the political aspects of this study – more than had been done hitherto.

Lowenthal, in analysing the English identity, within the European context, draws particular attention to the construction of archetypal, 'elite' views of the English rural landscape, actual or imagined, to form the visual foundations for a national identity which is found to be taken for granted, insular and vaguely defined but also, paradoxically, among the most confidently assumed in the world. Bassin, by contrast, shows how difficult it was to build a clear sense of national identity in that most continental of empires, Russia, in the nineteenth century. Its chronic uncertainty about its 'European' identity – a doubt reciprocated by most European nations – is shown to have been exacerbated by its role as a bearer of 'European enlightenment' to the 'East' (Siberia and Central Asia). Geographers and explorers played a major part in this endeavour but now again, with the collapse of the Empire, Russian as distinct from 'Soviet', national identity seems as elusive as ever.

On the European borderlands of Russia there are peoples and regions whose identities have been periodically submerged, often for long periods, but which have been kept alive and have burst into flame when the right conditions arose. The image of Poland, a 'transitionary' country whose boundaries have been shifted drastically while preserving a core, is discussed by Kristof in terms of the building up of concepts of Fatherland and Motherland, while Babicz contrasts the dimensions and characterizations put forward by two Polish geographers of the pre-communist period. Similarly Buttimer, through the eyes of the Estonian geographer Kant, who was 18 when his country gained its independence, shows how the study of *heimatkunde*, or home-country study, can successfully give form to a rejuvenated national identity. His evocation of a wider regional unity, Balto-Skandia, so distinct from the Russia which had enveloped Estonia for two centuries, is now especially poignant and significant.

Eastern and Central Europe is replete with such rediscoveries of national identities today, none more startling than that of Ukraine, which was born of a blend of the ancient glories of Kiev Rus and the Cossack frontier of the seventeenth century and thrust full-blown, as large as any other European country, into the ranks of independent states. Stebelsky has chronicled its evolving identity and shown how geographers and historians have attempted to define and illuminate it. The disintegration of Yugoslavia is by all reckonings the most dramatic and tortured to date, and Velikonja describes the identity of Slovenia, the least bloody, perhaps because the most homogeneous of its seceding successor states.

The ferment in Eastern Europe has encouraged several of the regional identities of the old states of Western Europe to aspire towards greater, if not complete, independence. Catalonia is the most prominent of these, and García-Ramon (with Nogué-Font) shows how, during the years of persecution under Franco, the popularity of hiking clubs, with scholarly studies, played an important role in defining and evoking the geographical basis of Catalan identity.

The case of Quebec as a beleaguered francophone outpost in an anglophone continent is described by Berdoulay. The role of epic stories about the early days of colonization and resistance, together with a rootedness to the environment of the lower St Lawrence, are shown to be essential to the formation of the distinctive identity which may still be parlayed into independence from, or within, Canada. The work of the local geographers, as well as those visiting from France, in formulating and conserving the identity images and features is shown to be significant. At the other end of the Western hemisphere, the invention of a distinctive identity in Argentina, based on vague beginnings in the nineteenth century, the settling of 'empty spaces', a more purely European population than is found in the rest of South America, and the belated role of the universities, is recounted by Reboratti *et al.* There are many parallels with the experience of Australia and New Zealand, as well as Siberia and western North America.

The Australian case is well represented in this book, owing to the fact that it was conceived in an international symposium held at the country town of Bundanoon, New South Wales, in 1988. Spate has contributed a reminiscing note about his impressions of Australia and its emptiness and strange identity, on arriving from Britain; while Powell and McCaskill have chronicled the way in which a succession of observant settlers or amateur geographers and explorers came to terms with new environments and formed new identities, ending up belatedly with the work of geographical educators.

European settlement of Australia, like its counterparts in other continents, rode roughshod over the lives, rights and environments of the aborigines, with their complete and inseparable identification with their land and their spiritual death when uprooted from it. In a similar vein Yoon writes about the 'geomentality' of the Maori people of New Zealand and their close identification with the elements of their natural environment. Recently the Maoris have become increasingly detached from their spiritual connection with the land and have taken on other identities. Their kinsmen, the Polynesians of the scattered islands of the South Pacific, have in fact acquired multiple identities – even more so, paradoxically, when they have gained their independence from the colonial powers, as Crocombe has shown. The island nations, while each has a recognizable local identity are also identified with their particular metropolitan nationality, Australian, American, British and so on, as well as having a 'Pacific' identity.

In the days of the multinational empires, there was a drive to create an overarching identity, with a world *lingua franca*. The survival of the (British) Commonwealth and also the French Community, each with a still valued cultural bond, bears witness to this. Whether the (Russian) Commonwealth of Independent States will persist in its present form is doubtful, but again the imposed *lingua franca*, alongside the established economic links will continue to be a functional and to some extent a cultural bond. However, the circumstances of the recent break-up, and the pent-up nationalist sentiments which have been smouldering under repression, have made national and regional identities in the former Soviet Union very inflamed and volatile. The USSR, in setting itself up as an explicitly federal state on ethnic lines in 1922, was also sowing the seeds of its own dissolution two generations on. Meanwhile however, the republics, especially in the non-Slavic southern fringes, were often deliberately gerrymandered by Stalin in a classical 'divide-and-rule' way, with the result that many potential Yugoslav-type problems, of which Nagorno-Karabakh is only one, are waiting in the wings, as in post-colonial Africa. New regional associations, based not only on ethnic considerations but also common ecological problems and economic interests, can be expected to crystallize out, sometimes in association with neighbouring foreign countries, in the increasingly anarchical conditions of the early 1990s.

While there are many parallels between the ex-Soviet empire and the still intact multinational empire of China, there are important differences. As Husmann points out, the Chinese have long maintained the 'inalienable' unity of their great 'family' of peoples, encompassing not only the Han (over nine-tenths of the population) but also the Mongols, Turkic-speaking Moslems and Tibetans, within 'natural boundaries' of almost celestial – and now ideological – legitimacy. However, the rapid changes in formerly Soviet Central Asia and Mongolia among linguistic, cultural and religious kinsmen of the western Chinese 'minorities', coupled with the possibility, at least, of further internal political convulsions in China in the context of the world-wide evaporation of communist power, could open the door to the final realization of long-nurtured dreams of independence among the non-Han (id)entities of imperial China.

Although China is often described, especially in the aftermath of the collapse of the Soviet Union, as the last of the multinational empires, several others can be recognized in the Third World (an obsolete and always ambiguous category which is nevertheless still curiously meaningful). A prime example is Indonesia, the sprawling successor state to the old Dutch East Indies, which has been consolidating its hold on the extraordinarily diverse peoples within its empire, as well as expanding into neighbouring areas. As Ragaz demonstrates,

Indonesia through its state ideology, *Pancasila*, is trying to achieve unity out of the extraordinarily heterogeneous collection of peoples, just as the Dutch did. Disintegration is an ever-present threat, since fissiparous tendencies are endemic and the territory is naturally fragmented, with fragile and tortuous communications. As in the case of China, the achievement of long-term unity and stability – held together by force to a significant extent – depends, as she indicates, on the success of economic development which in turn depends on the continuing influx of global capital. Modern Indonesian nation-building, based on the deliberate creation and nurturing of an overarching national identity, including Islam through *Pancasila* and the active promotion of the *lingua franca*, Bahasa Indonesia, or 'market Malay', may yet prove viable and take on a life of its own, obviating the need for a continuation of forceful means. However, it is likely to be at risk from centrifugal forces, in a context of an infectious atmosphere of latent nationalist fervour world-wide.

A polar opposite to this situation in East Asia is that of Japan, which is a famously homogeneous, naturally unitary and close-knit culture – one of the very few virtually true 'nation-states' in the world today. As Takeuchi shows, its emergence from isolation after 1860 involved an intensive examination and building up of Japanese identity in the subsequent age of imperialism and in the previously unaccustomed world view. Britain was the obvious parallel to Japan in the late nineteenth century, but American and German influence was also strong, as he indicates, and influential geographers contributed to the recognition and definition of Japan's internal and external identities and aspirations. Following the collapse of its imperial ambitions, there is no near-equivalent of the (British) Commonwealth among its short-lived former colonies, but the economic links in the Asian-Pacific sphere are, of course, pervading and irresistible.

The range of the case studies in this book, broad as it is, is necessarily constrained by the availability and willingness of scholars as much as by the distribution of objectively existing problem areas. Take, for instance, two absurdly contrasting cases for which it ultimately proved impossible to obtain scholarly studies: the Palestine–Israel identity conflict and the national identity of the USA. The endowment of religious symbolism upon a piece of land, with unprecedented conflict precipitated by the establishment of Israel after the Second World War – brought about immediately by Hitler's brutality – alongside the long-time Moslem religious significance of the area, will make that tiny piece of land a tortured example of multiple overlapping national identities for a long time.

The US identity, bolstered by a fortunate history of immigration into a melting pot which has only recently begun to harden and a dominant world-power position in the twentieth century, seems to have been comparatively secure since the Civil War. Identity questions do not dominate life there as they do in some countries, but it would be naive to think the USA will continue to be immune from those problems evident in all corners of the world. In this context, the multicultural movements, in creative tension with a strongly ideological supernationalism, which reveals itself especially whenever there is a war or a perceived threat, should be increasingly interesting to follow, given the end of the Cold War and the fundamental change in the ethnic make-up of US immigration since Johnson's removal of the restrictive quotas which prevailed from 1920 to 1965. It still contains many true 'nations' embedded in its fabric, most notably, of course, the deeply felt identities of the various native American 'nations' which increasingly reassert their integrity and, above all, their deep attachment to their land.

In conclusion, it may be repeated that we have in this book a sampling of the interplay of two of the most powerful movements of our time: our continuing sense of group identity, distinct from others of our species, and our consciousness of connections with our natural environments and the global context, both essentially geographical concepts which are at the same time of immemorial antiquity and of present and future urgency. The fact that nations as important as, say, India and those of the Middle East and Africa are unrepresented here is deeply regretted. There is plenty left to think about in the *increasingly* colourful, individual and yet fragile and unpredictable system of states and nations which make up our world. Geography and identity are inextricably tied up with each other and this is likely to be more, rather than less, true in the foreseeable future.

PART I

Long-established Imperial Identities

1

European and English Landscapes as National Symbols

David Lowenthal

THE NATIONAL EMPHASIS

National identity is a leitmotiv of modern politics, geography and history the bulwarks of national identity. Ancient credentials and links with sacred terrain buttress the sovereignty of peoples the world over. Nations reshape geographical images and historical memory for present self-respect and future hope. These cherished scenes and memories need not be precise and could not be complete. But they must be detailed and vivid enough to serve as prideful heritage.

Such icons of identity are not confined to nation-states. They are just as crucial to ethnic and other groups whose autonomy is partial or residual – Scots and Welsh, Basques and Bretons. And political subordination often makes such regional identity claims especially fervent.

In most sovereign states, however, national patriotism reduces regional and local identities to virtual impotence. The frustrations thus engendered have latterly fragmented much of eastern Europe. But despite what is termed 'an orgy of regional self-discovery' in matters linguistic,[1] the West (Spain excepted) remains wedded to political geographies that are overwhelmingly national.

Centrism is especially evident in France and Britain. After two centuries of French federalism, *départements* remain puppet regimes ruled from Paris, the pre-Revolutionary provinces bywords of nostalgic reaction; everything of consequence happens at the level of '*la France une et indivisible*'.

British centrism ignores or belittles outlying regions: 'national' is normatively English. As Prime Minister Rosebery warned his fellow Scots, the Celtic lands were to the English but 'lesser gems' in the British diadem.[2] 'When God wants a hard thing done, He tells it, not to His Britons, but to His Englishmen', rejoiced Prime Minister Baldwin. The lexicographer Fowler considered 'English' the only geographical term 'in tune with patriotic emotion'.[3]

A prime marker of centrism is school history, everywhere a focus of civic pride. Shaped and censored by stewards of state identity, this past is not just pro-national but almost exclusively national. Matters above and below that level – continental and global, regional and local – get short shrift; those of other nations are neglected. In French schools history is still the national saga immortalized by Michelet and sanctified by Lavisse; it is not about Europe, much less Picardy or Normandy, certainly not Alsace or the Midi.[4] Even the *Annales*, which initially arose out of a felt need for comparative analysis, now deals almost entirely with French themes.

In British schools history is British or English, not European, Welsh, East Anglian, Liverpudlian, or even Imperial or Commonwealth. The past Britons acclaim and preserve is the *national* past. For much of this century, says a rare exception, 'history has been taught and written along national lines, and hence tied, often unconsciously, to national ideologies and nation-building'.[5] School and university curricula scorn global and continental pasts as subversive pap. Children fed 'on a diet of watered-down, vague world history' would learn only to decry our own past, explains a Tory education spokesman. 'Society will be destroyed unless our children are taught the unique qualities of British history and institutions.'[6] Britain is the only member of the European Community not planning to publish the composite anti-nationalistic new *Histoire de l'Europe*, on the somewhat blinkered ground that 'this country's schools are traditionally used in the dissemination of knowledge and not for propaganda'.[7]

British school history is anyway preponderantly English. 'However glibly we talk about "British history" and ourselves as "British historians"', an eminent historian concluded in 1975, 'no worthwhile "History of Britain" has in fact been written' since William Camden. As late as 1985 one Cambridge college 'steered students away from Scottish, Irish or Welsh history should they show unhealthy signs of interest in such peripheral topics; ... British history was English history'.[8]

The national level reigns among all realms of British heritage. In schools the regional past scarcely figures. The salient architectural past is shown through National Trust and English Heritage not local lenses. Archaeological relics gravitate to the national centre; little accrues to local and provincial museums.[9]

Centrism prevails even in federal states that have murky national pasts. History in Germany is still mainly German history, though the traumatic Nazi hiatus shifts some weight to Europe and to Germany's constituent *Länder*. American history is largely all-American, though Old World antecedents have required attention to European pasts, now broadening to African and Asian. But today's multicultural vogue is ethnic, not regional: American school history deals not with New England or the South (save as fonts of origin), or even (save in perfunctory homage) with the states which use and pay for the texts; its chief concern is the USA entire.

Citizens of every nation-state receive heavily centralized versions of their heritage. From the start they learn that history and geography are mainly about the origins and growth, triumphs and vicissitudes of the nation-state against internal dissidents and external enemies. Such national identity claims embrace every aspect of life, from language and literature to politics and sports. New nations are held to have won their freedom inspired by love borne for 'my country, my forefathers, the geography, the music and language from which they stem'.[10] These reflect three main sources: the character of the land, the nature of its people, and the history of their joint careers.

From a European perspective I focus on the first of these – how geography constructs and reveals national identities.

EUROPEAN NATIONAL LANDSCAPES

Explicit or implicit in every people's favoured heritage are geographical traits felt integral to national identity. European talismans of space and place are age-old; nationalism intensifies landscape feeling.[11] 'Hills and rivers and woods cease to be merely familiar; they become ideological' as sites of shrines, battles and birthplaces.[12] So in England 'almost every rural locality contained or abutted on a field, hill, river or ruin which it associated with a saint or local hero or ... memorable event'.[13]

In long-settled Irish landscapes 'to run off the family names was to call to vision certain hills, rivers, plains; to recollect the place-names was to remember ancient tribes and deeds'.[14] And to crush the losers' national spirit invaders violate the vanquished locale, as Ireland again attests: the British Ordnance Survey's mapping and renaming of Erse features was a campaign of calculated rapine memorably limned in Brian Friel's play *Translations*.[15] Meanwhile the emigrant or exile mourns the auld sod and yearns to leave his bones in it.

Landscapes have much in common, but each nation treasures certain geographical features and elements of its own. Such criteria were once felt to be tainted by a geographical determinism that left inhabitants helpless to shape their own destinies; but these supposed fixities have once again become popular cultural icons. Countries commonly depict themselves in landscape terms; they hallow traits they fancy uniquely theirs. Every national anthem praises special scenic splendours or nature's unique bounties.

Diverse geographical virtues have long been felt to mirror various national characters. Ever since 1729, when Albrecht von Haller celebrated the Arcadian purity of Alpine altitude and air, echoed in Rousseau's primitivism and in the Romantic Sublime, the Swiss have ascribed their sturdy independence to frugal, communitarian Alpine life.[16] Five centuries of English have attributed their stable continuities to insular isolation. Farmsteads set in fields, birch groves, and fir forests – the traditional *glits* landscape – continue to symbolize the beauties and virtues of Latvia, even though Latvians are now overwhelmingly urban or seafaring.[17]

Nationalist impulses reveal the novel charms of otherwise unsung landscape features. The striking success of the Irish peat-digging industry's Bogland Express Tour stems in part from the mystique that much 'of the most cherished material culture of Ireland was found in a bog'; the poet Seamus Heaney celebrates 'an idea of bog as the memory of the landscape, or as a landscape that remembered everything that happened in and to it'.[18] Nor is the passion for bogs Irish alone; the multicoloured carpet of scrub and fern at Shropshire's

Whixall and Fenns Moss is touted as one of Britain's 'most unexpected and hauntingly beautiful landscapes'.[19]

Geographical traits serve non-sovereign Europeans too. Mountain fastness safeguards Basque purity from degenerate Castile, buttressing Basque egalitarian faith against the cosmopolitan nomads – bastardized Celts, decadent Latins, corrupted Moors – of the rest of Spain.[20] Cultural autonomy sustains island communities from Shetland to Sardinia, Corsica to Crete.

Among ex-Europeans overseas geography substitutes for felt lack of history or compensates too sad a saga. Tasmanians say, 'We have a history of which none of us can be proud – but the most beautiful country in the world.'[21] Americans found their unique identity in wilderness; a natural saga older and purer than Europe's human history became the national mystique. Americans needed no 'artificial' palaces and cathedrals, declared the frontier historian Turner; 'we have giant cathedrals, whose spires are moss-clad pines, whose frescoes are painted on the sky and mountain wall, and whose music surges through the leafy aisles in the deep toned base of cataracts'. As a visual and mythic force, the Wild West is America's archetypal icon. 'Americans never tire of pointing out the moral lessons' – freedom, exuberance, optimism – they 'derive from the [unity and immensity] of their landscape'.[22] By contrast, European scenes swamped in historic features often strike Americans as heterogeneous and diffuse.

Laudatory foreign views focus on landscape traits felt to epitomize the visited country by contrast with their own. Thus to the French litterateur Taine the hills on the English horizon seemed 'all drowned in that luminous vapour which melts colour into colour and gives the whole countryside an expression of tender happiness';[23] Taine implies the lack of such landscapes in France. New World visitors praised Britain as 'the only country in the world that is all finished, ... the rubbish picked up, ... no odds and ends lying around', 'the whole country look[ing] ... swept and dusted that morning'.[24] America's premier landscape gardener was struck both by the 'clean and careful cultivation and general tidiness of agriculture' and the order even of English trees, 'as if the face of each leaf was more nearly parallel with all the others near it, and as if all were more equally lighted than in our foliage'.[25]

Not all European landscapes are nationally distinctive, to be sure; many political boundaries straddle scenic traits. Normandy and Picardy merge without a break into Belgian Flanders, French Savoy into the Val d'Aosta; the Alps are indifferent to the national flags of Germany, Austria, France, Switzerland, Italy.[26] Some nations feel the want of prideful icons; stuck with windmills and tulips, the sub-sea-level Dutch may still cherish but nowadays can hardly celebrate the threatened compage of water meadows, copses, dairy holdings and meandering rivers immortalized by Rembrandt and van Ruysdael.[27] Other lands accord regional traits national status: Jutland's heaths hardly typify the Danish landscape, yet they symbolize Danish love of nature.[28]

Territorial shifts or historical traumas make some traditional national scenery less salient than before. Of a picture book on prewar Germany, the writer Stefan Heym recently remarked, 'all those German landscapes weren't German any more. They'd been lost by Hitler.' Heym meant Breslau and Danzig, now

Polish. But also Chemnitz and Dresden and other places erased by war. And beyond, a larger landscape still stained by Nazi memories.[29]

These memories do not deny the still popular view that German landscape virtues are racially determined, that harmony with nature is genealogical. 'No truly German tribe', asserts a landscape authority, 'does not think in terms of plants, trees, and green landscapes'; 'a German village must always and only be a green village'; 'the more green the more German'. Long-settled old Germany was an ecological paradise of fertile farmyards, fields, gardens, of carefully planned village aggregations – outcomes not of the natural fundament but of the German race.[30]

France celebrates its sheer multiplicity of admired features. Scholars and patriots alike laud the Jardin de France's natural riches for French superiority. Here was *'douce France'*, *'la belle France'*, the *'genius loci'* that prepared our existence as a nation', in Braudel's words; 'the beauty of the country, the fertility of the soil, the salubrity of the air' held by a fifteenth-century chancellor to 'efface all other countries'.[31]

The French landscape is held incomparable on several counts. First, a unique range of Mediterranean, Atlantic and continental climes lends the agricultural mosaic exceptional variety; celebrants since Michelet seize on diversity as the 'word which best characterizes France'. Second, seven millennia of unbroken occupance confer an unmatched depth of country life and lore. Third, intense family allegiance attaches to land the Revolution conferred on peasant proprietors. Fourth, the landscape attests total and continual fructification, scarcely an inch until lately left untended. Fifth, peasant memory is still fresh and public sentiment rural, France having remained mainly agricultural long after Britain or Germany. Half the French at the turn of the century, a third as late as 1954, still made their living from the soil, and peasants persist as France's principal proprietors.[32]

Farmers are indeed France's new musketeers, viewed as paragons of ecological virtue guarding sacred national values. Artichokes and rapeseed oil are subsidized not simply against GATT machinations but for French culture. Every major political figure has close family ties with the countryside, and lavish subsidies protect this relic of the national patrimony. The popular hit of 1992 was 'Les mariés de Vendée', a peasant wedding folk song featuring seventeenth-century smocks and farm gear with electric guitars.[33]

To sociologists like Le Play and geographers like Vidal de Blache, the heterogeneity of French landscapes and ways of life meant strengths surpassing those of any other nation. That others might claim comparable diversity made no difference; no other country felt so happy about it. Vaunted diversity underscores manifold inherited excellences: the infinite rich variety of French wines, cheeses, cuisine, customs, dialects.[34]

Yet this diversity in no way detracts from French unity. Alone in western Europe the ancient territorial state coincides with the national community (Germany and Italy were till lately fragmented, Britain and Spain remain composite in culture and language). Predestined to engross a God-given realm, France was foreshadowed in a millennium of regal conquests; even now, mediaeval maps in French history texts show embryo regions 'not yet reunited' to the nation-to-be. The Revolution further sacralized national unity: eradicating provincial bonds, republicans sought to make France homogeneous. And France's natural frontiers (Alps, Pyrenees, Rhine) became a fixture of national ideology. A previsioned France shaped diverse peoples into common views and allegiances – a patriotism praised for forgetting discordant Occitan, Breton and Huguenot ways.[35]

Full of nature's grandeurs, the French physical heritage is also a supreme artifice. As royal circuits used to be inscribed within Euclidean geometries, so the primary-school ideal of France since the 1880s is of a land 'symmetrical, proportioned, and regular'.[36] Timeless France is a six-sided ideal: the national estate as reified hexagon.[37] Contemplating this Gallic geographic triumph, an American is reminded of his own land's long love-affair with its Manifest Destiny, the imperial reach 'from sea to shining sea'.

ENGLISH LANDSCAPE AND IDENTITY

The manifold particularities of European geography echo in the diversity of attachments to landscape and place, attachments formed by ethnicity and history as well as by locale. Most of them share, in varying degrees, a strong attachment to some primordial natural fundament. In this Britain – more precisely, England – diverges from most of the rest. More than anywhere else in the Old World, the geography celebrated in England is what has been made and remade by many centuries of native folk.

One literary historian claims for Britain the world's 'most strongly defined sense of national identity'.[38] But that identity no longer needs to be itemized; 'this country has been around for so long that the British simply *feel* rather than think British. Reluctant to define Britishness', as *The Times* puts it, the British are 'simply less interested in their national identity' than most Europeans, let alone aggressively hyphenated Americans.[39]

So robust is British (or English) identity, it allegedly needs no symbolic trappings. 'The most anomalous thing about England [compared] with all other European nations', reflects a professor of English, 'is that it doesn't have the formal marks of national identity acquired even by Iceland or Finland, Luxembourg or Albania. It has no national anthem. It has no national dress.' Few know its national flag. England's identity 'does not depend on the kind of props thought normal by everyone else. Not to need a national anthem/flag/dress/history/myth of origins' is for him 'a sign of maturity and inner self confidence'.[40]

To foreigners amazed at the seeming absence of archetypal symbols, British spokesmen explain that felt national continuities make them redundant. But where are they talking? In a Tudor country house deep in the Sussex Downs. Here, in colloquia punctuated by croquet and tea and country walks, continental Europeans get steeped in the *scenic* essence of British identity.[41]

This geographical icon has a profoundly English cast. Nowhere else is landscape so freighted as legacy. Nowhere else does the very term suggest not just scenery and *genres de vie*, but quintessential national virtues. The historian

Butterfield reified that 'inescapable heritage of Englishmen', the Whig interpretation of history, as 'part of the landscape of English life, like our country
lanes or our November mists or historic inns'. Landscape is the leitmotiv of
the 'solid breakfasts and gloomy Sundays' of Orwell's England set in 'smoky
towns and winding roads, green fields and red pillar-boxes'.[42] The glossy
magazine *Heritage* is subtitled 'British Life and Countryside'. But the countryside is not British; it is English. Incoming Romans and Saxons found England
'such a precious spot of ground', wrote a seventeenth-century chronicler, 'they
fenced it in like a *Garden-Plot* with a mighty *Wall* and a monsterous *Dike*' to
keep out the Scots and the Welsh.[43]

The countryside has been England's supreme communal creation since pre-
history. Even early megalithic civilization is lauded for adding 'a new dignity
to Nature' and leaving 'English country ... more beautiful than they found
her'.[44] A fabric woven and kept by 50 generations of landlords and labourers
became the acme of nature. Culture inheres not just in these lineaments, but
in a millennium of celebrants' verse and song. Peopled and storied rural England
is endlessly hailed as a wonder of the world.

Landscape is not just the setting; it is the centre of the scene limned 60 years
ago by G. K. Chesterton, home from a transatlantic tour. For all America's
'vast magnificence', Chesterton was appalled that two centuries of white
Christianity there had engendered 'nothing like a village that was fit to look at'.
By contrast,

> the English village was ... like the relic of a great saint. That ... the roofs
> and walls seemed to mingle naturally with the fields and the trees; ...
> the naturalness of the inn, of the cross-roads, of the market cross [made it]
> a very precious possession, ... in a real sense the Crown Jewels. These
> were the national, the normal, the English, the unreplacable things.[45]

Archetypal England had inspired poets of the Great War; 'what men were
fighting and dying for was some very green meadow with a stream running
through it and willows on its banks'; to Stephen Spender 'freedom' was 'a
feeling for the English landscape'.[46]

The now hallowed visual cliché – the patchwork of meadow and pasture,
hedgerows and copses, immaculate villages nestling among small tilled fields
– is in fact quite recent; only after the pre-Raphaelites did the recognizably
'English' landscape become a mediaeval vision, all fertile, secure, small-
scale.[47] Eighteenth- and nineteenth-century landscapes looked and were lauded
differently; but general praise of some man-made geographical palimpsest goes
back half a millennium.[48]

Four special traits link English landscape with national ethos and imprint
its heritage role: insularity, artifice, stability and order.

Insularity

Insularity sets Britain off from all other European nations save Iceland and
Ireland. Atavistic loyalties are insular: 'on these shores', 'this sceptr'd isle',

'the defence needs of these islands'. Yet insularity also 'utterly maroons Britain from all the world'; echoing Virgil, Gore Vidal foresees this 'green and pleasant land drifting off to join those other islands that history has forgot – the Isle of Wight, the Isle of Man, the Isle of Dogs – Staten Island too'.[49]

But in native eyes insularity is a great merit. Auden's island is a 'little reef / The mole between all Europe and the exile-crowded sea'. Insular Britain is a precious fortress; the sea is laden with rootless transients envying our home, in Kermode's phrase, 'our known and loved bit of territory, the garden in which we walk in peace'.[50]

More bluntly chauvinist is a Conservative Party leader's insularity:

> Our Continental neighbours use 'insular' as a term of abuse, but we in Britain have every reason to be thankful for our insularity. Our boundaries (that troublesome one in Ireland apart) are drawn by the sea – some might say by Providence. Unlike those of most other nations they have not been drawn, rubbed out and redrawn time and again. ... The blessing of insularity has long protected us against rabid dogs and dictators alike.[51]

The sea limits size, marks boundaries, purges continental contaminants. Because the insular English largely 'escaped foreign influence' until the mid-eighteenth century, the historian Buckle judged that more than any other people they had 'worked out their civilization entirely by themselves'.[52]

Islandness also graces 'this other Eden' as 'a paradise of pleasure and the garden of God'.[53] Like its mediaeval prototype, the English garden fences out a menacing wilderness. Hence the enduring English reluctance to cede any part of English sovereignty to the European Community, and the scary front-page headline after the Channel Tunnel breakthrough: 'Britain is no longer an island.'[54]

Artifice

Also like the archetypal sacred garden, the English landscape is not natural but crafted. Other nations extol untouched nature: the Swiss revere edelweiss and erratic boulders, Americans the forest primeval.[55] Not to the 'aloof and unsympathetic beauty of glaciers and coral reefs and tropical forests' is English music said to owe its genius, but rather to 'the intimate and personal beauty of our own fields and lanes'. Englishmen tame and adorn nature; in Emerson's phrase, 'nothing is left as it was made; rivers, hills, valleys, the sea itself, feel the hand of a master'.[56] England's landscape is its consummate artefact – not just the locus of heritage but its mainstay.

The English see themselves as beneficent stewards of an enduring yet evolving legacy. The humanized scene attests prolonged loving guidance. The writer Quiller-Couch linked true patriotism to growing up in some 'green nook ... where the folk are slow, but there is seed-time and harvest'; unlike superficial foreigners, we have deep well-springs in the landscape of Home.[57]

The fabric of English countryside subsumes all creatures; they are domestic even if not domesticated. To an archetypal Tory the gravest Continental menace is the threat to 'our heritage of British country sports by countries [devoid of] respect for wildlife and conservation'.[58] Rural stewardship spells hunting. 'Britain continues to have the best managed wildlife in Europe', claims the editor of the conservative *Daily Telegraph*, 'because [it] is in the hands of private landowners who conduct their affairs sensibly and responsibly.'[59] And conserving hedges, copses and woods as cover for game – a source of landed profit, a sign of rural status – is praised for staving off arable monotony. Animal rightists' failed to end fox- and stag-hunting over National Trust lands in large measure because it was feared this 'would alienate the Trust from many rural communities, cause it to be seen as an antagonist of the countryside and its needs' and forfeit future legacies.[60] And urbanites (translation: animal rightists) are held not to 'realise the importance of the social side of hunting to tight-knit rural communities'.[61]

Stewardship tempers rural exploitation with concern for natural harmonies. But not for nature in the raw. 'If you could get through the bogs and jungles and the thickets [that covered] this country one million years ago, you would say, "What a dreadful place this is" ', an Environment Minister admonished Green primitivists.

> The valleys were mosquito-ridden swamps; the mountains were covered in hideous oak thickets and there were just a few shacks, where miserable people attempted to live. Now this is a country full of wonderful landscape, full of beautiful buildings, superb cities and towns, all built by man, [and] we are constantly enhancing it.[62]

The 'we' who are enhancing it, of course, are the landed elite. Landscape control is rural paternalism. An elite supposedly sensitive to community needs has long set its stamp not just on house and grounds but field and village.[63]

The social bent of this minutely managed landscape emerges in an eighteenth-century commentary. Anna Seward deplored the picturesque 'Jacobinism of taste that indulged nature as well as man in that uncurbed and wild luxuriance, which must soon render our landscape-island rank, weedy, damp, and unwholesome as the incultivate savannas of America'. She shuddered at 'living in tangled forests and amongst men unchecked by those guardian laws that bind the various orders of society in one common interest; the lawns [she trod should] be smoothed by healthful industry'.[64] To rural improvers uncultivated land betokened 'uncivilized nations, where nature pursues her own course'; enclosure in England not only aided agriculture but made the countryside safe and familiar.[65] As late as the 1940s one writer saw noble parklands 'integral to the landscape as to civilisation' – virtues threatened by equality and democracy.[66]

English landscapes are compages of datable acts ascribed to ancestral precursors. They are infused with memorable human processes, desires, decisions, tastes. Few lowland features lack embedded links with those who have held and tenanted and tilled them, 'every acre observed, considered, valued, reckoned, pondered over, owned, bought, sold, hedged'. In Dorset 'virtually every tree, every clump, and every hollow' is said to have its story.[67] The National Trust has exhaustively surveyed Great Langdale's 'every last Hogg Hole'. Specific remembrances suffuse the whole countryside – Henry James's 'connecting touch' in every corner and 'an impression in every bush'.[68]

Myth and art add auras that reach from Arthurian Wessex to Lakeland's Peter Rabbit. A woodwind figure in the rondo of Elgar's Second Symphony is said to 'breathe the scent of Severnside to those who know it'. Wordsworth graces the Lake District, Constable decorates Suffolk, Hardy enhances Dorset; every hillock accretes attachments that humanize 'the whole island [as] a *campo santo*'. Emerson reflected that 'it is a long way from a cromlech to York minster, yet all the intermediate steps may still be traced in this all-preserving island'.[69]

These scenes still cradle some communities. Where else would defenders of an ancient right of way get a 112-year-old to attest that he'd walked a country footpath for a whole century? Rural England is historical more in the mode of old men and ancient buildings than of geological strata. Ancestral trees offer favoured analogies, but most such trees are artefacts, intended as icons of noble continuity.[70]

Stability

Because permanence is felt admirable, the English landscape is pervasively antiquated. Most familiar features were created for purposes now outdated, for needs now redundant, by means no longer to hand, in circumstances that no longer hold. The landscape is England's prime anachronism – a vast museumized ruin. It is used, much of it intensively. But these uses bear less on home and workplace than relics that cradle heritage and proclaim national identity.

Love of landscape attests the demise of previous functions. As agriculture ceases to be viable farmers become scenic stewards for tourism. Stonehenge is barred to bona fide but uncouth worshippers, while English Heritage and the National Trust tidy it up for less *engagé* – and less troublesome – tourists.[71] As gentrification turfs rural folk out of cottages, urbanites retire there to contrive Constablesque visions. Huts that Constable felt too disgusting to enter in 1820 and that H. Rider Haggard lamented as deserted in 1901 are now 'desirable'.[72]

Landscape-as-heritage fosters nostalgic myth: childhood memories 'of sunlit fields where we could play all day without fear, of picturesque villages un-shaken by foreign juggernauts, of peaceful beaches with the English enjoying themselves in their own quiet, time-honoured fashion'.[73] But this myth is said to prevail especially among outsiders 'smitten by blind Anglophilia'; as Chris Hall and Kate Ashbrook see it, 'some Americans lose all sense of reality when confronted by a thatched cottage or a peer of the realm'.[74]

Hall and Ashbrook explicitly decry a gushy tribute to 'Nether Burton'. This 'tidy old [Chiltern] village' charmed American journalist Edwin Yoder as 'still the center of a working life and a plain but vibrant agricultural economy, now, as for centuries past joining country people busy with their crops and animals and machinery, with the great bustling worlds of London and Oxford'. How fitting that Yoder's host, 'Baron Omnium', 'undoubtedly knows the farmers, the butcher, the fellow at the pub and the pleasant man ... who fixes lawn mowers'. In Nether Burton 'an unrushed patience, a care for the landscape and the past, a respect for privacy' betoken 'ongoing strengths of the once and ... future England'.[75]

Nonsense, retort Hall and Ashbrook. These champions of public access also live in Nether Burton (actually Turville, near Henley-on-Thames). They term it 'a rural suburb for the affluent' with outside incomes; few village-born can afford housing; the school has closed for lack of children; 'vibrant' agriculture employs fewer workers each year. It is not England's 'ongoing strengths' that Nether Burton explains, but 'the destruction of those strengths and the countryside by Thatcherist greed'.[76]

Bleak realism? No. Hall and Ashbrook simply displace the ideal landscape back into a mythical past, Raymond Williams's Golden Age fount. They ignore its shaping by previous elites, no less greedy and callous then than now, who would destroy or relocate entire villages to make way for country-house parklands. We 'talk of order in those cleared estates and those landscaped parks, but what was being moved about and rearranged was not only earth and water but men'.[77]

Other English landscape devotees combine Yoder's delusion that the rural past is yet alive with Hall and Ashbrook's ancestral harmonies. They hunger for 'the hierarchical certainty of the old England, that amalgam of faith, diligence, loyalty, independence and authority' Trollope apotheosized.[78] A countryside spokesman concludes that 'England still has quiet villages, peaceful homes and pleasing prospects, bearing the stamp of centuries of builders, farmers, gardeners, the village blacksmith, the rich wool merchant, the parson, the squire and the yeoman'.[79] These quaint relics tread the Trollopian stage with Baron Omnium.

Similarly 'historic' is a former Environment Minister's England. An inveterate tree-planter, Heseltine meekly claims to 'only copy where others pioneered':

My yews were their yews, the beech their foresight, the chestnuts their commitment. There is a tiny church. Eight centuries of England lie buried around it. Of course now there are roads, cars, planes, television and living standards beyond yesterday's imagination. But there is, too, an England as she was: changeless in our fast-changing world.[80]

Defenders of Egdon Heath, a fast-changing tract implausibly immortalized for Thomas Hardy, proclaim it 'from prehistoric time as unaltered as the stars overhead'.[81] As a rural sociologist concludes, 'the countryside reassures us that not everything these days is superficial and transitory; that some things remain stable, permanent and enduring'.[82]

To be rural sanctions any status quo. Mythologizing his backwater roots, Prime Minister Baldwin preened himself as not 'the man in the street even, but a man in a field-path, a much simpler person steeped in tradition and impervious to new ideas'.[83] Where else would a head of state invoke the national landscape to praise atavistic inflexibility?

Order

Orderly control is the landscape's fourth touchstone. It is an English creed that all land needs care. Far from knowing best, nature requires vigilant guidance. Cheddar Gorge is periodically trimmed and shaved lest it 'lose its uniqueness and become just another ordinary, wooded valley'.[84]

The impending withdrawal of millions of arable acres arouses custodial alarm. Elsewhere such land is simply left to itself; but scraggly briar and bramble, motley ground cover, the untended seedlings are intolerable here. Unmanaged wasteland was utterly repugnant even to so staunch a populist as Raymond Williams. Praising the hedging and ditching, the parks and gardens still cared for as a matter of course, he deplored 'the urban blindness to all this work that actually produces and preserves much of the "nature" that visitors come to see'; there would be 'too much [wilderness] for most tastes if this kind of tending stopped'.[85]

'Left to themselves', a journalist imagines, 'the fields would fur over with weeds, waist-high and then head-high', and eventually with scrubby forest.

> Much of this land would revert to waterlogged swamp, as field drainage broke down. It would be good for birds, but also for rats, mosquitoes and accumulations of weed pollen to make the nation sneeze and to cover its gardens. In the dimness of the tangled undergrowth would lurk ... mounds of abandoned cars, refrigerators, and ... agricultural machines. This prospect terrifies the government, the planners, and the environmentalists. That is why ... they have invented this new idea of the farmer as museum custodian, whose duty is to preserve the look of the rural 'heritage'.[86]

A zoologist thinks most people 'would be appalled by any wholesale reversion to the impenetrable wildwood that 5000 years ago swathed our land'. Even the encroachment of scrub is spurned as 'untidy'. Vacated fields become not just weedy and scruffy but degenerate, lawless. A leading landscape architect warns that 'every piece of land removed from the care of nature must be adopted by someone if it is not to become derelict'.[87]

Adopted by *someone*, yes; but not just *any*one. 'That peasant and nature alike are intrinsically barbaric unless subject to stern paternalistic discipline has been the rationale of the landed elite for centuries', notes a countryside advocate.[88] The aristocracy and gentry alone are fit for this nurturing task.

The dominance of Britain's old landed elite is unique in Europe. Only here do repute and identity normatively equate with rural residence; as Howard Newby says, 'the English have always regarded the countryside as the proper place for proper people to live in'. Only here are social life and agriculture so wedded to cults of hunting and shooting. Only here are great estates still the norm; farm holdings average three to five times those in France and Germany; the whole country is held by a tiny fraction of its folk.[89] Heritage guardians commend this concentration. 'From end to end of England, wherever you meet seemly villages and a countryside that speaks of understanding and affection', says a National Trust spokesman, 'the chances are you will be on a large estate.'[90]

A once-imperilled rural oligarchy now thrives anew. 'Many superficially benign peers like to put in a "hard man" as land agent, and this is what they have in Mrs Thatcher', wrote *Harpers & Queen*'s editor. Refuting one historian's chronicle of elite decline, a Tory leader finds 'the great thing about the British aristocracy [is] the way they have hung on to their social and political influence'.[91]

An axiom of such control is that the elite knows best; it can be trusted to cherish splendour and stability. Coupling landscape heritage with elite taste is not new: Wordsworth termed the Lake District 'a sort of national property, in which every man has a right and interest who has an eye to perceive' – but such an eye was formed only at the court of cultivated taste. Having earlier sought to preserve the Lakes *for* visitors (then a few gentry) Wordsworth ended by protecting them *against* visitors (whose wanton incursions he likened to a 'child's cutting up his drum to learn where the sound comes from').[92]

Reiterating 'That Good Taste is not Natural but Acquired', a staunch countryside defender of the 1940s wanted every urbanite who hadn't passed an 'examination in country lore and country manners to wear an "L" [for learner] upon his back when he walked abroad in the country'. But if the previous century had not taught the masses country manners, neither, landowners charge, has the half century since.[93]

A recent Environment Minister advised old elites too broke to keep up the great houses 'their ancestors had often bought, stolen or married into' to sell out to *nouveaux riches*. Lord Saye and Sele reacted with hauteur: 'Do you think that by removing us from [Broughton Castle] and installing a *nouveau riche* family, the heritage would be maintained in the way that we maintain it? Would they want to open their Peter Jones-furnished rooms to the public?' Only old 'families who cherish them and can give them continuity' look after historic estates properly. Lord Saye and Sele feels he is not so much 'the owner of Broughton as the temporary custodian'.[94] The editor of *Country Life* hails 'the majority of dukes as responsible, committed and influential men who recognize their duty to maintain their estates and houses for posterity'.[95]

Noblesse oblige slithers into egalitarian self-delusion. As Lady Saye and Sele hangs her washing on the battlements and makes instant coffee for art students, her consort 'like[s] to think we're the same as other people except for the fact that we live in a castle'. At least Lady Saye and Sele gets hefty fees for letting film-makers use the castle – and smiles when they intersperse her roses with plastic flowers. Some chatelaines fare worse: 'It is galling to be told that your dinner guests cannot park in the courtyard, or to be charged 6p by the gardener every time you pick a sprig of parsley.'[96] But not even such indignities daunt aristocratic former owners kept on as tenants for the National Trust ideal of living continuity.

Snobbery bolsters both heritage and tourism: truly philistine gentry subject house guests to nostalgic discomforts. As in Emerson's day, 'perfunctory hospitality puts no sweetness into their unaccommodating manners'. Today's genuine English country house is a chilly, comfortless place, inhabited by rude upper-class people who complain endlessly about the cracks in the gunroom ceiling. You eat lumps of mashed parsnips and home-grown grouse that is full of lead shot, while six smelly dogs drop dead voles into your lap.[97]

Such ordeals at least lend authenticity to arcadian stereotypes. But often the country-house door – along with the countryside itself – is firmly shut. The National Trust is accused of 'underwriting comfortable existences, with nothing asked in return beyond opening a room or two once in a blue moon'. Secretive about its holdings, it protects tenants and lessees by making historical gems and landscapes off limits or unfindable. Thousands of 'ghost' properties, unmarked on maps, are omitted from Trust guides. As seen by admirers, the Trust deploys elite expertise to public benefit; as seen by Open Space crusaders, the Trust is 'an elitist club of art experts dedicated to preserving dinosaur country houses, [owing to] anti-public sentiments picked up along with the feudal values of the land it has bought and inherited'.[98]

Nostalgia for rural order harks back to times when boundaries were firmly marked. Everything was in its place; people too knew their place. Place and order lent moral certitudes now lacking, in a hierarchy consensually united despite the blatant inequality of squire and serf, master and servant. Populist progress now lets the serfs' descendants wander in wonderment round the former masters' mansions, but these gawkish interlopers no longer keep to or even know where they belong.[99]

The confusion infects most scenic cynosures. Look at Blenheim, writes the poet Peter Levi, that 'perfect ducal paradise: how beautiful it was once, before it was full of people. How strange and sinister a world built it. How lucky we are still to have it.' Many feel 'baroque triumph or a pastoral exaltation', as intended; others recall Winston Churchill; some show envy. More mingle 'puritanic disapproval of the past, furious resentment of the old, complacent upper classes, [with] the odd illusion all these landscapes and lakes and palaces are "our inheritance", belong to us and represent our creative powers'.[100]

The old order's outstanding legacy is the image of controlled harmony. While Americans learned to defy authority and negate order in the free forests and on the open frontier, obsession with order is still patent in preferred English landscapes. Seemliness and propriety are *de rigueur*; mess is ill-mannered and offensive; hedges and walls mark clear-cut bounds. Neatness, 'a passion for tidiness, for trimmed edges, for mucklessness, mysterylessness, boringness, homogeneity' is the National Trust aesthetic. 'You are in a world of good taste, of discretion, of no incongruity. It is a scraped world.'[101]

'The love of order', felt the poet Crabbe, ranked 'with all that's low, degrading, mean, and base'. But among 'pompous men in public seats obey'd' tidiness still prevails. Within 'an ancient order of immemorial subordination' this ideal extends from mansion to cottage. 'The people which has made such homes as these is distinguished above all things', now as in Gissing's day, 'by its love of order.'[102]

Nostalgia for order rues landscape linkages severed for at least a century. To the casual eye rural England looks peopled, but few live in it; fewer still are at work there. Though four-fifths cultivated, the rural scene is largely an empty shell, shunned even by its stewards. Other Britons view England's landscape more and more as outsiders. Access may improve, interpretation expand, history come alive, the Countryside Code intone reciprocal rights and duties; but for most urbanites the rural scene is strictly of the past. It may be the more treasured *because* of this; it is none the less remote.[103]

The English rural heritage is now much imagined. City dwellers' countryside contacts are scenic or sportive; the landscape is as shallow as a Tussaud wax-work, as exotic as the Elgin Marbles. Visual primacy is at once pampered – 'a private British Rail view for one of the best shows on earth – the English countryside' (an Oxfordshire field) – and panned – 'we can't have rural policy being dictated from the car window'.[104] The heritage landscape is less and less England, more and more 'England-land', Europe's offshore theme park.[105]

Heeding reiterated rural perils – midges, ticks, bracken spores, bramble wounds, poison and pollution – landscape lovers may instead opt for the 'virtual reality' of Center Parcs resembling Orwell's prophetic paradise where one is 'never within sight of wild vegetation or natural objects'.[106] Or else, 'rather than leaving the car-park to hazard the real thing, users could lie back and think of England as they want it to be, looking like a Constable, smelling of roses, the undergrowth debugged, desnaked and dethorned'.[107]

Actual or imagined, rural England will in any case continue to supply a geography suitable for a national identity uniquely devoted to insularity, artifice, stability and order. It is questionable whether any other Europeans find greater contentment in or commitment to the dwindling stock of natural features with which they more generally identify.

DOMESTIC LANDSCAPES AND NATIONAL FOCI

While English attachments to nature tamed are most intense, they are not unique. Verdant, cultivated, inhabited scenes are beloved all over Europe. The wild or the ruinous may charm aesthetes and tourists, but domestic rural locales rank first in popular affection.

To be sure, early nineteenth-century national spokesmen extolled Nature's pristine splendours, and by the century's end wild scenery featured in most national logos. Rugged mountains, dense forests, deep lakes, storm-scarred coasts and cliffs suited nascent and militant chauvinisms. But domestic emblems endured along with dramatic vistas. And European national icons today stress intimate, humanized chocolate-box scenes – a figured landscape, whose traces of cultural heritage are now often embellished. Photographers' props in Sweden enrich the classic Dalecarlian lakeland scene: a collapsible wooden fence, a model in folk costume, a replica birch-bark horn, a couple of goats.[108]

Against these national affections, the current cult of regional and local geographies seems more touristic than popular. National landscape tastes have largely rural roots and a populist, democratizing bent. In landscape as in language, regionalism in continental Europe still conjures up memories of privilege, injustice and landless bondage. By contrast, national landscape icons, wild and tamed alike, bespeak liberation and freedom: liberation from subjugations both foreign and domestic, freedom for farmers empowered by sovereign statehood.

Nationhood for most Europeans came when they were still rural, linking peasant emancipation with national sovereignty. Except in England, where agricultural labourers remained landless, rural folk tended to welcome the centralized state as a release from ancient inherited provincial and local bondages, a stimulus to peasant proprietorship. That is why Europe's landscapes, however lightly inhabited, remain compelling icons of *national* identity. In Britain, English landscapes inspire affection for a realm that long ago ceased to be a nation – England. Geography is still valued largely through national lenses.

NOTES

1 Neal Ascherson, 'No regional governments, please, we're English', *Independent on Sunday*, 24 November 1991.

2 Lord Rosebery, 'The patriotism of the Scot' (1882), in his *Miscellanies: literary and historical*, 2 vols, London: Hodder & Stoughton, 1921, vol. 2, pp. 111–12.

3 Stanley Baldwin, 'England' (1924), in his *On England and Other Addresses*, London: Philip Allan, 1927, p. 1; H.W. Fowler, *A Dictionary of Modern English Usage* (1926), 2nd edn, rev. Ernest Gowers, Oxford: Clarendon Press, 1965: 'England, English(man)'. Irish revolt and the Great War made the English uncomfortably aware of the UK's Celtic adjuncts. 'Englishmen!' exhorted a *Times* ad in 1914, 'Please use "Britain", "British", and "Briton", when the United Kingdom or the Empire is in question – at least during the war' (quoted in H.R. Hanham, *Scottish Nationalism*, London: Faber & Faber, 1969, p. 130).

4 Suzanne Citron, *Le Mythe national: l'histoire de France en question*, 2nd edn, Paris: Edn Ouvrières, 1991; Pierre Nora, 'Lavisse, instituteur national', in his (ed.) *Les Lieux de mémoire*, I: *La République*, Paris: Gallimard, 1984, pp. 247–89.

5 Hugh Kearney, *The British Isles: a history of four nations*, Cambridge: Cambridge University Press, 1989, p. 3. At school in England after the Second World War, a French-Norwegian boy learned 'that England had won every damned battle ever fought against the French', still portrayed (1985 *Oxford Junior History*) as beret-wearing, backward peasants (Frédéric Delouche, quoted in Julian Nundy, 'History leaves Britain behind', *Independent on Sunday*, 19 January 1992, 15).

6 Rhodes Boyson, *Centre Forward: a radical Conservative programme*, London, 1978, pp. 188–9.

7 Teddy Taylor MP, quoted in Matthew d'Ancona, 'Publishers shun EC view of history', *The Times*, 20 January 1992, 7. *The Times* likewise ridicules this 'dubious concept of a pan-Europeanism (chief exponents, Charlemagne, Napoleon, Hitler?)' ('History lessons', 20 January 1992, 13).

8 J. G. A. Pocock, 'England', in Orest Ranum (ed.), *National Consciousness, History, and Political Culture in Early-Modern Europe*, Baltimore, Md: Johns Hopkins University Press, 1975, p. 100; Kearney, *British Isles*, p. 213. See also J. G. A. Pocock, 'British history: a plea for a new subject', *Journal of Modern History*, 47 (1975), 601–21.

9 Kearney, *British Isles*, p. 3; Robin Shapp, 'Lindow person', *New Scientist*, 11 June 1987, 68; on the neglect of regional history, David Hackett Fischer, *Albion's Seed: four British folkways in America*, New York: Oxford University Press, 1989, pp. 788–9.

10 Harry Morgan (letter), 'The fight for racial identity', *Observer*, 20 October 1991, 50.

11 On national and other landscapes, see my 'Finding valued landscapes', *Progress in Human Geography*, 2 (1978), 375–418.

12 Joshua Fishman, 'Nationality-nationalism and nation-nationalism', in Fishman, Charles A. Ferguson and J. D. Gupta (eds), *Language Problems of Developing Nations*, New York: John Wiley, 1968, p. 41.

13 D.R. Woolf, 'The "common voice": history, folklore and oral tradition in early modern England', *Past & Present*, no. 120 (1988), 26–52, quotation on 31.

14 Daniel Corkery, *The Hidden Ireland: a study of Gaelic Munster in the eighteenth century* (1924), Dublin: Gill & Macmillan, 1970, pp. 64–6.

15 Brian Friel, *Translations*, London: Faber & Faber, 1981.

16 Albrecht von Haller, *Die Alpen* (1729), Lausanne: Edn André Gonin, 1943; Souren Melikian, 'Switzerland revealed: how foreign artists changed the national image', *International Herald Tribune*, 24–5 August 1991, 8.

17 Edmunds Bunkśe, 'Landscape symbolism in the Latvian drive for independence', *Geografisker Notiser*, no. 4 (1990), 171–2.

18 Seamus Heaney, 'Feeling into words' (1974), in his *Preoccupations: selected prose 1968–1978*, London: Faber & Faber, 1980, p. 54. See Mary Tubridy, 'The final episode in the story of (almost) all bogs', in her (ed.) *The Heritage of Clanmacnoise*, Dublin: Environmental Sciences Unit, Trinity College, 1987, pp. 83–93.

19 Edward Pilkington, 'Our backyard: threat to peat bogs fuels unrest', *Observer Magazine*, 25 March 1990, 7. On concern for England's vanishing lowland raised bogs, see Polly Ghazi, 'Heseltine prepares to wade into peat bog exploiters', *Observer*, 20 October 1991.

20 Marianne Heiberg, *The Making of the Basque Nation*, Cambridge: Cambridge University Press, 1989, p. 48.

21 Remark of former premier William A. Neilson, Agent-General of Tasmania in London, introducing a film (*c.* 1978) about the extirpation of the Aborigines.

22 Frederick Jackson Turner (1884), quoted in William Coleman, 'Science and symbol in the Turner hypothesis', *American Historical Review*, 61 (1966), 46; Wilbur Zelinsky, *Nation into State: the shifting symbolic foundations of American nationalism*, Chapel Hill, N.C.: University of North Carolina Press, 1988, pp. 161–2. See David Lowenthal, 'The place of the past in the American landscape', in Lowenthal and Martyn J. Bowden (eds), *Geographies of the Mind*, New York: Oxford University Press, 1975, pp. 99–117; Lillian B. Miller, 'Paintings, sculpture, and the national character, 1815–1860', *Journal of American History*, 53 (1967), 696–707.

23 Hippolyte Taine, *Notes on England* (1872), London: Thames & Hudson, 1957, p. 188. For similar American responses to English landscape, see note 75 below.

24 Allison Lockwood, *Passionate Pilgrims: the American traveler in Great Britain. 1800–1914*, New York: Cornwall Books, 1981, p. 444.

25 Frederick Law Olmsted, *Walks and Talks of an American Farmer in England* (1859), Ann Arbor, Mich.: University of Michigan Press, 1967, pp. 228–9.

26 Luigi Barzini, *The Europeans*, Harmondsworth, Middx: Penguin, 1984, p. 12.

27 Mark Fuller, 'Flood defences mar Old Master landscape', *The Times*, 29 June 1991, 10.

28 Kenneth Olwig, *Nature's Ideological Landscape: a literary and geographic perspective*

on its development and preservation in Denmark's Jutland heath, London: George Allen & Unwin, 1984; Niels Hørlück Jessen, 'A conservation scheme on the west coast of Denmark', Council of Europe colloquy on Management of Public Access to the Heritage Landscape, Dublin, 16 September 1991.

29 Stefan Heym, quoted in Günter Grass, *Two States – One Nation? The case against German reunification*, London: Secker & Warburg, 1990, pp. 40–3.

30 Heinrich Friedrich Wiepking-Jürgensmann (1940), professor of landscape architecture during and after National Socialism, quoted in Gert Gröning, 'The feeling for landscape – a German example', *Landscape Research*, 17 (1992), 108–15.

31 Citron, *Mythe national*, p. 136.

32 Armand Frémont, 'La Terre', in Pierra Nora (ed.), *Les Lieux de mémoire*, III: *Les France*, Paris: Gallimard, 1992, vol. 2, pp. 18–55.

33 Charles Bremner, 'Hit song captures Gallic nostalgia for peasant roots', *The Times*, 11 November 1992, 11; 'Behind farm crisis: French fear the loss of a "way of life" ', *International Herald Tribune*, 27 November 1992, 2.

34 Thierry Gasnier, 'Le Local: une et indivisible', in Nora, III: *Les France*, vol. 2, pp. 462–525.

35 Citron, *Mythe national*, pp. 19, 34, 85, 264–6; Ernest Renan, 'What is a nation?' (1882), in Homi K. Bhabha (ed.), *Nation and Narration*, London: Routledge, 1990, pp. 8–22.

36 Ferdinand Buisson (1887), quoted in Georges Vigarello, 'Le Tour de France', in Nora, III: *Les France*, vol. 2, pp. 884–925, quotation on p. 886.

37 Richard Bernstein, *A Fragile Glory: a portrait of France and the French*, London: Bodley Head, 1991, pp. 105–6; Eugen Weber, *My France*, Cambridge, Mass.: Harvard University Press, 1991, pp. 154–5.

38 Ian Ousby, *The Englishman's England: taste, travel and the rise of tourism*, Cambridge: Cambridge University Press, 1990, p. 2. What follows is based on my 'British national identity and the English landscape', *Rural History*, 2 (1991), 205–30.

39 'Rabbit droppings', *The Times*, 10 November 1990, 13.

40 Tom Shippey, 'Footpaths', *London Review of Books*, 26 July 1990, 14.

41 Tom Nairn, *The Break-up of Britain: crisis and neo-nationalism*, 2nd edn, London: Verso, 1981, pp. 270, 291–3.

42 Herbert Butterfield, *The Englishman and his History*, Cambridge: Cambridge University Press, 1945, p. 2; George Orwell, *The Lion and the Unicorn* (1941), in his *England, Your England, and Other Essays*, London: Secker & Warburg, 1954, pp. 193–4.

43 Edward Chamberlayne, *Angliae Notitia; or, The Present State of England*, 5th edn, 2 vols, London: 1671, vol. 1, p. 5.

44 H.J. Massingham, *The Heritage of Man*, London: Jonathan Cape, 1929, p. 294.

45 G.K. Chesterton, quoted in 'Preservation of rural England', *The Times*, 30 April 1931, 11.

46 Stephen Spender, *Love–Hate Relations: a study of Anglo-American sensibilities*, London: Hamish Hamilton, 1974, p. 140. On the countryside images of Rupert Brooke, E. V. Lucas and Siegfried Sassoon, see Alun Howkins, 'The discovery of rural England', in Robert Colls and Philip Dodd (eds), *Englishness: politics and culture*, London: Croom Helm, pp. 80–1.

47 Louis James, 'Landscape in 19th-century literature', in G. E. Mingay (ed.), *The Victorian Countryside*, 2 vols, London: Routledge & Kegan Paul, 1981, vol. 1, pp. 150–65; Alex Potts, ' "Constable country" between the wars', in Raphael Samuel (ed.), *Patriotism: the making and unmaking of British national identity*, vol. 3: *National Fictions*, London: Routledge, 1989, pp. 166–7.

48 The rambling English villages enjoyed by early nineteenth-century aesthetes were quite unlike the immaculate ones of today; damp thatched cottages, some deliberately dilapidated, reflected views that equated poverty with rural felicity (Tom Williamson and Liz Bellamy, *Property and Landscape: a social history of landownership and the English countryside*, London: George Philip, 1987, p. 169).

49 'Penitus toto divisos orbe Britannos' (Virgil, *Eclogues*, i.66); Gore Vidal, 'A new world on prime-time television', *Observer*, 31 December 1989.

50 W. H. Auden, 'O Love, the interest itself in thoughtless Heaven' (1932), in his *Selected Poems*, London: Faber & Faber, 1979, pp. 25–6; Frank Kermode, *History and Value*, Oxford: Clarendon Press, 1989, pp. 75–6. On England as a garden, see Christopher Mulvey, *Anglo-American Landscapes: a study of nineteenth-century Anglo-American travel literature*, Cambridge: Cambridge University Press, 1983, pp. 126–7.

51 Norman Tebbit, 'Being British, what it means to me: time we learned to be insular', *The Field*, 272 (May 1990), 76–8.

52 Henry Thomas Buckle, *History of Civilization in England* (1857), rev. edn, 3 vols, London: Longmans, Green, 1873, vol. 1, pp. 232–5.

53 E.S. Nadal, *Impressions of London Social Life, with Other Papers* ..., London: Macmillan, 1875, p. 172.

54 'Britain is no longer an island', *The Sunday Times*, 2 December 1990.

55 François Walter, 'Attitudes toward the environment in Switzerland 1880–1914', *Journal of Historical Geography*, 15 (1989), 279–99; Roderick Nash, *Wilderness and the American Mind*, New Haven, Conn.: Yale University Press, 1967.

56 Ralph Vaughan Williams, 'What have we learnt from Elgar?', *Music and Letters*, 16 (1935), 13–19, quotation on 16; Ralph Waldo Emerson, 'English traits' (1856), in *The Portable Emerson*, New York: Viking, 1946, pp. 353–488, quotation on p. 353. The English landscape – misty, green, moist, rich – is similarly said to pervade Vaughan Williams's music (Herbert Foss, *Ralph Vaughan Williams: a study*, London: Harrap, 1950, p. 66). See Jeremy Crump, 'The identity of English music: the reception of Elgar, 1898–1935', in Colls and Dodd, *Englishness*, p. 181.

57 Arthur Quiller-Couch, 'Patriotism in literature, I', in *Studies in Literature*, Cambridge: Cambridge University Press, 1918, pp. 290–306, references on pp. 301, 306; Herbert Read, 'Introduction', in his *The English Vision: an anthology*, London: Eyre & Spottiswoode, 1933, p. x.

58 Tebbit, 'Being British', 78.

59 Max Hastings, 'Diary', *Spectator*, 3 November 1990, 7.

60 John Young, 'Move to ban hunting on Trust land', *The Times*, 18 August 1990, 12; National Trust, 'Annual General Meeting 1990' notice, pp. 13–14. For the general background, see Williamson and Bellamy, *Property and Landscape*, pp. 201–5, 219.

61 Peter Barnard, 'Fear and loathing in the National Trust', *The Times, Weekend*, 26 October 1991, 1.

62 Nicholas Ridley, in *The Future of the Public Heritage*, Cubitt Trust Panel conference, 15 October 1986, London: Royal Society of Arts, 1987. p. 92.

63 F.M.L. Thompson, *English Landed Society in the Nineteenth Century*, London: Routledge & Kegan Paul, 1963, ch. 7; *idem*, 'Landowners and the rural community', in G. E. Mingay (ed.), *The Unquiet Countryside*, London: Routledge, 1989, pp. 80–98; D.C. Moore, 'The landed aristocracy', and 'The gentry', in Mingay, *Victorian Countryside*, vol. 2, pp. 367–98; Howard Newby, *Country Life: a social history of rural England*, London: Weidenfeld & Nicolson, 1987, pp. 56–8.

64 To J. Johnson, 20 September 1794. *Letters of Anna Seward, written between the Years 1784 and 1807*, 3rd edn, 6 vols, Edinburgh, 1811, vol. 4, pp. 10–11.

65 Abraham and William Driver, *General View of the Agriculture of the County of Hants, with Observations on the Means of Its Improvement*, London, 1794, p. 10. For encomia of order, see John Barrell, *The Idea of Landscape and the Sense of Place, 1730–1840: an approach to the poetry of John Clare*, Cambridge: Cambridge University Press, 1972, pp. 94–5; Keith Thomas, *Man and the Natural World: changing attitudes in England, 1500–1800*, London: Allen Lane, 1983, pp. 254–7.

66 William Beach Thomas (1938), quoted in Malcolm Chase, 'This is no claptrap: this is our heritage', in Christopher Shaw and Malcolm Chase (eds), *The Imagined Past: history and nostalgia*, Manchester: Manchester University Press, 1989, pp. 128–46, reference on p. 138. 'Whatever we feel as to the desirability of reducing the inequality of fortune between man and man, we must realise that we have to pay a heavy price in natural beauty for a more democratic ideal' (Arthur Gardner, *Britain's Mountain Heritage and its Preservation as National Parks*, London: Batsford, 1942, pp. 1–2).

67 Fay Weldon, 'Letter to Laura', in Richard Mabey (ed.), *Second Nature*, London: Jonathan Cape, 1984, p. 68; Sarah Lonsdale, 'Natives rage as Hardy's heath waits to be blasted', *Observer*, 14 July 1991, 4.

68 Michael Hall, 'Every last hogg hole', *Country Life*, 2 May 1991, 82–5; Henry James, 'Old Suffolk' (1897), in his *English Hours*, New York: Orion Press, 1960, p. 196.

69 Diana MacVeagh, *Edward Elgar: his life and music*, London: Dent, 1955, p. 166; on the *campo santo*, Oliver Wendell Holmes, *Our Hundred Days in Europe*, London: Sampson, Low, 1887, p. 279; Emerson, 'English traits', p. 356.

70 'Memory lane', *The Times*, 4 October 1989, 3; Thomas, *Man and the Natural World*, pp. 217–23. This is not to gainsay British fascination with the geological and palaeontological heritage. See David Elliston Allen, *The Naturalist in Britain: a social history*, London: Allen Lane, 1976; Peter J. Bowler, *The Invention of Progress: the Victorians and the past*, Oxford: Basil Blackwell, 1989; Ousby, *Englishman's England*; Thomas, *Man and the Natural World*, pp. 269–71, 280–4. Robert Foulke, 'A conversation with John Fowles', *Salmagundi*, nos 68–9 (1986), 367–84, privileges natural over human history.

71 Christopher Chippindale et al., *Who Owns Stonehenge?*, London: Batsford, 1990.

72 Ronald Blythe, 'The dangerous idyll', in his *From the Headlands*, London: Chatto & Windus, 1982, p. 161. Weekend commuters 'change their clothes … when they get down to the country; join appeals and campaigns to keep one last bit of England green and unspoilt; and then go back, spiritually refreshed, to invest in the smoke and the spoil' (Raymond Williams, 'Ideas of nature' (1971), in his *Problems in Materialism and Culture*, London: Verso, 1980, p. 81).

73 V.P. Springett (letter), 'Recollections of a golden past', *The Times*, 11 March 1989, 11.

74 ' "Nether Burton" revisited', *International Herald Tribune*, 26 July 1990.

75 Edwin M. Yoder, Jr, 'In praise of Baron Omnium and his old English village', *International Herald Tribune*, 20 July 1990. Yoder's idyll has old American antecedents. English relics and place-names evoked 'deep-rooted sympathies; … a suppositious pedigree, a silver mug, [were] potent enough to turn the brain of many an honest republican' (Nathaniel Hawthorne, 'Consular experiences' (1863), in his *Our Old Home, Works*, Ohio: State University Press, 1962–80, vol. 5, pp. 19–20). Americans steeped in Shakespeare and Tennyson pre-empted English heritage as their ancestral own, and in England enjoyed aristocratic privilege anathema in America; delighted by 'contented' English domestics, a Yankee parson who never spoke of 'servants' at home saw 'no harm in it where it is customary,

especially as [it is] abundantly sanctioned in the Scriptures' (Heman Humphrey (1838), quoted in Lockwood, *Passionate Pilgrims*, p. 137; see also p. 448). The Great House crowned the English landscape, preserving not only the visual heritage and the owner's wealth but 'an ancient and honourable pattern of human relationships' – notably those of master and servant – 'lost to the modern, democratic, and American order of things' (Mulvey, *Anglo-American Landscapes*, p. 127). 'The well-appointed, well-administered, well-filled country house [is] the most perfect, the most characteristic ... of all the great things the English have invented and made part of the national character', but Henry James set against these glories the grim workhouse and orphanage idiots he saw the same day ('An English New Year' (1879), in his *English Hours*, pp. 170–1).

It was the 'latent preparedness of the American mind' for the English scene that made James's devotion 'total and sacred' ('A passionate pilgrim' (1875), in his *The Reverberator and Other Stories*, London: Macmillan, 1909, p. 335). Indeed, only an American could truly savour historic England; so oblivious to hoary antiquity seemed England's natives that Hawthorne suggested they all be removed 'to some convenient wilderness in the Great West' and replaced by Americans ('Leamington Spa' (1862), in *Our Old Home*, vol. 5, p. 64). But the 'phlegmatic' English response to their heritage concealed strong attachments; natives, after all, could not pass the whole of life in tourist euphoria (Mulvey, *Anglo-American Landscapes*, p. 62).

76 ' "Nether Burton" revisited'. Hall heads the Ramblers' Association, Ashbrook the Open Spaces Society. Their English past is likewise mythical: old-time Turville was anything but peaceful to runaway serfs hunted by Chiltern Hundred stewards and stagecoach passengers held up by highwaymen.

77 Williams, 'Ideas of nature', p. 80; *idem*, *The Country and the City*, London: Chatto & Windus, 1973, pp. 74–9. The miseries of dispossession limned in Oliver Goldsmith's 'The Deserted Village' (1770) were quite real; the unsightly villagers of 'Auburn' were tidied away from Earl Harcourt's landscape at Nuneham Courtenay, Oxfordshire. See Paul Coones and John Patten, *The Penguin Guide to the Landscape of England and Wales*, Harmondsworth, Middx: Penguin, 1986, pp. 247–8; John Barrell, 'The golden age of labour', in Mabey, *Second Nature*, pp. 177–95; Newby, *Country Life*, pp. 78–91; Williamson and Bellamy, *Property and Landscape*, pp. 137–8.

78 Jan Morris, 'Barchester lives on', in Ronald Blythe, *Places: an anthology of Britain*, Oxford University Press, 1981, p. 146.

79 Reg Gammon, 'Our country, all earthly things above, as always', *The Field*, 272 (May 1990), 82–3. See also his *One Man's Furrow: ninety years of country living*, Exeter: Webb & Bower, 1990, p. 176.

80 Michael Heseltine, 'Is it at risk, this England?', *The Field*, 272 (May 1990), 78–9.

81 Lonsdale, 'Natives rage'.

82 Howard Newby, 'Revitalizing the countryside: the opportunities and pitfalls of counter-urban trends', *Royal Society of Arts Journal*, 138 (1990), 630–6. On the harm to rural society caused by faith in its enduring stability, see Newby, *Country Life*, pp. 219–24; Chase, 'This is no claptrap', p. 133.

83 Stanley Baldwin, 'The classics' (1926), in his *On England*, p. 101; see also Dennis Smith, 'Englishness and the liberal inheritance after 1886', in Colls and Dodd, *Englishness*, p. 264.

84 Geordie Greig, 'Which Cheddar Gorge do you like best?', *The Sunday Times*, 7 February 1988.

85 Raymond Williams, 'Between country and city', in Mabey, *Second Nature*, p. 218.

86 Neal Ascherson, 'The land belongs to the people', *Observer*, 25 January 1987, 9.

87 Michael Brooke, 'A day in the country', *New Scientist*, 7 April 1990, 68; Sylvia Crowe, *Tomorrow's Landscape*, London: Architectural Press, 1956, p. 137.
88 Richard Mabey, 'Strange vision of a promised land', *Observer*, 1 February 1987, 26.
89 Newby, 'Revitalizing the countryside', p. 631; 1983 European Community figures in Marion Shoard, *This Land is Our Land*, London: Paladin/Grafton, 1987, pp. 144–5.
90 Robin Fedden, *The National Trust: past and present*, rev. edn, London: Jonathan Cape, 1974, p. 98. Fedden's view is widely shared. Against his own bias, H. E. Bates acclaimed the great landowners: 'I doubt if the poor have ever beautified the English landscape. It is the rich and the prosperous who have left on it the hall-marks of beauty' ('The hedge chequer work', in H. J. Massingham (ed.), *The English Countryside: a survey of its chief features*, 3rd edn, London: Batsford, 1951, p. 50).
    A dissenting view contrasted the 'peasants' country' of south Germany, the Touraine, the Midi, with typical English 'landlords' country': 'the open woods, the large grass fields and wide hedges, the ample demesnes which signify a country given up less to industry than to opulence and dignified ease; ... sparsely cultivated, but convenient for hunting and shooting' (C. F. G. Masterman, *The Condition of England*, London: Methuen, 1909, pp. 201–2).
91 Nicholas Coleridge, 'Why the Lords love the lady', *Spectator*, 22 October 1988, 9–11; Lord St John of Fawsley, quoted in Andrew Alderson, 'Study of aristocratic decline makes the blue-bloods see red', *The Sunday Times*, 2 September 1990, sect. 1, 7. David Cannadine, *The Decline and Fall of the British Aristocracy*, London: Yale University Press, 1990, traces the elite's dwindling political power; Shoard, *This Land is Our Land*, pp. 127–43, and Jeremy Paxman, *Friends in High Places: who runs Britain?*, London: Michael Joseph, 1990, pp. 39–46, document continuing elite control of land and social institutions.
92 William Wordsworth, 'A guide through the district of the Lakes' (5th edn, 1835), and 'Kendal and Windermere Railway (1844), in *The Prose Works of William Wordsworth*, 3 vols, Oxford: Clarendon Press, 1974, vol. 2, p. 225; vol. 3, pp. 346, 355.
93 C. E. M. Joad, *The Untutored Townsman's Invasion of the Countryside*, London: Faber & Faber, 1946, p. 219; *idem*, 'The people's claim', in Clough Williams-Ellis (ed.), *Britain and the Beast*, London: Readers' Union, 1938, pp. 64–85, reference on p. 80. Today's townsmen 'play football on a hay field; pick our fruit; run their dogs off the lead and let them chase stock; they sunbathe, picnic and copulate in the fields; they scatter rubbish and ... leave gates open' ('The rights and wrongs of way' (letter), *Observer*, 9 September 1990, 44). On elitist assumptions, see W. J. Keith, *The Rural Tradition: a study of non-fiction prose writers of the English countryside*, Toronto: University of Toronto Press, 1974, p. 11; Thomas, *Man and the Natural World*, p. 267; Ousby, *Englishman's England*, pp. 189–92.
94 Nicholas Ridley at Historic Houses Association, 22 November 1988, in Martin Fletcher, 'Sell stately homes to nouveaux riches, says Ridley', *The Times*, 23 November 1988, 1; Marcus Binney, 'Mr Ridley's bad house-keeping', *The Times*, 24 November 1988; Lord Saye and Sele, in Sally Brompton, 'Family castle not for sale', *The Times*, 17 December 1988. The custodial sentiment is an old cliché: Charles Trevelyan 'considered Wallington not so much *owned* by him as entrusted to him by inheritance for the benefit of the public' (Martin Drury, 'The early houses of the country houses scheme', *National Trust Magazine*, no. 52 (Autumn 1987), 33).

95 Clive Aslet, quoted in Richard Duce, 'Prince loses green battle of the dukes', *The Times*, 5 September 1991, 5.

96 Brompton, 'Family castle not for sale'; Lady Saye and Sele, in Andrew Lycett, 'Saved in the last reel', *The Times*, 13 August 1990, 16; John Martin Robinson, 'Holding the fort for 50 years: the National Trust's country house scheme', *Spectator*, 18 April 1987, 35–6.

97 Emerson, 'English traits', p. 484; Kate Saunders, 'Modern manors', *The Sunday Times*, 26 August 1990, sect. 5, 1. Unlike country-house hotels, 'real English country houses have an essential shabbiness ... They have draughts and mouse-holes and untuned pianos and mangy old dogs' (Sue Arnold, 'Country kitsch by the yard', *Observer*, 27 January 1990, 51).

The National Trust concedes past indignities (Adrian Tinniswoode, *A History of Country House Visiting: five centuries of tourism and taste*, Oxford: Basil Blackwell/National Trust, 1989, ch. 5), but it heeds complaints *by* today's visitors less than complaints *about* them; aristocratic tenants find them 'frightfully inconvenient' (Trust chairman Jennifer Jenkins, quoted in Valerie Grove, 'Keeping Britain's earls and toads in [Germolene] pink', *The Sunday Times*, 19 June 1988, B4). See Patrick Wright, 'Brideshead and the tower blocks', *London Review of Books*, 2 June 1988, 3–7.

98 Sarah Lonsdale and Michael Prestage, 'National Trust gems closed to the public', and Gillian Darley, 'It's open season on the Trust', *Observer*, 28 October 1990, 4, 20; Rodney Legg, quoted in Lin Jenkins, 'National Trust rejects an open and shut case', *The Times*, 23 October 1990, 1, 22.

99 Michael Billig, 'Collective memory, ideology and the British Royal family', in David Middleton and Derek Edwards (eds), *Collective Remembering*, London: Sage, 1990, pp. 60–80; Ousby, *Englishman's England*, p. 60. Raphael Samuel notes that the newly populist and democratic heritage, despite radical and egalitarian roots, 'feeds on a nostalgia for visible social differences. "The World We Have Lost" is one where people knew where they stood, where classes were classes, localities localities, and the British an indigenous people' ('Introduction: exciting to be English', in his (ed.) *Patriotism*, vol. 1, p. xlix).

100 Peter Levi, 'Knowing a place', in Mabey, *Second Nature*, p. 41.

101 Adam Nicolson, 'Tidiness and the Trust', *National Trust Magazine*, no. 58 (Autumn 1989), 37–9. Hugh Prince and I treat love of order in 'English landscape tastes', *Geographical Review*, 55 (1965), 186–222. John Bayley sees order as British, not English; today 'Englishness' conjures up sturdy folk liberties and radical protest, while 'Britishness has come to represent law and order and the orthodoxy of established power' ('Embarrassments of the national past', *The Times Literary Supplement*, 23 February – 1 March 1990, 188).

102 George Crabbe, 'The learned boy' (1812), in *Poetical Works of George Crabbe*, London: Oxford University Press, 1908, p. 334; George Gissing, *The Private Papers of Henry Ryecroft* (1903), New York: Modern Library, pp. 214–17; 'Heaven's order' is in Alexander Pope, *Essay on Man* (1733–4), Epistle IV, line 49.

103 'The common image of the country is now an image of the past' (Williams, *The Country and the City*, p. 297). See Martin J. Wiener, *English Culture and the Decline of the Industrial Spirit, 1850–1980*, Cambridge: Cambridge University Press, 1981. On how the emptiness of the English countryside has affected the ruralist mystique, see John Lucas, *Modern English Poetry from Hardy to Hughes*, London: Batsford, 1986, pp. 50–69, and his *England and Englishness: ideas of nationhood in English poetry, 1688–1900*, London: Hogarth Press, 1990, p. 204.

104 John Hopkinson, director, British Field Sports Society, quoted in John Young,

' "Green" policies may harm wildlife', *The Times*, 18 August 1990, 7.

105 Stephen Haseler, 'How a nation has slipped into its dotage', *The Sunday Times*, 25 February 1990, C6; Simon Hoggart, 'Thinking of England-land', *Observer Supplement*, 12 August 1990, 5.

106 George Orwell, 'Pleasure spots', cited in Angela Lambert, 'It's the vulnerable who feel the pinch', *Independent on Sunday*, 1 September 1991, 21.

107 Nigel Hawkes, 'Ardours of Arcadia', *The Times*, 31 August 1991, 11.

108 Orvar Löfgren, 'Landscapes and mindscapes', *Folk*, 31 (1989), 183–208.

2

From Michelet to Braudel: Personality, Identity and Organization of France

Paul Claval

THE ORIGINS OF AN IDENTITY CRISIS

During the nineteenth century, French intellectuals were very worried about French identity. It was a consequence of the rise of nationalism, the French Revolution and the ensuing events: they had deeply affected French society.

The Ancien Régime *identities*

In *Ancien Régime* France, national identity was based on a simple principle:[1] all that belonged to the King was considered French. This type of answer was not considered satisfactory by the enlightened part of the population. At the end of the sixteenth century, for instance, humanists often wondered about the genesis and nature of their country. The origins of the French monarchy, at the time of Clovis and the Frankish invasion, had ignored the Gallic and Gallo-Roman past of the people. It was rediscovered by the scholars of the 'Perfect History' group.[2] They were glad to dissociate French identity from faithfulness to a prince who was at that time prosecuting some of his subjects.

The lesson of the theoreticians of 'Perfect History' was essential. It glossed over the mediaeval centuries of differentiation in focusing on a lost unity. But in order to be convincing, such a conception also involved unity in the future. From that time on, French identity became peculiar because it was based not on the similarity between present Frenchmen but on the idea that they derived from the same distant past and had to build their future together. At no time did France exist except in those places where the project inherent in her nature began to be realized, that is to say, around the Crown in Paris or, later, at Versailles.

Such a conception was not welcomed by the French monarchy since it ignored the Franks' contribution to French identity and the role of Clovis, but the idea of France as a project was not foreign to the kings. There was always something Utopian in the kings' assignment. Was not France the eldest daughter of the Roman Catholic Church? Was it not a universal commitment present in the idea of the Sun King? Was it not the duty of the King and of his agents to spread everywhere a particular ideal of civilization and greatness?

The impact of the French Revolution

The Revolution destroyed the traditional roots of French identity and erased the provincial identities traditionally associated with it. Until 1789 somebody living in Provence, for instance, thought of himself first as a *provençal*. He was a Frenchman because of the annexation of Provence to the French kingdom on the death of King René.

The Federation celebration on 14 July 1790 was organized to provide Frenchmen with a new form of identity, contractual this time. Delegates representing all the French departments laid down new foundations for France at a time when its unity was threatened by the dynamics of the Revolution.

To rebuild an identity was not an easy task. At that time, the idea that French society had a dual identity weighed heavily on many minds. Since Father Le Laboureur[3] during the reign of Louis XIV, the nobility was thought of as stemming from the German conquerors and the lower classes as the descendants of the defeated Gallo-Romans. Had not Father Sieyès stated that this was the origin of the *Tiers-Etat* in his pamphlet *Qu'est-ce que le Tiers-Etat?* The theme of race struggle – from which came later the idea of class struggle – prevented the perception of France as a united society.

But the French Revolution had other consequences. The *Ancien Régime* provinces varied widely. This was accepted as normal since the kingdom was a mosaic of lands collected together by reason of a complicated history, each with its own traditions. Allegiance to the King was the common link. The Constituent Assembly made all Frenchmen equal and did away with the *Ancien Régime* provinces. Had Frenchmen suddenly developed a new homogeneity? Not at all. The evidence of diversity was more acute than ever. *Ancien Régime* France had been governed through a small French-speaking elite which adhered to French culture and values and which linked regional societies to the nation state. The men who took over the country during the Revolution came from lower classes. Even if they all spoke French, it was with local accents. The recruits, more numerous than ever because of conscription, were discovering their differences. Historians date from the Revolution the time when people began to use the term *Midi* for Southern France.[4] Every country has a north, a south, an east and a west; from the time of the Revolution France was different because it had no south, but a *Midi*; the word shows that everybody was conscious of the profound differences in this part of France, where people spoke French with a strong regional accent and had attitudes and behaved in ways which puzzled people from northern France.

There are many signs to show that the new political organization of France sharpened opposing views instead of obliterating them.[5] Almost everywhere, details of vernacular architecture, styles of furniture, regional costumes still differentiated between regions. At a time when increased mobility made contacts easier and more frequent, Frenchmen reacted by stressing their local identities.

Nineteenth-century Frenchmen identified themselves on two levels. They were first and foremost members of a local and regional community, a fact that was significant in shaping their personalities. Inasmuch as they felt themselves full members of these local cells, they felt themselves to be French. This conferred on everyone something which was shared by all. This something was thought essential, at least by the educated part of the population; but to take advantage of it, it was first necessary to be integrated into regional society. Those for whom national reality was directly experienced constituted only a small group, except in Paris.

Evidence of the growing acceptance during the first half of the nineteenth century of this way of experiencing national realities is provided by many factors: the strength of regional identities; the new interest in regional novels;[6] the growing taste for travel, and for guidebooks which helped to prepare them for the journey and showed them the diversity of the country.

A new situation: general agreement on the present boundaries of France

Revolution and Empire had other consequences. At the end of the *Ancien Régime* France was satisfied with its boundaries. Along the Austrian Netherlands, borders which were no more than a century old had been widely accepted. In Lorraine, Alsace and Franche-Comté the French expansion had been supported by the local populations, and no one thought it necessary to go further. The independence of Switzerland was recognized. On the south-east border, Savoy was a feudal state where French was officially used (as it was in Piémont, on the other side of the Alps): France had no territorial ambitions on this side.

The policy of expansion and annexation during the period of the Revolution and the Empire had been unsuccessful. At the Congress of Vienna the European nations were wise enough to give back to France its 1789 boundaries, and to accept the integration of former enclaves such as Avignon, Montbéliard and Mulhouse. As a result, from that time nobody in France looked for territorial expansion in Europe. The boundaries of France were considered fixed. Nobody resented the fact that they did not exactly match the linguistic boundaries. Wallonia, Savoy, Aosta and Nice escaped France. At the same time, there were within French boundaries Flemings to the north, German-speaking people to the east, Italian-speaking Corsicans, and Catalans and Basques on the Spanish border. France was not conceived of as an ethnic unit. It had been built through history from a variety of groups, and the limits which had been reached during the eighteenth century were considered well-fitted to the national will, since everyone who wished to build a common future as Frenchmen lived within the same state. When there was an opportunity, France was ready to integrate regions of French language and culture, as with the referenda which gave it Savoy and Nice. But French opinion was not passionately involved in these problems: Aosta was given to Italy against the wish of its inhabitants because it gave Victor-Emmanuel a well-liked hunting ground.

French reality, as it was experienced during the nineteenth century, relied on a two-level territorial membership: local or regional and national. National boundaries were considered permanent. If there was an identity problem, it did not grow out of French limits; it grew out of the nature of France itself. What was the justification, after the end of the monarchy, for the territorial construction which resulted from the thousand-year-long policy of the Capetian kings?

MICHELET'S INTERPRETATION: FRANCE AS A PERSON

The majority of historians contemporary with Michelet considered that French society incorporated two races, Franks and Gallo-Romans, but he refused to accept this view. He thought that the history of France started later: 'The history of France starts with the French language. The language is the main sign of a nationality. The first monument of our language is the oath dictated by Charles the Bald to his brother, at the treaty of 843.'[7] In this way Michelet got rid of the quarrels that had engrossed scholars for more than a century. For him, a country was basically founded on a language and the territory in which it was used. With such a perspective, the diversity of France was more striking: 'It was within the fifty years which followed [the Verdun Treaty in 843] that a feudal dynasty rose in each of the various parts of France until then merged into an obscure and vague unity.'[8] The country was transformed into a mosaic. Michelet spoke of 'the infinite variety of the feudal world, the multiplicity of objects through which it first exhausted view and attention'.[9] But mixture did not mean disorder: 'The country appears in its local diversity, drawn by its mountains and rivers. Instead of confusion and chaos, as someone said, it offers order, inescapable and fatal regularity.'[10]

For Michelet, France existed because it occupied a particular space: 'History is first completely geography. We can only tell the feudal or *provincial* period (the last term fits it as well) without a view of the main characteristics of each of its provinces. But it is not enough to draw the geographical patterns of each of these diverse regions. It is mainly through the fruits which they bore that they can be explained – I mean the men and events that their history must offer.'[11]

Michelet had received a kind of training which differed from that of contemporary historians. Like them, he was acquainted with the value of documents and the necessity to rely on archives, but his craft was completed by a conception of history which he owed to reading the works of Gian-Battisto Vico and of Herder. For the latter, history concerns both groups and the environment where they settled and which they had mastered. The best way to understand the destiny of a people is to consider it in the context of the place where it lives. When speaking of the French provinces, Michelet wrote: 'From our point of view, we shall predict what each of them must do and produce, we shall mark their destiny, we shall provide each of them with a dowry from the cradle.'[12]

What were the results of the approach that Michelet developed? The idea of a complementarity of the French regions welded around Paris: 'Concerning the centre of the centre. Paris, Ile-de-France ... they received and produced the national spirit.'[13] 'How has this big and wholesome symbol of the country been encapsulated in a city? ... The Parisian genius is at the same time the most complex and highest form of France. It would seem that something which resulted from the annihilation of all local spirit, of all provinciality, had to be purely negative. It is not true. Out of this negation of material, local, particular ideas a living generality, a positive thing, a live force result.'[14]

France was made up of diversity, but of a transcended diversity: a diversity transcended through its centre. The conception of France developed by Michelet was new, but it stuck to the idea that the country existed only at the point where all local sluggishness had been eliminated and the spiritual forces which shape the country asserted. For Michelet as for the Perfect History group or the theoreticians of monarchy, France had to be structured around a shared project.

The geography of France expressed this unity which stemmed from variety:

It is a great and marvellous spectacle to cast an eye from its centre to its extremities and to take in at a glance this extended and powerful organism where the different parts are so deftly combined, the strong with the weak, the negative with the positive ... Considered along longitudes, France is built as a double organic system, just as the human body is double, gastric and cerebrospinal. On one side, the provinces of Normandy, Brittany, Poitou, Auvergne and Guyenne; on the other one, those of Languedoc and Provence, Burgundy and Champagne, and finally those of Picardy and Flanders, where the two systems merge. Paris is the sensory centre.

The force and beauty of the whole consist in mutual reciprocity, the solidarity of parts, the distribution of functions, the division of social labour. The resisting and military strength, the virtue of action lies in the extremities, intelligence is at the centre. The centre knows itself and knows all the other parts ... The centre, protected from war, thinks, innovates in industry, science, politics. It transforms all the things that it receives. It drinks in wild life and transfigures it.[15]

The idea of an organic solidarity was central to Michelet's idea of France. It covered economics, but also concerned the fields of military strength and innovation. It gave France its originality:

But France must not be taken part by part, it has to be seized as a whole. It is just because centralization is powerful, and the common life strong and energetic, that local life is weak. I would even say that it is the beauty of our country. It does not have the head of England, monstrously strong with industry and wealth, but neither does it have the desert of the Highlands of Scotland, the cancer of Ireland. You will not find, as in Germany or in Italy, twenty centres of science and art: there is only one, one of social life. England is an Empire: Germany, a race. France is a person.[16]

And Michelet gave a more precise form to his thought through an analogy:

> Personality, unity, it is through this that a being ranks high in the chain
> of beings ... Nations can be ranked just like animals. The common use
> of a great number of parts, the mutual solidarity of these parts, the
> reciprocity of the functions that they assume one for the other, it is there
> that social superiority lies.[17]

The movement through which the unity of France was achieved meant also
the rise of spirit out of the bonds of matter:

> In this way the general and universal spirit of the country was shaped;
> the local spirit disappeared day by day; the influence of soil, climate,
> race has retreated before political and social action. The fatality of
> places has been defeated, man has escaped the tyranny of material cir-
> cumstances ... Society, freedom have mastered nature, history has rubbed
> out geography. In this marvellous transformation, spirit has won over
> matter, the general over the particular, and idea over contingencies. The
> individual man is materialistic, he willingly adheres to local and private
> interest: human society is spiritualist, it constantly tends to overcome the
> miseries of local existence and to reach the high and abstract unity of
> homeland.[18]

Michelet had understood Herder. People bear the mark of the country
where they live. Slowly, they manage to escape the constraints of their environment, they form a nation and begin, in this world, to give substance to
the reign of the ideal. Such an interpretation was close to the philosophies of
his time – it had Hegelian overtones – and it fitted the concerns of nineteenth-
century Frenchmen, who experienced the nation at two levels, the local and
the general, except maybe for the Parisians who reached the upper one directly.
Nothing was left of the old theme of opposition between Franks and Gallo-
Romans. The influence of the Celts, of Rome, the Iberian imprint or the
Germanic mark were presented in their normal place, when speaking of the
provinces where they were the strongest. They constituted one of the dimensions of French diversity, of this diversity which was progressively overcome
by the building of the nation. Unity was seen as a process, as an ongoing
construction, and not as an everlasting reality. Contemporary boundaries seemed
satisfactory for a correct functioning of the whole and for the harmonious
development of France as a person. Lastly, the nation appeared as a spiritual
adventure. It did not testify to an egoist withdrawal into oneself, but to an effort
to overcome local pettiness. The French nation owes to its very foundations
its vocation for progress and its role as an engine for the progressive transformation of humanity.

Were there weaknesses in such an analysis? Yes, evidently. Michelet did not
try to go very deep in his analysis of provinces. He used the old Hippocratic
theory of humours applied to the different parts of the nation. Was not he
speaking of the 'eloquent and winy Burgundy, ... the ironic innocence of

Champagne, the critical, polemical and warlike toughness of Franche-Comté and Lorraine, ... the Languedocian fanaticism, the Provençal lightness and the Gascon indifference'?[19] What a beautiful collection of stereotypes! But as soon as he developed his analysis, brilliant comments redeemed it. For instance, presenting the general pattern of Provence: 'In Provence, all life is on the border.'[20] On its climate: 'And this powerful sun, the ordinary feast of this land of feasts, hurts the head violently when one of its rays transforms winter into summer. It gives life to the tree and burns it. And frosts burn as well. More often, storms, and rivulets become rivers. The farmer picks up his soil down the hill, or looks at it as it flows downstream, giving wealth to the land of his neighbours. Capricious, passionate, quick-tempered and charming nature!'[21] But where men are concerned, simplification won: 'Here the genius of Basse Provence, violent, noisy, barbarian, but with grace.'[22] History is solicited in order to speak of ethical problems or to judge institutions: 'This spirit of equality cannot surprise in this land of Republics, amongst Greek cities and Roman municipes. Even in rural areas, serfdom never weighd so heavily in other parts of France. These peasants were their own liberators and victors against the Moors.'[23]

The vivid picture, the sovereign view did not suffice to give consistency to these regional descriptions. The overview is seductive, but the presentation of the mosaic of provinces is too impressionistic to be scientific. It was the main weakness of an interpretation which had all the other qualities demanded by Frenchmen of the time.

VIDAL DE LA BLACHE, OR THE SCIENTIFIC APPROACH TO THE IDEA OF PERSONALITY

The fin de siècle *atmosphere*

When geography really began to prosper, during the third part of the nineteenth century, it had to cover the problem of French identity. Nationalism was growing stronger every day. The young discipline had to provide children and young men with a clear image of their homeland and of the problems which confronted it throughout the world. The problem was made more acute in France by the 1870 defeat by Prussia. The country had lost provinces which were considered essential. Their inhabitants had not been consulted. Alsace and a part of Lorraine were speaking a German dialect, but it was not an acceptable argument for Frenchmen accustomed to consider their country as a unifying mosaic. The Germans had not even respected the only justification that they could invoke: they had added to the German-speaking part of Lorraine the city of Metz, where French had always been used. It was such a stronghold that Bismarck did not want to see it on the French side of the border.

Intellectual life was a bustling and restless one during the last decades of the nineteenth century. All opinions were represented in France, from the ultra-left of the anarchists (Elisée Reclus was one of its members), or the Marxists (with Jules Guesde), to the ultra-right which was taking on new features, since the old themes of counter-revolutionary thought of the early nineteenth century had been interwoven with new fashionable ideas. The constraints which limit human freedom ceased to be considered as an expression of God's will. They were thought of as resulting from the inescapable laws of evolution, of racial inheritance more particularly. The dominant mood was more moderate. In spite of the harsh quarrel between Republicans and Roman Catholics, these two groups agreed on some attitudes. All of them supported strong patriotic attitudes, and adhered to idealist philosophies – it was the time of Kantian renewal.[24]

Geography, developing in such an atmosphere, gave an important place to France and its colonies. It was still a fragile discipline in the 1870s. During the 1880s it acquired strong foundations. Its influence grew rapidly during the next twenty years.

The genesis of regional interpretations: Vidal de la Blache

The chief merit of Vidal de la Blache was to personify this long and patient work of organization and reflection. He always kept in mind the main questions that geography had to answer, and would succeed in solving, he hoped. Hence the attention that he displayed to national problems, as shown in his book *Etats et nations d'Europe. Autour de la France.*[25] The nations he depicted were not presented as homogeneous areas but as articulated wholes. Was he speaking of Germany? The convergence of navigable rivers made Berlin a place that was more convenient to dominate the great northern European plain than people were generally prepared to admit. Italy? Transhumance and temporary migrations triggered by the limited natural resources of the mountains and by the size of the population gave the peninsula a measure of unity which was not evident to those who considered only the fragmented nature of its topography. Vidal was conducting simultaneously an analysis of phenomena that were occurring at different scales, regional and national.

Vidal was also relying on the spatial reflection of geologists. They had developed an early interest in the diversity of France and had been the first to note the coincidence of the names given to the *pays* with the small natural areas unified by the nature of their underlying rocks. Brie, Beauce and many other areas had been carefully studied by them. During the 1840s. Elie de Beaumont and Dufresnoy synthesized decades of former research. In their book, *L'Explication de la carte géologique de France,*[26] they showed how the geology and the topography of the country were structured into two concentric sets, one built around a high centre, an area from which water flows in all directions, the Central Massif, and the other characterized by the convergence of rivers towards a common focus, Paris. The formation of France could be easily explained in this way: the Parisian nexus, the steady unification of provinces around Ile-de-France, and the decisive inroads which were conducive to the control of the Central Massif and of the lower districts which allow for easy movement on its periphery: Poitou, Lauraguais, Burgundy. Vidal learnt two lessons from the geologists: geographers had to rely on field-work and examine the details of topography and landscapes; they also had to give to the elements they discovered their place in broader contexts, at a different scale. These were the dominant themes at the time when he wrote his paper, *Des divisions fondamentales du sols français.*[27]

For Vidal, geography was a discipline based on field evidence, sight and spatial dialectics. The regional approach was the only piece missing in his thesis: the study of Southern and Mediterranean Europe provided him with it.[28] This area always interested him. His inaugural lecture at the University of Nancy dealt with it:[29] the opening of the Suez Canal had restored its former strategic significance. At that time. Vidal's curiosity was twofold. It was first motivated by the feeling of estrangement and wonder which people commonly experienced when confronted with milieux where our civilization was born but which looked so strange when living in Atlantic Europe. It was also linked with geostrategic actuality.

How to understand the Mediterranean realities? Vidal reviewed the different answers to this question in his article on 'Des rapports de la population et du climat sur les bords européens de la Méditerranée'.[30] Could the clemency of the climate, the purity of the light and their psychological effects on the inhabitants serve as an explanation? Vidal was tempted by this solution for a short while, but considered it insufficient. Another solution was more fruitful. It had been suggested to Vidal by Fischer, the great German geographer who had been the first to understand that the summer drought was the key factor in Mediterranean originality and Mediterranean life. Mediterranean nature is not plentiful, as was generally thought, but difficult and avaricious. Earth is fertile only when it is irrigated, in the hills below the mountains, or on the plains – but marshes often make the latter unhealthy. Hills are healthy but their soil is poor, and without trees which can withstand the long summer droughts crops would be even scarcer. Mountains offer a favourable environment for a few months each year during the summer, but winter is harsh because food is lacking both for people and for cattle and sheep.

Is the Mediterranean world uniform? Not at all. Summer drought is present everywhere, but its impact is not the same everywhere, this leads to many human responses. Man has imagined ways of life to fit each specific environment. He can also escape some of the environmental constraints through trade: mountain dwellers sell the honey, the wool, the skins and the cheese which they produce and buy from nearby plains the corn that they lack. The mobility of men and cattle completes the exchange of goods: herds feed during winter on the pastures of the plains where the temperature remains mild. Mountain-dwellers work during summer on the big farms of the depopulated plains and in the cities where they are appreciated because they accept all types of jobs and are good workers.

The analysis of ways of life was in this sense complementary to that of natural features and aptitudes. It showed the diversity of the forms of social life that men have developed. As a result, Vidalian geography incorporated a social dimension and was as keen to understand local history as to explore natural endowments.

The identity of France according to Vidal de la Blache

It was while writing his *Tableau de la géographie de la France*[31] that Vidal had to explain the identity of France. He was commissioned to write this book by the historian Ernest Lavisse who wished to open his monumental *Histoire de France* with a geographical introduction, just as had been the case 70 years earlier in the treatise of Michelet. Vidal could not ignore his famous predecessor. He had to write a geographical introduction to the history of a nation. The problem was not to set an anonymous stage on which a standard play would be performed. It was to show how the stage and the players were dominated by the same inner logic, and how each part of the country was a necessary component of the whole. The identity of France did not result from one factor, such as the presence of an ethnic group. It grew out of a long process of elaboration. It was an historical construction.

Vidal de la Blache accepted the main theses of Michelet. He wrote, for instance: 'We willingly repeat this sentence of Michelet: "France is a person." '[32] He added to this theme an older one, the idea of the global harmony of French territory – a commonplace since Strabo had praised 'the correspondence which exists there between rivers and sea, the Mediterranean and the Ocean'.[33] Vidal added: 'These rivers facilitate the relations between the seas; this correspondence, which is so exceptional all around the Mediterranean sea, and which here is present, suggested [to Strabo] the idea of an organism composed at will, "just as the result of an intelligent anticipation".'[34]

Was it enough to define the personality of a country? No:

> A geographical identity does not result only from considerations of geology and climate. It is not something which is proposed ready-made by nature. People have to start from the idea that a country is a reservoir where energies the germ of which had been laid down by nature are sleeping, but the use of which depends on Man. It is he who, when curbing them to its use, brings its individuality to light. He connects scattered features; for the incoherent effects of local circumstances he substitutes a systematic array of forces. It is then that a country becomes increasingly precise and differentiated, and at length like a medal stamped in the image of a people.[35]

L'Histoire de France by Lavisse stopped at the end of the *Ancien Régime*. Vidal was thus asked to present traditional France and to show how its spatial divisions were instrumental in the birth of the French nation. He had to resolve the same problem as Michelet. The way in which the personalities of provinces were described in *Tableau de la France* did not satisfy him. The old theory of humours and the connection between climate and psychological disposition which are its geographic expression were obsolete. Vidal, who had tried to mobilize these old interpretations in the case of the Mediterranean world and had experienced their limitations, knew that. He could have used the ideas of Elie de Beaumont and Dufresnoy, as in *Les Divisions fondamentales du sol français*,[36] and stressed in this way the harmony of French territory, relying on themes imagined during the nineteenth century but close to those of Strabo. But did the environment play so functional a role in the development of a country? Vidal did not try to show how France was working or how history had shaped it. He wished to grasp what was responsible for her identity – or, to use Michelet's expression, what made her a person.

The analysis of ways of life offered him the solution: France was not only the organic assembly of parts linked by reciprocal relations. She grew out of the complementarity of their genius. Hence the curious plan chosen for the book, which started with northern France, a country of villages where economic relations developed early and social life is very strong; it then dealt with the large area between the Alps and the Atlantic Ocean, where human settlements are smaller and social life less active, continued with western France where dispersion is total, all types of relations difficult and life is contained within very restricted milieux. The south, the *Midi*, constituted the fourth part, differing in many of its features, but where the climate is milder and types of behaviour often unique.

Vidal noted:

What strikes one first within this entire physiognomy is the magnitude of differences. In an area which covers only the eighteenth part of Europe, we see regions such as Flanders and Normandy on one side, Béarn, Roussillon and Provence on the other, regions the similarities of which are with Lower Germany or England, or with Asturias and Greece. No other country of the same size contains such diversities. For what reasons did these contrasts not appear as focuses of centrifugal action ...?

It is because in between these contrasting poles, the nature of France develops a range of transitions which does not exist ... elsewhere.[37]

What was the proof of this proposition? The mixture of north and south which is closer here than under other skies. Hence the frequency of good regions and the fertility of land. 'What a Frenchman sees in France, as is proved by his regrets when he leaves Her, is the quality of the land, and the pleasure to live on it. [France] is for him the quintessential land, that is to say something which is intimately linked with the instinctive ideal of what is life for him.'[38] Nature is preparing in this way the unity of France.

The ways of life which took advantage of the diversity of the country and made its wealth did not remain foreign one to the other: 'A surrounding atmosphere, inspiring ways of feeling, expressions, locutions, a peculiar type of sociability, had enveloped the diverse populations that fate had gathered on the land of France.'[39] And Vidal continued: 'There is thus a beneficent force, a *genius loci*, which prepared our national existence and gives it something healthy. It is something which flutters above regional differences. It balances them and combines them into a whole; nevertheless, these varieties subsist, and remain alive.[40]

The identity of France is thus linked, as Michelet said, to her diversity, a diversity expressed by the modes of spatial organization and the ways of life described with many details in *Le Tableau de la géographie de la France*. This diversity is tempered by the overlapping of contraries and the multiplicity of transitions. It owes much to the very pattern of environments which is less conducive to tensions than elsewhere. But it is, finally, the way in which her forms of social life combine which gives France her personality: 'This word personality', noted Vidal, '... corresponds to a soon advanced level of development of general relations.'

'This degree has been reached early by France. Earlier than other countries, ours developed from this imprecise and rudimentary state where the geographic aptitudes and resources of a country remained latent, where nothing that characterized a living personality could still be detected. It was one of the countries which took shape early.'[41]

The identity of France according to Vidal de la Blache and Michelet

Here there is an interpretation of the identity of France very close to Michelet's. For Vidal, as for Michelet, the nation expressed the triumph of 'the general life' or 'general relations'. But for Vidal, diversity was rooted in the environment and not in an adventurous psychology of peoples. It could be read in the varied strategies developed by groups to master the environments where they settled. The existence of transitions and of recurring features certainly favoured bringing extremes closer together. Was it necessarily conducive to a centralized construction? Vidal de la Blache showed the part played by necessity and what resulted from history:

> A weight thrown on the scales disturbed, in France, the balance of geographic forces. Some natural affinities were exaggerated. It is no more pure geography but history which can be seen in this concentrated, inner-orientated organism jealous to bring back to a focus the life scattered all over the country and to retain it. A more condensed individuality has replaced the one which was expressed in the previous network. The system has been nationalized ...[42]

> As a result, of the energies embedded in the homeland, part had been obliterated when others had been brought to the fore and sometimes their consequences overdone. Our history obeys a kind of logic which insists on some geographic aptitudes while it subordinates others and maintains them in the background.[43]

Michelet was speaking of a person: for him France was a being, the existence of which had taken forms prepared by geography. Vidal used instead the term personality: French space favours unity because it develops complementarities and solidarities between its components. These features give her a personality, but without forcing upon her a destiny written from the beginning. The part of choices and of political mistakes must not be neglected. Vidal de la Blache was the heir, through Michelet (and through the German geographers), of Herder, but he did not accept the teleological views which made up the strength and the weakness of the German philosopher.

The Vidalian interpretation of France seduced public opinion in the late nineteenth and early twentieth nineteenth centuries. To be French was experienced at two levels: one was a member of a local or regional community and, because of that, was integrated into the national whole. Diversity appeared as a necessary component of global identity and was thus accepted and valued.

The idea that France was in the making, that she had potential, with a future to be determined, remained. At the same time, Vidalian interpretation escaped criticism. The forms of social life which combined had nothing to do with the theory of humours or with the direct influence of climate. To the nationalists who spoke of a French race, *Tableau de la géographie de la France* reminded them that France was the result of the fusion of different peoples. To the anarchists who deprecated the idea of the nation, Vidal explained that the French nation was not artificial, even though its political and administrative organization could have taken other forms.

National identity, French geography and German geography

With this answer to questions on French identity, Vidal de la Blache imposed some specific orientations on French geography. Geography as a science of sight, as a field discipline? French geographers were not the only ones to make this discovery. At the time they received this idea from geologists, German geographers were also introduced to it, but from natural scientists with broader views on what constitutes a milieu than in France. This led them to a more synthetic vision of environment. The dialectics of scales? It existed everywhere: it was present in Ratzel's works even if he did not use it with a subtlety equal to Vidal's. Was there any point in which French geographical practice differed from German? Yes: French geographers concentrated more on social groups and ways of life, and German geographers on landscapes. For what reason? Because the problem of national identity was not identical in the two countries. The problem of boundaries was not central to the French case: they appeared as a fact; everyone accepted the existing ones. The problem was to understand what gave unity to the territorial being thus defined. The German situation was different. What pre-existed there was the ethnic fact. Vidal noted, for instance: 'Germany represents mainly, for German people, an ethnic idea.'[44] It was the state that had to be delimited. Geographers were asked to fix its boundaries. Hence the attention devoted to landscape. Is it not marked by the peoples who shaped it? Geography, in Germany, had no need to dwell on the way in which social groups develop the capability to exploit their environment through the creation of adapted ways of life. Its credit was higher when it appeared just as a natural science of landscapes: it was the best way for it to fulfil its national mission.

Problems of national identity were central to the differentiation of French and German geographies at the end of the nineteenth century. They shared many things: first, their sources, Herder in the background and, more directly, Ritter. They were dominated by the same nationalist ambiance. But the answer given in France to the problem of national identity was, since Michelet and in conformity with the commonly experienced feeling of membership on two levels, conceived as the historical outcome of the passage from a primitive identity to a broader feeling of integration. What was at stake was the nature of the nation, its content, and not its boundaries. In Germany, the problem was not to say what was the meaning of being German, but to determine where the country stopped: hence the prevalent natural sciences bias of the approach.

TOWARDS A NEW PERCEPTION OF THE IDENTITY OF FRANCE: BRAUDEL'S
QUESTIONING

The success of the Vidalian interpretation

The success of *Tableau de la géographie de la France* was considerable. Emmanuel
de Martonne described the reactions to the publication of the book thus:

> it arouses enthusiasm. P. Vidal de la Blache is made a member of the
> Institute of France (Academy of Moral and Political Sciences), is ap-
> pointed as chairman of the Commission of Geography in the Ministry
> of Public Instruction, becomes a member of the Consultative Committee
> of Higher Education and receives the sash of Commander of the *Légion
> d'honneur* order. His name is from that time considered in France and
> abroad as the head of the French school of Geography.[45]

The success was total: it came to the man, to the school that he had created
and to the ideas which he had sown. The way in which he was conceiving
of France fitted perfectly with the concerns of his time. It was neither racist
nor prone to revenge, but showed why Alsace and Lorraine were part of
France; Vidal de la Blache made his position clearer during the First World
War when he wrote *La France de l'Est*.[46] The image which he gave of the
country was that of a voluntary construction rooted in real complementarities: France was an adult nation, sure of its values and without expansionist
ambitions.

French transformations

The problem of French identity ceased to be set in the same terms after the
Versailles treaty. All territorial ambitions had disappeared, France had boundaries
which fitted it. The slaughter of the First World War diverted attention from
the nationalist cult of *patrie*. The conception of French identity developed by
Vidal de la Blache remained widely received but without any real involvement.
Geography suffered from this evolution, losing its status because it was deprived
of its former political significance.

Did the conception of France remain stable and the interpretation developed
after Michelet retain all its validity? No; but people had to wait some time
before becoming aware of the ongoing changes. During the years after the
Second World War, the ambiance did not favour this kind of questioning: all
energies were devoted to economic growth. Few people were really interested
in the real nature of France. Contests could have come only from regionalist
or independence movements: because they had compromised with Germans
and fascism at the time of Vichy, they had lost their credibility.

The atmosphere changed during the 1960s. The national idea was eroded
in various ways. The criticism of colonialism introduced doubts about the
civilizing mission of Western countries, and of France in particular. The building of Europe and the bringing closer of European cultures which it entailed
made people who had long been enemies conscious of what they shared. With

the focus now directed towards European genius, the genius of France began to appear less distinctive. The very idea of culture was evolving. Until the Second World War Frenchmen had stuck to the word civilization.[47] They began to adopt the term used by other European languages in the early 1950s. This was a very significant change: the sets of means and ideas which characterize people ceased to be ordered hierarchically. Each group creates a culture which is equal in dignity to any other and has the same right to be respected.

France was not conceived of as an ethnic unity. What it was carrying was a civilization – Civilization itself, in fact, for the majority of people in the eighteenth as well as in the nineteenth centuries. Renouncing the idea of civilization meant a lowering of the idea of France and a new lease of life for the long-forgotten regional cultures. Regionalism was stimulated by this change.

The country is nevertheless becoming more homogeneous than ever: the school system has worked, the French language is known and used at home even in the most remote rural areas. Newspapers, radio and TV complete the cultural standardization. The mobility of a population which has ceased to be a peasantry during the 30 years of high growth (which are called in France *les Trente Glorieuses*) is bringing people closer together. What is the significance of the petty differences of soil which have been perpetuated through the names of *pays* at a time when agriculture has mastered many of the constraints of yesterday? In an era of high speed, what is the role of the very tiny territorial units that geographers liked to describe? Cars travel too quickly to allow awareness of what makes the unity of a landscape and what distinguishes it from the previous or following ones. The shared perception of the French nation has ceased to be a two-level one. Nothing is more significant in this field than the less-frequent use of the term *Midi*. France today is losing her *Midi* and gaining a south, just like any other country, and an Occitanie: the national dream of a small minority. The idea of a two-tier identity, which made one *Méridional* and French, *Méridional* then French, French because *Méridional* – or *Normand*, *Savoyard* or *Lorrain* – is disappearing.

Braudel's reflection

What is France? This question is again pertinent at the end of this century. Braudel concentrated on it during the last years of his life. His death interrupted the impressive work that he had undertaken, *L'Identité de la France*,[48] but some conclusions were already evident. The first is the calling into question of France as a land of diversity: Braudel started from this idea, just like all his predecessors, and then became critical of it, as he explained in the second volume of the second part of his work:

> But why this hetereogeneity, this perpetual, obsessing skipping, which appears to Frenchmen as a major characteristic of their country, but which they are wrong to consider without equivalent through the World? In 1982, at the University of Göttingen, I was speaking without remorse of this unequalled diversity, which proves that, without showing any critical attitude, people have a natural tendency to repeat shared ideas and to

conform to the pretences, even innocent, of their fellow countrymen. When the discussion started, the floor began to exclaim with insistence and even amusement that Germany was too diversity, and to offer proofs of it. I knew it for all that, and also that Italy, Spain, Poland, England were diversity.[49]

The interpretation commonly shared since Michelet was in this way called into question. It did not mean that variety did not exist – Braudel devoted to it much discussion. But to consider it, he felt, as the cornerstone of French identity does not give a fair perspective on the most specific features of our country. As a result, he tried to cover all aspects of French originality.

The influence of milieu, of geography?[50] It was real, but became more significant when expressed in the homogeneous settlement and spatial organization pattern, built on villages, towns and a very early regular urban network than by the singularities of topography and the ease of relations. This last factor was indeed significant: the French isthmus does exist. The wealth and the density of settlement in the Parisian basin gave many opportunities to Paris. But it was mainly through the constant threat to the continental and maritime boundaries of the country, and in the patient task of fortifying its continental borders and of opening and at the same time protecting its coasts that the history of France was geared to environment.

If geography was not finally a major factor in the identity of France, was it not in history that an explanation must be sought? The next two volumes of *L'Identité de la France*[51] were devoted to this question: Braudel showed the slow evolution of settlement patterns, the early high densities, the crises, but also the slowness of nineteenth century and early twentieth century growth and industrialization. What history mainly shows is the central role of a society and an economy which remained predominantly peasant until the end of the Second World War. Such was the basic infrastructure[52] of French society. It clearly had 'superstructures' too,[53] but cities, transport and manufactures had little success in transforming this weighty and solid base – the change only began to be perceptible during the last third of the nineteenth century, but people still had to wait until the last 40 years to observe substantial changes.

Did Braudel succeed in epitomizing in this way the identity of France? He managed to isolate some of its features and permanencies. But he knew better than anyone else that the identity of France could not come down just to the regularity of its settlement pattern, to its peasant basis and to the mediocrity of the capitalism which she had produced. To take it further, Braudel had planned two other volumes: one devoted to *Etat, culture, société* would have explored the results in France of politics, anthropology and sociology. The other would have dealt with *France outside France*, and would have 'overpass[ed] the testimonies of the international relations and [would] have served as a conclusion to the whole enterprise'.[54]

We shall never know exactly what ideas Braudel was developing about the identity of France. What would have been the results of his work if it had not been interrupted by death? No doubt, he would have delineated other features to be added to those that he had already noted: the early strength of the state, the taste for a divided society perhaps, and a culture in which anxiety is expressed more in the search for rigour and in the quest for order than in the fascination with death. Was it sufficient to define the identity of a geographical being? We do not think so.

The relation between Braudel and geography was always ambiguous. Our discipline had given him many insights, as he willingly acknowledged, but for him its fecundity was over. It had showed him, it is true, how to capture the originality of the Mediterranean world at the time of Philip II, and had made him conscious of the significance of long-term trends, to which he had devoted so much of his time. But he refused to keep abreast of the development of our discipline. That is why, in his search for the identity of France, he had to review the results of all the social sciences. What he missed was an integrative point of view. Cultural geography would have provided it for him. It shows, in fact, how, from the numerous combinations of which a culture is made, some are articulated into systems and are surprisingly long-lived. To consider geographical entities as 'persons' and to define their 'personalities' is not futile. The instruments to do it correctly are currently being brought up-to-date.

A culture is made of tools, techniques, know-how and ideas. What distinguishes the study of fragmented societies of the past is the strength of the feeling of rootedness which characterized them. People were grateful to the land on which they lived, even if it was poor, since it offered them what they needed to live and also simple pleasures, the pleasures of eating and drinking. The cultures of traditional France were, perhaps more than others, imbued with a familiar reverence towards the environment. Their ideal was expressed in a down-to-earth way, in daily life: it was made of an aspiration towards tranquillity which did not exclude moral rigour.

All local cultures share many features in their fundamental attitudes even if their know-how differs widely, as we have seen. But they owe to their ingrained ideas their aptitude to resist the possibilities of openness and mobility, and the appeal of large spaces. In order to overcome this tendency towards withdrawal, better transport and communication facilities are not sufficient. Strong motives are needed in order to create the permanent field of tension which is needed to overcome this resistance: it was this field that French civilization provided over local cultures. It did not contradict them; it accepted the feeling that nature is a benevolent force in daily life, but proposed alternatives to give meaning to over-confined existences through the pursuit of some shared aims: the Christian faith, or progress, charity, liberty and equality.

Was Michelet wrong when he examined the identity of France? We do not think so. France would not exist without that permanent tension between a divided present dominated by egoism and localism, and an open and generous future. There is, in the idea of France, a share of Utopia which perhaps makes destiny here more tasty than elsewhere – but generally makes Frenchmen insufferable to other people.

Is this fragile balance of tension destined to last? Is not the erosion of local societies which served as a counterpoint to idealist aspirations threatening the equilibrium upon which French identity rested? Are not collective destinies evolving increasingly towards supranational scales? Such are the real problems which confront Frenchmen today.

NOTES

1 Xavier de Planhol, *Géographie historique de la France*, Paris: Flammarion, 1988.
2 George Huppert, *L'Idée de l'histoire parfaite*, Paris: Flammarion, 1973; American original edition, Urbana, Ill.: University of Illinois Press, 1970.
3 'At the origin of germanic traditionalism, there was an unpublished work of Abbé Le Laboureur, commissioned on the 13 March 1664 by the peers of France in order to discover in history proofs of the rights and privileges attached to their blood' (Raymond Aron, *Les Étapes de la sociologie*, Paris: Gallimard, 1967, p. 73).
4 O. Carbonell, 'Midi, histoire d'un mot'. *Midi: Revue de Sciences Humaines et de Littérature de la France du Sud*, 2 (1987) 11–15.
5 Planhol, *Géographie historique de la France*.
6 Paul Claval, 'Le thème régional dans la littérature française', *L'Espace Géographique*, 16(1) (1987), 60–73.
7 Jules Michelet, *Tableau de la France*. In *Histoire de France*, vol. 2, Paris, 1833. Quotations made in Hermès-Pierre Waleffe edition, Paris, 1966, conform to the revised text of the 1861 edition.
8 Ibid., p. 11.
9 Ibid.
10 Ibid.
11 Ibid., p. 12.
12 Ibid.
13 Ibid., p. 138.
14 Ibid., p. 140.
15 Ibid., p. 141.
16 Ibid., pp. 148–9.
17 Ibid., p. 149.
18 Ibid., p. 154.
19 Ibid., p. 141.
20 Ibid., p. 75.
21 Ibid., p. 78.
22 Ibid.
23 Ibid.
24 On the intellectual context at the end of the nineteenth century and its influence on French geography, see Vincent Berdoulay, *La Formation de l'école française de géographie (1870–1914)*, Paris: Bibliothèque Nationale, 1981.
25 Paul Vidal de la Blache, *Etats et nations d'Europe. Autour de la France*, Paris: Delagrave, 1886.
26 O.P. Dufresnoy, L. Elie de Beaumont, *L'Explication de la carte géologique de France*, vol. 1, Paris, 1841.
27 Paul Vidal de la Blache, 'Des divisions fondamentales du sol français', *Bulletin littéraire*, 2 (1888–9), 1–7, 49–57.
28 Paul Claval, 'Les géographes français et le monde méditerranéen', *Annales de géographie*, 97(542) (1988), 385–403.
29 Paul Vidal de la Blache, *La Péninsule européenne, l'Océan et la Méditerranée*, Nancy: Berger-Levrault, 1873.

30 Paul Vidal de la Blache, 'Des rapports entre la population et le climat sur les bords européens de la Méditerranée'. *Revue de géographie*, 10 (1886), 402–19.

31 Paul Vidal de la Blache, *Tableau de la géographie de la France*, Paris: Hachette, 1903; Paris: Tallandier, 1979.

32 Ibid., p. 7.

33 Strabo, quoted in Ibid., p. 11.

34 Ibid.

35 Ibid., p. 8.

36 Vidal de la Blache, 'Des divisions fondamentales du sol français'.

37 Vidal de la Blache, *Tableau de la géographie de la France*, p. 49.

38 Ibid., p. 50.

39 Ibid., p. 51.

40 Ibid., pp. 51–2.

41 Ibid., p. 8.

42 Ibid., p. 381.

43 Ibid., p. 382.

44 Ibid., p. 50.

45 Emmanuel de Martonne, 'Vidal de la Blache', *Association amicale de secours des anciens élèves de l'Ecole Normale Supérieure* (1919) pp. 28–33, esp. p. 33.

46 Paul Vidal de la Blache, *La France de l'Est*, Paris: A. Colin, 1917.

47 Lucien Febvre, *Civilisation. Le Mot, la chose*, Paris: La Renaissance du divre, 1930.

48 Fernand Braudel, *L'Identité de la France*, 2 vols. Paris: Arthaud-Flammarion, 1986, vol. 1: *Espace et histoire*; vol. 2: *Les Hommes et les choses*.

49 Braudel, *Les Hommes et les choses*, p. 423.

50 Braudel, *Espace et histoire*.

51 Braudel, *Les Hommes et les choses*.

52 Ibid., pp. 9–182.

53 Ibid., pp. 183–421.

54 Braudel, *Espace et histoire*, p. 19.

3

National Identity in Vidal's *Tableau de la géographie de la France*: From Political Geography to Human Geography

Marie-Claire Robic

The Ecole française de géographie founded 'human geography' at the beginning of the twentieth century through a rejection of 'political geography', both an obsolete genre and a new discipline created by F. Ratzel. This innovation was based on biological references taken from ecology, a new science at the time. *Tableau de la géographie de la France* by Vidal de la Blache (1903) is a good example of the contradictions of this school of thought in its early period. Indeed, the *Tableau*, as part and parcel of a collective effort to legitimize French national identity in the early period of the Third Republic, tends to negate the internal and external differences against which this entity is constructed, in an effort to establish French identity on geographic principles. The insistence on structural spatial contrasts, the refusal to posit clear-cut boundaries, the defence of diversity and fusion bear the mark of the universalistic trend distinguishing French national ideology from the German one.

Many recent historiographic studies give new insights into the reality of the French nation and its representations at the beginning of the twentieth century, and insist on the originality of Vidal's approach to the notion of nation. The general process of elaboration of geography, of which the *Tableau* is a by-product, is described in this chapter. Vidal's work on human geography is studied in its relationship to Ratzel's political geography, to historical geography and French romantic and critical historiography. Paradoxically, the naturalistic justification for human geography was put to use in the definition of a national identity.

INTRODUCTION[1]

Published in 1903 as an introduction to Lavisse's *Histoire de France, des origines à la Révolution*, the *Tableau de la géographie de la France* has always been the object of contradictory appraisals. For Lacoste (1979), it can be contrasted with *La France de l'Est*, published at the end of the First World War, as the very prototype of apolitical geography and as a good example of the evolution of academic geography at the end of the nineteenth century. On the contrary, Sanguin (1988) insisted on the permanency of Vidal de la Blache's interest for political geography and the continuity of his work in this field. In his preface to a recent edition of the book, Claval (1979) stressed the relevance and originality of his ideas concerning human geography. Lastly, several historians demonstrated that Vidal de la Blache played a part in the construction of a new national identity, but that he did not adhere to the idea of a collective political memory reduced to the chronology of dynasties, great men and important events (Nora, 1984–6).

The pattern of ideas underlying the *Tableau* is ambiguous, with its hesitations between 'political' and 'human' geography (*géographie humaine*).[2] From the start, Vidal adopts the point of view of human geography when, after Michelet, in order to clarify the notion of a 'geographic personality of France', he writes that the word personality belongs to the field and vocabulary of human geography. But he goes on to examine a question bordering on politics since, to quote his own words, the problem is to determine 'how a fragment of the surface of the earth that is neither a peninsula nor an island and that should not properly be considered as a unit by geography became a political entity and then a country (*patrie*)' (p. 6).[3] Along the same lines, Vidal apparently assimilates human geography and political geography (pp. 16–17). However, in the very year the *Tableau* was published, Vidal wrote an article for a historical journal on the nature of human geography, a decisive innovation in the field of geography, according to him. I shall try to substantiate the claim that it is possible to interpret the *Tableau* in several ways. This fundamental ambiguity corresponds to the replacement of the traditional political geography that Vidal rejected by a new human geography that was to become an independent branch of geography – a development that parallels the invention of ecology by naturalists a few years earlier. This replacement was still in progress in French academic geography at the turn of the century. My investigation will be conducted in the light of three kinds of evidence: new insights derived from historical work on the objective state of French national unity and on the propagation of nationalistic ideas in the early years of the Third Republic; a new reading of the *Tableau* taking into account the draft versions included in Vidal's Notebooks;[4] a confrontation of the original aspects of the *Tableau* with the founding principles of human geography as stated in 1902–3 at the very moment the *Tableau* was published.

THE *TABLEAU DE LA GÉOGRAPHIE DE LA FRANCE*: A DESCRIPTION OF A
POLITICAL ENTITY

A contribution using the concepts of political geography

It is evident that the *Tableau* corresponds to research in political geography.
But it differs from previous work on the subject in political, administrative
or historical geography.[5] The importance of political issues appears in various
ways, but nevertheless it has not always been clearly seen. For instance, Claval
(1979) has shown that Vidal distinguishes two main divisions in France on the
basis of the organization of local life: the north with its villages and country
towns, its sociable population, as opposed to the west, an agricultural region
of isolated farms. But, beyond this basic contrast, Vidal proposes a typology
of organizing units that is more politically based than derived from social
considerations. It ranges from the '*pays*', with its autonomous collective life,
to the region on a global scale (Europe for instance) and includes the political
provinces, the major cities and the states. The Saône-Rhône corridor is a
typical case with the Jura '*val*' at one end of the scale, Burgundy and the city
of Lyon as intermediate units, and the Lyon region as a potential unit at the
other end (Loi et al., 1988). Apart from this typology, the *Tableau* gives a
number of laws, or at least principles, governing political geography. They
mainly concern the size of the units, their situation and – an original idea
not included in Ratzel's work – what might be called their structural organization: 'Harmonious proportions are required for political structures to grow'
(p. 233); 'This corridor is not vast enough to support an independent state'
(p. 233); 'A degree of subordination of the parts to the whole is necessary
to the formation of a state' (p. 85). The analysis of every region could be
seen as a confirmation of the existence of these laws. The description of France
as a whole is itself informed by these considerations, and in particular by the
principle that 'power concentrates in places where antagonism induces effort'
(p. 60).

In a less explicit way, the general plan of the *Tableau* seems to be dependent on principles derived from political geography. The order of treatment
of the various regions in the descriptive part of the book and their importance[6]
seem to be a direct function of their role in the making of France as a territorial unit. Guiomar (1986), too, notes that the privileged treatment of the
northern part of France over the western and southern parts has its source in
the prominent role played by this region in the historical process of unification. This choice is explicitly justified in the text and in the drafts included
in Vidal's notes. Constant reference is made to the anteriority and historical
dynamism of the north as opposed to the southern and western margins:
'French unity is a consequence of internal development and resulted in the
fusion of its various regions. Our history has reached the shores of the Atlantic
ocean and of the Mediterranean sea only recently.' Finally, this hypothesis is
corroborated by various tentative plans and titles proposed around 1900 with
reference to notions derived from political geography: 'The French soil (*sol*):
a study in political geography' becomes 'The French land (*pays*): a study
in political geography', a paragraph including explicit development of the idea

(Notebook 20); projects that are not very different from the final version make reference to the expansion of 'germanism' (*germanisme*) as a driving force in the formation of European states (Notebook 16).

The identity of France: a paradoxical unit

Indeed, this political orientation is toned down in the titles of the final version. It went unnoticed at the time and is difficult to perceive for contemporary analysts because what is in question is a nation-state that no longer seems problematic. It will be shown that the contribution of university teaching to this state of affairs is not negligible. However, two original aspects of Vidal's representation are worth mentioning. One is that it tends to minimize French idiosyncracies and at the same time to exalt them, an apparent contradiction. The other is that it proposes an original image of the nation, an image that does not conform to the stereotypes that were current at the time.

Contrary to many geographic presentations of France, mainly those found in school textbooks, France, in Vidal's eyes, is an open territory. No reference is made to a specific shape except in the famous expression of the 'French isthmus' (*isthme français*) derived from France's fundamental role of a bridge between two seas and of a meeting point between two continental regions. But at a time when France is variously described as an octagon, an hexagon or a pentagon, especially in textbook imagery, Vidal does not resort to geometry to symbolize its harmony (Smith, 1969; Weber, 1986; Robic, 1989).

In accordance with his insistence on France's open character, Vidal downplays the role of borders and boundaries. They are rarely described in a precise fashion and are no more than fuzzy lines in the cultural or geomorphological regions through which they are supposed to run. The treatment of borders is even worse: they are often ignored or criticized because they cut through linguistic or social units or because they hinder exchanges. Ironically enough, Vidal's most radical criticism of the arbitrary nature of borders concerns their disruptive influence on pastoral communities in mountain regions. In the case of Alsace-Lorraine – the lost province! – this treatment of borders is used to integrate this region in the *Tableau* without any further discussion, as well as to include it in a wider eastern region, a crossroads region that extends well into Germany.

Far from insisting on the difference between France and other nations, Vidal shows that its peculiarity lies in its microcosmic quality, in the fact that it is a miniature of the world. France's originality lies not so much in its diversity, its nuances[7] as in its fusion of diverse entities. This idea appears repeatedly in the Notebooks, together with the idea of early development (*précocité*). One can even go one step further and maintain that the unity of France is based on spatial contrasts, that its structure is 'fractal' in nature, in that, at all levels, it is based on pairs of opposite notions: sea versus ocean, climatic opposition between north and south, geological dualism, contact between mediterranean and continental populations, and so on. This pattern of territorial contrasts, favouring the appearance of exchanges at the general as well as at the local level, could perhaps account for the specificity of France (Robic, 1992a).

Unlike the positivist historians who contributed to the same series, the image of France's national heritage proposed by Vidal is not based on famous figures or politics but on a mixture of concrete images. They include landscapes as symbols of the permanency and reality of the country, material aspects of life epitomized in specific modes of existence (pp. 49–50) mainly in relation to food: France is primarily a country living on white bread, the country of good food; 'see Lenain's paintings representing peasants', he adds (p. 51). The French way of life, the general organization of society, too, are good clues to the identity of France, an identity which is best analysed in the long run and in the population as a whole. He pays little attention to the 'powers that be' and prefers to base his analyses on country life because it corresponds to a structured social and economic set-up, a 'closely knit world' with its peasants, its rural bourgeoisie (*bourgeois agriculteur*), its nobility living on their estates (p. 384): a rural world rooted in crops, with well-established legal traditions and relationships firmly grounded in everyday life. The peasant is the epitome, the best witness, a good sample of all this. But it will also be seen that he is a vestige, a relic of the past and therefore more reliable than texts for studying history.

Indeed, national identity exists in time and is subject to change. Vidal stresses the unequal development of the various potential resources of France. He insists on the close dependence of spatial organization on systems of government. For instance, he compares the systems of communication they generate, a grid pattern for the Roman Empire opposed to a centralized road system corresponding to the French monarchy. New potentialities are opened up by scientific discoveries and variations in the situation of France in the world. As a consequence of the great importance Vidal gives to communication, the basic turning point in the history of France is not the Great Revolution of 1789, but the revolution in transportation during the nineteenth century. It introduces generalized interdependence so that, for a geographer, to plan the future within a framework of worldwide competition is a national priority. The modern world will be dominated by economic problems.

BETWEEN A SCIENTIFIC CHALLENGE AND A NATIONAL CHALLENGE

The contents of the *Tableau* are more political than is usually thought. If it cannot be reduced to a treatise of regional and descriptive geography, it is much more than the book on general human geography that Claval describes in his preface to the new edition. The latter analyses the Ritterian concern for human geography seen as 'a study of social groups in their day to day existence, in their adaptation to their environment and their aspirations' (p. xv). He highlights Vidal's original choices and reflections in general geography, his distinction of 'two levels of social organization: the level of local relationships and the level of general trends' (p. xv). According to him, Vidal concentrated on the local level for purely editorial reasons: the pre-eminence of this level is a characteristic of pre-revolutionary France and the subject of Lavisse's series. But his method, derived from his readings of Ritter and applied as early as 1889 in *Etats et nations de l'Europe* was, in all essentials, based on a 'taste for the analysis of the relative position of sites, topographic configurations and their role in relational life' (p. xv).

A critique of Vidal's predecessors and contemporary authors in political geography

Obviously Vidal took his inspiration from Ritter, who is abundantly referred to in his work of the period and especially in his reviews of Ratzel's work. However, Vidal's critique is more fundamental in nature. His work leads to a rejection of classical historical geography, even in its most advanced form, as exemplified by the work of his predecessor in the Sorbonne (Berdoulay, 1981). In the very year that Vidal wrote his notes for the *Tableau*, he succeeded Himly, the chair of historical geography becoming a chair of general geography. The tribute included in his inaugural address was very critical: 'M. Himly's study of Central Europe is geographically based but deals with history from an historical point of view' (Vidal de la Blache, 1899, p. 99). However, in his work, Vidal kept Himly's overall view of the part played by 'germanism' in the formation of European states. The *Tableau* is a reassessment of the relationship between geography and history. According to him, geography is neither a background element nor one of the founding factors which, according to Michelet, are 'erased' by history once unification around a capital city has been accomplished. Geography has a permanent influence on history.

This work on France also stands as a test of Vidal's reflections on Ratzel's work on political geography and on the political organization of space. If the sheer number of reviews and general papers written between 1896 and 1898 is any indication, it seems that he is more interested in *Politische Geographie* than in *Anthropogeographie*. His notes clearly demonstrate that the case of France is a constant reference, either as an illustration of the laws posited by Ratzel or as a counter-example. For instance, in his analysis of an article by Ratzel concerning the growth of states:

> 3rd law: the growth of states has its origin in the agglomeration of smaller parts, the result being fusion yielding a closer connexion between the people and their land: examples derived mainly from the formation of the Roman Empire. The example of France would have been more topical. I don't know of a better example. (Vidal's drafts)

In the manuscript we also find: 'the best example, not referred to in the text, would be the formation of France. But did the United States, and more generally colonial states, grow along the same lines?' (Vidal's drafts). This dialogue with Ratzel is also found in the correspondence established between Ratzel's ideas rejected by Vidal and his own leitmotiv about France. The notion of virtuality concerning a country, closely connected to what L. Febvre called 'possibilism', is a good case in point. The famous passages of the *Tableau* on the 'capability', the potential energies of countries – 'a country is a pool where nature has deposited the seeds of energies that remain dormant until man puts them to use' (p. 8) – are echoed by notes on the relationships between state and space analysed by Ratzel. For instance, the following remark, 'the organic character of states is such that growth in one domain results in growth for the whole territory and its inhabitants and in increased possibilities', is taken up in the final version of the review: 'States are integrated in their territory: on the other hand, the corresponding territory, under the influence of the state it personifies, acquires new geographic properties that were hitherto latent' (Bibliographie de 1896, no. 113).

An adaptation of a monographic study to a general reflection on the nature of human geography

The *Tableau* is also part of a study of the relationship between general geography and regional geography. This is clearly shown in the draft versions found in the Notebooks. To be more precise, the *Tableau* is an element in the development of Vidal's criticism of Ratzel's political geography. He deplores its excessive dogmatism and its lack of clear methodological statements. But first and foremost, he finds it wanting in a coherent definition of systematic relationships between the various branches of geography. When it is examined in the light of the draft versions of the articles on Ratzel, Vidal's work published between 1896 and 1903 can be viewed as a constant effort to define the place of human geography within the domain of general geography. 'Human geography', a term which appeared for the first time in Vidal's work in 1898, was gradually to replace political geography in the so-called French school of geography (Robic, 1991 and 1993). The construction of a coherent general frame goes with the replacement of the opposition between political and physical geography by an opposition between human and natural geography. The analysis of this transformation is beyond the scope of the present chapter but let us mention in passing that this duality is transcended by the higher unity of the 'geography of life'. Human geography takes its models from the new scientific ecology, which has its origins in the work of Warming and Schimper at the turn of the century. Their work was rapidly available to the geographers of the *Annales de géographie* group through the French specialists in botanic geography such as Flahaut and other neo-Lamarckian scientists (Robic, 1990 and 1992b).

The *Tableau* was written between the 1890s and 1903, a period during which Vidal went from a rather traditional conception of the human aspect of geography to a new conception viewing the relationship between men and their environment as part of a kind of biogeography. Apparently, the change took place between 1898 and 1902–3. When he started writing the *Tableau*, the categories Vidal used conformed to the tradition that was prevalent in France at the time, including physical, political, historical and economic geography. These categories are used, for instance, in his *Atlas général* (1894) and in *La terre* (1883). Political geography is then defined as 'the study of cities, states and other human organizations in their relationship with the land'. In 1903, in an article devoted to 'Human geography and its relationship to the geography of life' (1903b), Vidal launches into an apology for human geography. It is described as a totally new point of view, whose adoption is justified by its inclusion in life-sciences: it shares their principles and their methods of comparative investigation. From that moment on, the study of ethnographic material already mentioned in 1899 becomes of considerable heuristic value. In a monographic work, it allows the reconstruction of the 'characteristics of a country', the 'signature of a people' (Vidal de la Blache, 1902) and, in general texts, it is the 'reflection of the geographic life of the earth in the social life of men'. Human geography is from then on a branch of general geography, informed by the ecological paradigm. In 1902 political geography still had an independent status but it was just mentioned in passing in the article published in 1903, in which Vidal asserts that the influence of political problems is diminishing and is going to be replaced by economic factors. This article, corresponding to the final stage in this process, draws the ultimate consequences of the inclusion of human geography as a branch of biogeography. It proposes definitions for the main concepts of the theory, namely ecology, environment,[8] adaptation, and insists on the importance of ethnographic museums and observation of the *'genres de vie'*[9] in order to characterize the relationships between civilizations and their natural surroundings.

The *Tableau*, published in 1903, is certainly a good example of the new status of human geography as a part of biogeography. This conclusion is further justified in Vidal's explicit comparisons between the migrations of human beings and the spread of natural species: 'Naturalists analyse the growth of vegetal and animal life in islands and on continents in a different way. These notions can be applied to the study of the facts of human geography' (pp. 25–6). The analogy with biogeography also accounts for the fact that Vidal relies heavily on ethnographic material, exploiting it in an original manner It has to be noted that he pays more attention to living conditions and to housing than to dress – always a favourite with folklore specialists – because they fit better his naturalist conceptions of the relationship between man, as a biological species, and his environment.

French nationalism after the defeat of 1870

Political experts have shown in a precise way that the growth of French universities as research and teaching institutions is central to the conception of democracy shared by *'Républicains opportunistes'* (Nicolet, 1982; Weisz, 1979). The authors of *Les Lieux de mémoire* insist on the part played by republican historiography in the spreading of a new version of French nationalism (Nora, 1984–6). Vidal's *Tableau* is included in a historical series written exclusively by university professors, under the leadership of E. Lavisse. The latter saw it as the glorification of the continuity of a nation that had come of age and had been realized in the adoption of republican institutions. The defeat at the hands of Prussia changed the definition of France as a nation (Nora, 1986). So far, it had been based on an internal opposition between the *Ancien Régime* and modern France. But after 1870, one of the consequences of the defeat was to give a primary role to the opposition between the French nation, as a free association based on free will, and the German nation. The identity of France lies in a 'moral conscience' generated by the 'deep complexity of its history', to quote E. Renan's famous Sorbonne conference, 'Qu'est-ce qu'une Nation?' (1882).

Geography and nationalism

'The whole past is redeemed' (Nora, 1986). The *Tableau de la géographie de la France*, significantly the first volume in a series, is a perfect image of a long, and even very long, memory. But while positivist historians – with their reliance on a critical assessment of written documents – based their research exclusively on state archives, thus reducing the history of France to the history of political power,[10] Vidal, on the contrary, relied on field-work, like all naturalists. He hunted for ethnographic documents, used new methods based on observation, searching at the same time for the true face of France and its concrete features. He thus invented new methods of historical investigation – later adopted by the members of the *Ecole des annales* (Guiomar, 1986; Pomian, 1986; Roncayolo, 1986). To the conception of national history as the history of the state, the *Tableau* opposes another conception based on the peasant as a living memory and as a symbol of national identity. But the period corresponds to the end of regional differences[11] and the final establishment of national unity, a process that was completed only during the Third Republic through schooling, universal suffrage and military service, according to the somewhat unorthodox opinion of some Anglo-Saxon historians who investigated local ethnological realities (Ozouf, 1984; Zeldin, 1973–7; Weber, 1976).

To sum up, Vidal sees the legitimacy of the French nation not in ethnic principles nor in constraint but in the spontaneous fusion generated by the genius loci. This providential land is closely analysed in the *Tableau*. In Vidal's mind the *genius loci* of the French territory is due to its spatial pattern and its history can be viewed as a process leading from a natural spatial order to a human spatial organization (Robic, 1992a). It is why the dual nature of France outlined by Vidal, with its spatial distribution of contrasts that is typical of the French territory, is so important. This 'fractal' structure is a metaphor of the notion of universality that is a characteristic of the French brand of nationalism. Vidal's construction can be compared to Dumont's analysis of nationalism in his *Essais sur l'individualisme* (1983). He mentions the problems posed by the combination of two different conceptions of the nation 'as a collection of individuals, on the one hand, and as a collective individual confronted to other nations as individuals'. In this light, the microcosmic nature and the extraordinarily open quality of its map as described in the *Tableau* represent the genius of the French variety of the idea of nation. As opposed to German national ideology, it is characterized by the fact that the nation as a collective individual is not highly valued, so that, as a result, the differences between France and other nations are only marginally taken into account and the antagonism between nations minimized. Primarily, the nation, in its French variant, provides a frame for the emancipation of the individual. It is much less a factor of integration, a reference to common origins than the German nation. According to Dumont, the German variant of the idea of nation is

'I am a man because I am German', whereas 'for the French, I am a man by nature and French by accident'. One may wonder whether this is not the deep meaning of the naturalism pervading Vidal's geography of France. From a scientific point of view, the preference given to human geography over political geography, which is after all the main characteristic of the *Tableau* in its final version and the path taken by the French school of geography, would correspond to a choice that is typically French in nature.

CONCLUSIONS

To analyse the *Tableau* as 'a-political' would therefore be naive, except if one accepted a reification of the idea of nation. Gaudin (1979) has criticized Lacoste's analysis of Vidal's work as idealistic, demonstrating that the ideology of numerous geographers, characterized by conservative tendencies and the importance given to country people, was consistent with the compromise agreement between the industrial middle class and peasants against the working class after the Commune de Paris, a compromise represented by Méline and his followers. In the same way, Englund (1988) scoffed at the idealization of the concept of nation found in the books of Nora et al., and proposed a critical approach to this notion as a complex system of ideological discourse. The *Tableau* bears the mark of the nationalism and the conservatism that are the characteristics of the moderate republicanism of the 1900s. This republicanism is the ideology of the university teachers who were the pioneers of the school system of the Third Republic, and Vidal was one of them. His work is apparently a-political because, for them, politics was reduced to a desire to overcome the internal contradictions of the nation. The apology for national identity is in a sense a defence of the established order.

The *Tableau* also bears the traces of interplay between two distinct struggles for identity, one that was scientific in nature and another that was political. It coincides with a turning point in the process of invention of human geography. A certain degree of similarity between the French conception of national identity and Vidal's project concerning human geography favoured an osmotic *rapprochement* between two projects that were of a different nature: the scientific and the ideological projects. These similarities were the source of the new intuitions mentioned by the historiographers of nationalism: a wider scope for history, a richer national memory by the inclusion of material culture and folklore and the relative indeterminacy of history. On the other hand, this position entails a subordination of social and political questions to natural processes, thus giving precedence to spontaneous control over creative power. As far as Vidal's subsequent evolution is concerned, his growing preference for economic over political problems might well be a consequence of the diminished role of politics in the definition of national identity. As shown by Vidal's *La France de l'Est*, the trauma of the First World War was nevertheless to rekindle his interest in the political dimension of nationalism.

NOTES

1 English version translated by Alain Nicaise.
2 The translations of Vidal's quotations are original. Some of Vidal's terms have been kept in their primitive French form (in parentheses and/or italicized).
3 The reference comes from the new edition of the *Tableau* (1979), which is a photographic reproduction of the first edition with a preface of P. Claval.
4 See Joseph and Robic, 1987; and Loi et al., 1988.
5 These branches of geography were: a nomenclature of place-names, an inventory of administrative units (after the fashion of traditional regional statistics), and a description of the evolution of territories.
6 In terms of number of pages and of relative importance compared to their surface.
7 An idea stressed by many commentators, such as Braudel (cf. P. Claval's chapter in this book, Revel, 1989; and Kearns, 1991).
8 '*Milieu*' or even '*environnement*', a new acceptance of the English word in Vidal's text.
9 In this context, the term 'form of life' would certainly be a better approximation of Vidal's *genre de vie* than lifestyle or way of living.
10 A point of view that is after all no different from that adopted by the memorialists of the *Ancien Régime* (Nora, 1986).
11 The '*fin des terroirs*' to quote the French translation of Weber's title, 'Peasants into Frenchmen ...'.

REFERENCES

Berdoulay, V. 1981: *La Formation de l'école française de géographie (1870–1914)*. Paris: Bibliothèque Nationale.
Broc, N. 1977: 'La géographie française face à la science allemande (1870–1914)'. *Annales de géographie*, pp. 71–94.
Buttimer, A. 1971: *Society and Milieu in the French Geographical Tradition*. Chicago: Rand McNally (AAAG Monograph no. 6).
Canu, J. 1931: 'Les Tableaux de la France. Premiers essais: Michelet, Reclus, Vidal de la Blache, Jean Brunhes'. *Publications of the Modern Language Association*, 46, 554–604.
Chamboredon, J.-C. 1984: 'Emile Durkheim: le social, objet de science. Du moral au politique?' *Critique*, 445–6, 460–531.
Claval, P. 1979: 'Préface'. In P. Vidal de la Blache *Tableau de la géographie de la France*, Paris: Tallandier, pp. i–xxii.
—— 1992: 'Le rôle de Demangeon, de Brunhes et de Gallois dans la formation de l'Ecole française, 1905–1910'. In P. Claval (ed.), *Autour de Vidal de la Blache. La formation de l'école française de géographie*, Paris: CNRS Editions, Mémoires et documents de géographie, pp. 149–58.
—— 1993: 'From Michelet to Braudel: personality, identity and organization of France'. Chapter 2 in this volume.
Digeon, C. 1959: *La Crise allemande de la pensée française (1870–1914)*. Paris: PUF.
Dosse, F. 1987: *L'Histoire en miettes. Des Annales à la 'nouvelle histoire'*. Paris: La Découverte.
Dumont, L. 1983: *Essais sur l'individualisme*. Paris: Seuil.
Englund, S. 1988: 'L'Identité de la société française, De l'usage du mot "nation" par les historiens et réciproquement'. *Le Monde diplomatique*, March, 28–9.

Gaudin, J.-P. 1979: *L'Aménagement de la société. Politiques, savoirs, représentations sociales. La production de l'espace aux XIXè siècle et XXè siècle.* Paris: Anthropos.

Girardet, R. 1966: *Le Nationalisme français. Anthologie (1871–1914).* Paris: A. Colin.

Guiomar, J.-Y. 1986: 'Le *Tableau de la géographie de la France* de Vidal de la Blache'. In P. Nora (ed.), *Les Lieux de mémoire.* II: *La Nation,* 1. Paris: Gallimard, pp. 568–97.

Himly, A. 1876: *Histoire de la formation territoriale des états de l'Europe centrale,* 2 vols. Paris.

Joseph, B. and Robic, M.-C. 1987: 'Exploration d'archives: autour des papiers d'E. de Martonne'. *Acta Geographica,* 4th term, 37–65.

Kearns, G. 1991: 'Historical geography'. *Progress in Human Geography,* 15(1), 47–56.

Lacoste, Y. 1979: 'A bas Vidal ... Viva Vidal!'. *Hérodote,* 16, 68–81.

Langlois, C.V. and Seignobos, C. 1898: *Introduction aux études historiques.* Paris: Hachette.

Loi, D., Robic, M.-C. and Tissier, J.-L. 1988: 'Les carnets de Vidal de la Blache, esquisses du *Tableau?'. Bulletin de l'Association de géographes français,* 4, 297–311 (*Vidal de la Blache. Lecture et relectures*).

Nicolas, O.G. and Guanzini, C. 1987: 'Paul Vidal de la Blache; géographie et politique; espace, science et politique'. *Eratosthène-Méridien,* 1 (Lausanne).

Nicolet, C. 1982: *L'Idée républicaine en France (1789–1924). Essai d'histoire critique.* Paris: Gallimard.

Noiriel, G. 1991: 'La question nationale comme objet de l'histoire sociale'. *Genèses,* 4, 72–94.

Nora, P. (ed.) 1984–6: *Les Lieux de mémoire.* I: *La République*; II: *La Nation.* Paris: Gallimard.

Nora, P. 1986: 'L'*Histoire de France* de Lavisse. Pietas erga patriam'. In P. Nora (ed.), *Les Lieux de mémoire.* II: *La Nation,* 1. Paris: Gallimard.

Ozouf, M. 1984: 'Jules Ferry et l'unité française'. In *L'école de la France.* Paris: Gallimard, pp. 400–15.

Pomian, K. 1986: 'L'heure des *Annales.* La terre, les hommes, le monde'. In P. Nora (ed.), *Les Lieux de mémoire.* II: *La Nation,* 1. Paris: Gallimard, pp. 377–425.

Ratzel, F. 1987: *La Géographie politique. Les Concepts fondamentaux.* Paris: Fayard.

—— 1988: *Géographie politique.* Paris: Editions régionales européennes.

Renan, E. 1882: 'Qu' est-ce qu'une nation?' Conference at the Sorbonne, 11 March.

Revel, J. (ed.) 1989: *L'Espace français (Histoire de la France),* vol. 1. Paris: Seuil.

Rhein, C. 1982: 'La géographie, discipline scolaire et/ou science sociale? (1860–1920)'. *Revue française de sociologie,* XXIII, 223–51.

Richard-Pettier, P. 1993: 'Michelet géographie'. *L'Information géographique,* 2.

Robic, M.-C. 1989: 'Sur les formes de l'Hexagone'. *Mappemonde,* 4, 18–23.

—— 1990: 'La géographie humaine, science de la vie?'. *REED (Sretie Info),* 31, 6–9.

—— 1991a: 'La *Bibliographie géographique internationale* (1891–1991), témoin d'un siècle, de géographie: quelques enseignements d'une analyse formelle'. *Annales de géographie,* 3–4.

—— 1991b: 'L'invention de la géographie humaine au tournant des années 1900: les vidaliens et l'écologie'. *Actes du colloque Réflexions sur l'histoire de la géographie française au XIXè siècle et au début du XXè siècle, Paris, 1990.* Paris: CNRS (Mémoires et Documents) (forthcoming).

—— 1992a: 'De la distribution à la disposition ou la France matricielle (sur Vidal de la Blache)'. *Littérature et nation,* 9, 2nd series, 153–85.

—— 1992b: 'Géograhie et écologie végétale: le tournant de la Belle Epoque'. In M.-C. Robic (ed.), *Du milieu à l'environnement. Pratiques et répresentations du rapport homme/nature depuis la Renaissance,* Paris: Economica, pp. 125–65.

—— 1993: 'L'invention de la "géographie homaine" au tournant des années 1900: les Vidaliens et l'écologie'. In P. Claval (ed.), *Autour de Vidal de la Blache. La formation de l'école française de géographie*, Paris: CNRS Editions, Mémoires et documents de géographie, pp. 137–47.

Roncayolo, M. 1986: 'Le paysage du savant'. In P. Nora (ed.), *Les Lieux de mémoire.* II: *La Nation*, 1, Paris: Gallimard, pp. 487–528,

Sanguin, A.L. 1988: 'Vidal de la Blache et la géographie politique'. *Bulletin de l'Association de géographes français*, 4, 321–33 (*Vidal de la Blache. Lecture et relectures*).

Seignobos, C. 1901: *La Méthode historique appliquée aux sciences sociales.* Paris: Alcan.

Smith, N.B. 1969: 'The idea of the French hexagon'. *French Historical Studies*, VI, 2, 139–55.

Tissier, J.-L., Besse, J.-M., Loi, D. and Robic, M.-C. 1987: 'Lyon et ses possibles: la région lyonnaise dans le, *Tableau de la géographie de la France* de Paul Vidal de la Blache'. *Actes du 112° Congrès national des Sociétés savantes, Lyon, 1987, Géographie*, Paris: Comité des travaux historiques et scientifiques.

Vidal de la Blache, P. 1888: 'Les divisions fondamentales du sol français'. *Bulletin littéraire*, Nov., 49–57.

—— 1889: *Etats et nations de l'Europe. Autour de la France.* Paris: Delagrave.

—— 1896: 'Le principe de la géographie générale'. *Annales de géographie*, 129–42.

—— 1897: 'L'éducation des indigènes'. *Revue scientifique*, March, 353–60.

—— 1898: 'La géographie politique, à propos des écrits de M. F. Ratzel'. *Annales de géographie*, 97–111.

—— 1899: 'Leçon d'ouverture du cours de géographie'. *Annales de géographie*, 98–109.

—— 1902: 'Les conditions géographiques des faits sociaux'. *Annales de géographie*, 13–23.

—— 1903a: *Tableau de la géographie de la France.* Paris: Hachette.

—— 1903b: 'La géographie humaine, ses rapports avec la géographie de la vie'. *Revue de synthèse historique*, 219–40.

—— 1905: 'Sur l'internationalisme'. In *Libres entretiens*, 2è série, pp. 26–43.

—— 1913: 'Les régions françaises'. *Revue de Paris*, Dec., 841–9.

—— 1917: *La France de l'Est: Lorraine-Alsace.* Paris: A. Colin.

—— 1922: *Principes de géographie humaine.* Paris: A. Colin.

—— *Notebooks* and *drafts*, see Joseph, B. and Robic, M.-C. 1987.

—— Reviews of Ratzel's writings in the *Bibliographie annuelle* of the *Annales de géographie*: nos 201 and 202 (*Bibliographie de 1895*); nos 113, 114 and 115 (*Bibliographie de 1896*).

Weber, E. 1976: *Peasants into Frenchmen: the modernization of rural France, 1870–1914.* Stanford, Cal.: Stanford University Press (French edition: La Fin des terroirs. La modernisation de la France rurale, 1870–1914. Paris: Fayard, 1983).

—— 1986: 'L'Hexagone'. In P. Nora (ed.), *Les Lieux de mémoire.* II: *La Nation*, 2, Paris: Gallimard, pp. 97–116.

Weisz, G. 1979: 'L'Idéologie républicaine et les sciences sociales'. *Revue française de sociologie*, XX, 83–112.

Zeldin, T. 1973–77: *France 1848–1945.* London: Oxford University Press (French edition: *Histoire des passions françaises*, 4 vols. Paris: Seuil, 1978).

4

In Search of Identity: German Nationalism and Geography, 1871–1910

Gerhard Sandner

Within 8 years of the foundation of the German Reich in 1871, 11 chairs of geography were established at German universities, with 13 more up to 1910. The conceptional development of German geographical studies and the struggle to establish their legitimacy and to obtain institutional recognition were linked to the specific political sphere of the period. One decisive element in this context was *Bildungswert*, meaning the potential educational effect in schools and for the public in general. It was this element that linked the formative process of nationalism and imperialism with that of expanding geography.

This linkage is illustrated by some examples: (a) the fight for the expansion of geographical education in schools made use of the political spirit of the time and was also decisive in the methodological discussion; (b) maritime geography expanded under the impact of the new imperialist naval politics after 1900; (c) rural settlement geography was imbued in its formative period by nationalist and 'volkish' thinking; (d) colonial teaching and research is exemplified by the work of Siegfried Passarge and his direct political involvement.

In each case mentioned there is a specific combination of geographical education in schools, geographical research and institutionalization at the university level. The legitimization function of geography for German nationalism and imperialism in the 1871–1910 period has been overstressed. Developing in the specific political sphere of the period, German geography was primarily concerned with its own legitimization and identity, which included political opportunism and nationalistic adaptation. The adverse effects became visible much later.

INTRODUCTION

The search for identity and the discussion about the meaning and content of national consciousness have been constants in German history for about

150 years. The far-reaching controversies which in the 1980s characterized *Historikerstreit* and the *Revisionismus-Debatte* on the same issue (see Wehler, 1988; Sandner, 1988) are part of a continuum which has always included contradictions and inconsistencies. Many studies have contributed to a differentiated insight into the complex interrelations and the background of these fundamental themes of German self-reflection.

The problematic nature of German nationalism and national consciousness is partly rooted in the specific style and the contradictions within the long-lasting endeavour to create a German national state. Among the many constitutive elements of this legacy there is that insoluble problem of applying the concept of a territorial national state (represented by France, for example) to the vast region 'on the other side of the Limes', defined through history and by collective memory not in Cartesian terms of territory, boundary and unambiguous assignment, but in the context of 'horizons', of movement, configuration and ambiguity (Steger, 1987).

The 'problem character' of German nationalism, identity and national consciousness was not 'produced' by Bismarck's *kleindeutsch* solution in creating a unified ('national') German state which excluded millions of Germans living in adjacent states. It is due to the fundamental difficulty of operating with concepts of the national state ('nation', 'territory', 'boundary' and 'borderline', people and *Volk*) in a region traditionally styled by completely different concepts. On the fundaments of an extreme fragmentation in the territorial pattern of sovereignty and power, cultural identity could develop only as a complex identity within a field characterized not by boundaries but by overlying strata, by degrees of vicinity and affinity, and by reference to a set of traditions and bounds beyond language. So 'national' identity in the sense of a nation-state was necessarily and from the beginning incongruent with cultural identity, in a *kleindeutsch* as well as in a *grossdeutsch* solution, needing intensive use of public symbols and action from above to become viable. These actions from above would have been ineffective without the national dreams of the intelligentsia. From its beginning the process of nation-building in Germany was to a large degree the activity of an intelligent elite.

This is why the school system became so important during the last quarter of the nineteenth century. It was used to translate and transfer concepts, meaning and consciousness of national unity 'down' to the people. After 1871 the nationalist and patriotic role of the school was enlarged to serve as a bulwark against 'problems' related to rapid urbanization, industrialization and the transformation of society. In 1889 Wilhelm II declared that it had long been on his mind 'to use the school in its different levels in order to act against the expansion of socialist and communist ideas' (quoted from Filipp, 1987, p. 66).

These and other elements define the specific problematic nature of German national consciousness and nationalism. In this aspect I disagree with Bassin when he states: 'as with the nation-state principle, Germany shared basic elements of its late nineteenth-century imperialist thinking with parallel tendencies in western Europe' (Bassin, 1987b, p. 475).

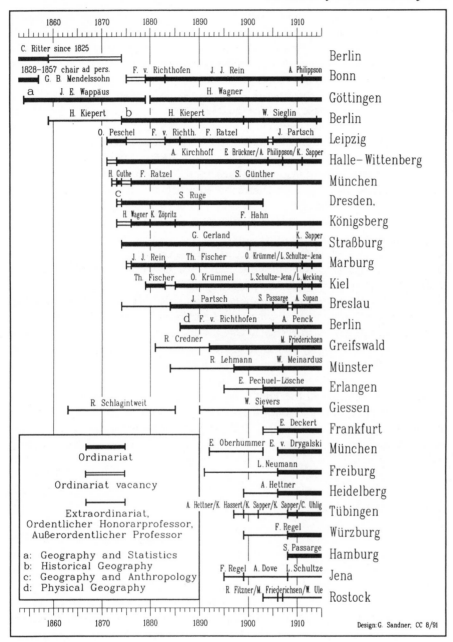

Figure 4.1 The establishment of geographical chairs in German universities, 1860–1914

Among the specific elements of German imperialism at the end of the nineteenth century was a constant interaction between self-assuring nationalism, which tried in vain to solve the contradictions mentioned, and expansionist concepts both within Europe and beyond, where the emerging *Weltpolitik* included colonialism. Another aspect is the specific role of patriotism and nationalism in producing confidence in a 'unity' beyond diversity, a process intensified by public symbols and an overheated propagation of supremacy. Excessive nationalism and patriotism can be seen as compensation for the congenital defects of the new German *Reich* and the specific problems of *Kaisertum* in the Bismarck and post-Bismarck eras, with its parliamentary system affected by polarized political parties. Immediately after the creation of the Reich in 1871, a series of organizations and movements began to shape public opinion and to claim political influence, among them the ultra-chauvinistic Pan-German League (for a comprehensive study, see Smith, 1986).

The rapid expansion of institutional geography at German universities between 1871 and 1880 (see figure 4.1) cannot be attributed exclusively to expectations coming from above that geography could contribute to nationalism and patriotism and to the goals of imperialist *Weltpolitik* and *Weltgeltung* (world politics and global role). In other countries there was a considerable expansion of geography as well, and in 1871 the first International Geographical Congress took place in Antwerp. On the other hand, the creation of so many chairs of geography was not a reward for scientific performance in the previous decades, much less a recognition of an established coherence as a scientific discipline legitimated by a set of aims, concepts, methods and advances. Of course, exploration and discovery, spectacular voyages and information about distant parts of the world were highly esteemed in the nineteenth century in Germany as elsewhere. But the period of explosive institutionalization as an act from above was motivated less by these aspects than by expectations related to the old fundamental element of the continuity and utility of geography, that is its *Bildungswert*, the potential educational value and impact in schools and for the public in general. It was this legitimative and functional element that linked the formative process of nationalism and imperialism with that of the expanding science of geography.

THE ROLE OF GEOGRAPHICAL EDUCATION

In a letter to H. Wuttke, J. E. Wappäus of Göttingen complained in 1871 that, according to his experience in examinations,

> geography at school must be without exception beneath contempt; considering the fact that people of this sort will be entrusted with geographical education at higher schools, we must lose all hope of improvement ... No doubt the governments are to be blamed for this evil. (This letter is reproduced in Engelmann, 1983, p. 154)

In his memorandum on the future development of geography at Berlin University of 1898, von Richthofen stated:

> According to a widely held opinion in neighbouring countries, Germany has for some time occupied the leading position in the recognition and

attention to geography. The excellent training of German officers in terrain evaluation and in the practical use of maps – which became obvious during the wars – may have contributed a great part to this positive assessment ... [But] it had not been noticed in other countries that geographical education at grammar schools is, with some exceptions, still in a very subordinate position, and general geographical education in Germany is not very common. (The memorandum is included in Engelmann, 1983, pp. 157–65, this passage at p. 161)

Indeed, geographical education in schools was subordinated to history, limited to the lower grades and dedicated mainly to *Staatenkunde* and the description of the topographical and regional arena of history. In the Prussian school system geography was introduced as a specific subject only in 1867. Most of the teachers had not been trained in geography, and geographers at the universities did not participate in the examination of candidates for the award of a degree. This situation changed rapidly after 1871.

In 1873 A. Kirchhoff was appointed to the chair of geography at Halle University, which he occupied until 1904. His qualifications were his teaching ability and two years of service teaching geography at the military academy in Berlin. In 1877 he succeeded in his first significant reform: geographers were included in the examination commissions for the graduation of students at the university. In 1881 the first congress of German geography in Berlin presented demands and declarations in favour of school geography from Kirchhoff and others. In 1882 a new curriculum was introduced, strengthening the position of geography. The reformed curriculum of 1892 included the subject 'geography of German colonies'; however, it was only after the reform of 1901 that the teaching of geography in schools was restricted to teachers trained in geography.

The fight to professionalize the teaching of geography in schools during the period 1871–1910 was not just a charitable act of university geography 'down' to the school level. In his critical commentary on the genesis and function of opportunism in geography, Hard reminds us of the fact that the expansionist period of German geography at the university level

between 1871 and 1906 took place against the advice and the resistance of the faculties. It was a political octroi, directed from the state towards a 'school-science' for the education of teachers, of the *Volk* and in general, carried through by (cultural) politicians due to (political) pedagogical motives and hopes. (Hard, 1983, p. 16)

This view is based on Schultz, who reminds us of a debate in the Prussian parliament in March 1875, where the primary interest of the government in the educational function becomes quite clear. Critical remarks of the budget commission and of Mommsen as a representative of the historical sciences were rejected by the representative of the government with the following remark: 'Since geography is a very important discipline for general education, particularly for the education of teachers, the government had no hesitation in creating geographical chairs extensively without asking the universities if they were inclined to incorporate a professor of geography' (quoted in Schultz, 1980, p. 66). The representative of school administration complained that 'geographical education at school is, in most of the secondary schools, nothing but ridiculous which is due to the lack of university trained teachers' (quoted in ibid., p. 67).

Here and on other occasions, for instance in the memorandum of 1883 in which the ministry demanded the expansion of geography to all universities, the priority was defined by the education of school teachers at the secondary level. *Bildungswert* (educational value) became the most important legitimating instrument in the institutional expansion of geography. The concept was far from neutral, being pervaded by the functional nationalism and patriotism of the time. In the foreword to his *Deutschland*, with the indicative subtitle *Einführung in die Heimatkunde*, published by Ratzel in 1898, the aims are clearly defined:

> First of all, the German should know what he has in his country. The present intent is based on the conviction that this purpose can only be achieved by showing that *Boden und Volk* [soil and people] belong together. May this book have a stimulating influence on the lessons of *Vaterlandskunde*. (Ratzel, 1898, p. v)

Hettner described the function of his voluminous regional geography of Europe, published in 1907, as follows:

> This book is not intended to be a textbook for lessons at school, nor a reference work for practical use; it is intended as a short scientific presentation of regional geography for teachers and students of geography and for all those coming from neighbouring fields who seek geographical instruction [*Belehrung*], and for all educated persons as well. (Hettner, 1907, p. iii)

Speaking about the unity of geography in science and in schools, he appealed to comparative regional geography in order to demonstrate geography's *Bildungswert*, stressing that only methodological concentration

> can win that position at school that geography needs to penetrate the education of our people ... Particularly in these dark days we have to consider the national task which geography has to fulfil. It teaches us to know and to understand our Fatherland, and thus it strengthens the love for our Fatherland. In teaching to know and to understand foreign countries, geography shows what we may learn, but also what treatment we may expect from them. (Hettner, 1919, p. 32)

On the same lines is Harm's textbook *Vaterländische Erdkunde* ('Patriotic Geography', with 17 editions up to 1926), dedicated 'to the strenghtening of national consciousness'.

At the first meeting of German geographers in Berlin in 1881, Kirchhoff was already claiming that geography should be strengthened in schools, referring to patriotism, love of the Fatherland and *Heimat*, so readily imparted by geography, 'that most German of all sciences' (*diese deutscheste aller Wissenschaften*, Kirchhoff, 1882a, p. 91). On this occasion a resolution was published in favour of geography in schools, but at the second meeting it had to be repeated, since nothing had improved. Now school geography feared that it was being overtaken by the progress of geography in universities. This impression led to even more expressive references to *Bildungswert* and patriotism. When the *Geographischer Anzeiger* was launched in 1900, Theobald Fischer initiated the journal with a short article on 'The German Reich in its actual boundaries: an ephemeral mayfly', welcoming the new journal as 'serving the national idea'. At the thirteenth meeting of German geographers in Breslau in 1901 a 'Central Commission for Geographical Education' in schools was founded on the initiative of H. Fischer, H. Wagner and A. Kirchhoff (Fischer, 1901).

Among the arguments to strengthen the cause of geographical education in school by appeals to nationalistic and patriotic *Bildungswert* was the emphasis on German colonies, that had been present since the late 1880s. The reformed curriculum for secondary schools introduced in 1892 included this subject within geography; repeated calls for *Kolonialkunde* as a subject in its own right were rejected by the administration. This reflects the relatively low awareness of the 'colonial question' by the public. The situation changed completely when the Reichstag initiated a more coherent and dynamic colonial policy in 1904, reflecting the growing economic importance of the colonies. Immediately, colonial education was intensified, combining economic and commercial geography with nationalist colonial politics. Due to the belief in the 'fundamental necessity of colonial possessions for world powers', this subject survived after the First World War and the loss of all German colonies (see Zimmer, 1978).

Another aspect to be mentioned in this context is the function of the school system for 'volkish' concern with Germans living outside the Reich. Düwell (1976) and Weidenfeller (1981) published detailed information and documents on this topic. Parallel to massive German emigration overseas, the motto *Erhaltung des Deutschtums im Ausland* ('preservation of Germandom abroad') had been present since the middle of the last century in all the discussions about nationalism and the nation-state. In 1881 the *Allgemeiner Deutscher Schulverein zur Erhaltung des Deutschtums im Ausland* was founded with the active participation of geographers. Its basic perspective was 'protection against the pressure of assimilation' to preserve language, *Volksgeist* and 'German culture'. One of its main functions was to strengthen German schools and teaching overseas, in which respect it can be compared with the 'Alliance Française'. In 1908 the association was transformed into the *Verein für Deutschtum im Ausland* (VDA, 'Association for Germandom Overseas'). Contrasts with radical nationalist associations such as the *Alldeutscher Verband* were reduced and the VDA became part of the *Volkstumskampf* with an aggressive missionary concept, working with cultural propaganda based on 'German

Supremacy'. In 1915 the VDA had about 57,000 members, mostly from the educated bourgeoisie.

Though not all the geographers occupying new chairs at universities were so intensively involved in activities to strengthen geographical education in schools as A. Kirchhoff and H. Wagner, both former school-teachers, or Th. Fischer, many used the reference to patriotic *Bildungswert* for the legitimation of further development of the subject. In fact eight new chairs were created at universities between 1902 and 1908 (figure 4.1).

Even more important for the development of geography as a science were the repercussions of geography teaching in schools upon the methodological discussions about the themes, methods and unity of geography, which in this period constituted the search for identity of scientific geography. The fundamental concepts of human and political geography introduced by Ratzel and others, the development of physical geography and the broadening knowledge accumulated during expeditions and field studies abroad became relevant for school geography much later. It was the discussion on *how* to teach geography in the context mentioned above, which intensified the dispute about *Landschaftskunde* and *Länderkunde*.

In his comprehensive study, Schultz analyses the role of 'school geography' as a source of innovation for 'university geography', concentrating on *Landschaftskunde* (Schultz, 1980, pp. 95–122). Disproving the opinion that geography teaching in schools lagged behind 'scientific geography' and received innovations from the university level, he points at the decisive role of Kirchhoff, who used *Landschaftskunde* as an element for integration and unity before 1900.

> The concept of *Landschaft* was created by the needs of schoolteaching ... it was used as a principle for the fight against the 'school-ghosts of geographical registers of names and numbers' ... Hettner's question 'What is geography' reads in the context of school geography 'How can geography best be taught?' (Schultz, 1980, p. 97)

The old controversy between geography as *Staatenkunde* (political geography) or as *Länderkunde* (regional geography), which survived the era of Ritter, was finally decided in favour of *Länderkunde*. The concept of *Land + Volk = Staat* persisted, but the concept *Natur + Mensch = Landschaft* became decisive (Schultz, 1980, p. 95). Hettner confessed he had based his concept of *Länderkunde* on the ideas of Kirchhoff. Recently, Ute Wardenga has uncovered the didactic function of Hettner's geography in general and his *Länderkunde* in particular:

> Hettner learned from experience how to work out a concept of geography that was based on separating research from teaching. He intended to create a programme for teaching, not for research. In his view of geographical research the following principles apply: interdisciplinarity, pluralism of methods, the principle of cause and effect, and the application of theories. ... Because of continuity of space, 'regions' [*Länder*]

do not exist in reality. The geographer constructs them for the purpose of teaching. Originally they are not subject of research. Therefore *Länderkunde* is a didactic problem. Questions concerning these problems are questions of how to organize the subject systematically. (Wardenga, 1987, p. 207)

In the context mentioned, 'teaching' cannot be reduced to teaching in schools. It includes transfer and mediation of insight and knowledge to the public in general and training in universities. On the other hand, the concepts orientated towards teaching, the public functions and the legitimation of geography, had a reverse influence on research and 'scientific' geography in universities. The majority of the geographers occupying new chairs at universities after 1890 had been trained in geography and became involved in the exploration and consolidation of research areas not linked directly to school teaching.

The developing subdisciplines of geography differed in their reference to the legitimative functions of nationalism and patriotic education in schools. Some, such as geomorphology, developed as academic subject areas based on worldwide research. Others, such as maritime geography, referred to nationalist and patriotic functions as an instrument for institutional consolidation in the period of take-off. Yet others were imbued in their formative period by nationalism and volkish thinking which persisted; an example is the development of German settlement geography on a historical-genetical basis.

LEGITIMATIVE ROOTS AND NATIONALIST FUNCTIONS OF DEVELOPING SUBDISCIPLINES: THE CASE OF MARITIME AND SETTLEMENT GEOGRAPHY

German regional research overseas, and particularly oceanographic exploration and research, had a long tradition of international co-operation before the founding of the Reich in 1871. Since the famous second voyage of James Cook with the *Resolution* in 1772–5, which included the two Forsters, many German scientists participated in British, Russian and other expeditions (Paffen and Kortum, 1984). International co-operation continued in the period of intensification and diversification of oceanographic research which followed the Challenger expedition of 1872–6. The international congresses of geography served as generators for joint ventures. That of Berlin in 1899 was decisive for the foundation of the *Conseil permanent international pour l'exploitation de la mer* in 1902, for the British–Swedish–German co-operation in Antarctic exploration and for the start of an international bathymetric chart, the forerunner of GEBCO.

The inclusion and involvement of geography in oceanographic research during the 1880s combined different elements: the demonstration of the 'importance' of geography in this respect; the legitimation of chairs and institutional expansion; the persistence of an image related to exploration and discovery; and participation in 'modern' techniques of natural sciences. In his famous address of April 1883, considered by Hettner as 'the fundamental programmatic statement relating to modern geography' (Hettner, 1929, p. 106), von Richthofen included oceanography as an integral, though specific, part of geography (von Richthofen 1883). In the same year Krümmel was appointed to a new chair at Kiel University that had special reference to oceanography. In Ratzel's series of geographical handbooks, a two-volume oceanography was published, the first volume by Bogulawski in 1884 (the author was a section head in the Admirality's Hydrographic Service), the second by Krümmel in 1887.

In the political sphere, the transition to resolute imperialism after the dismissal of Bismarck in 1890 led to an immediate response from oceanographic research and maritime activities. The Prussian naval expedition to Eastern Asia of 1860–2, which included von Richthofen, had already combined the elements of trade politics, the search for economically interesting outpost settlements and the stimulation for diplomatic action, which finally led to the acquisition of colonies. What changed after 1890 was the style in the presentation and institutionalization of Wilhelminian imperialist *Weltpolitik*, defined by reductionist concepts like the Kaiser's formula, 'Germany's future lies on the sea.' In 1897 Admiral Tirpitz became Secretary of State. His strategic concept of intimidation of the British naval power by means of a strong German fleet was decisive for the huge naval construction programme of 1898 and the *Flottengesetz* (law on the fleet) of 1900, both pushed through the Reichstag against opposition. Of course, neither the appeal to nationalistic 'eleation', to 'manifest destiny' and 'fateful necessity', nor the reference to the role of 'seapower in history and greatness' were specific to Germany at the end of the century (Mahan, 1890).

What matters in the context of this chapter, is the way geography used this political sphere as an instrument for development and consolidation. Ratzel's study on the sea as a source of national greatness (*Völkergrösse*, Ratzel, 1900) was the immediate response of geography to the *Flottengesetz* of the same year. This year also saw the foundation of the Berlin *Institut für Meereskunde*, with von Richthofen as president. Six years later the Kaiser himself inaugurated the Berlin *Museum für Meereskunde*. The directorate of both institutes was linked to the chair of geography until 1921 (von Richthofen, 1900–5; A. Penck, 1906–21; see Paffen and Kortum, 1984, pp. 96–9). Describing the new institutions, Penck referred explicitly to 'nationalistic concepts' and the 'personal interests of the Kaiser' (Penck, 1907, 1912).

The development of German oceanographic and maritime research cannot be subsumed under nationalism and patriotism. Under the influence of rapidly changing techniques, accumulation of data and knowledge, and international interchange, it developed into a highly differentiated research area, moving away from its strong links with geography.

As with oceanographic and maritime geography, the geography of rural settlements developed as a subdiscipline in the 1871–1910 period, but based on different and in some aspects deeper roots. Here, too, the 'spirit of time' was reflected in the topics and approaches. Unlike France with the influence of Vidal de la Blache's '*géographie humaine*' and his concept of '*genres de vie*', German settlement geography had a strong 'volkish' touch in its formative period. The early approaches were directed towards distributions, classification and natural determination, but then history and *Volkstum* were included, integrating different approaches. Arnold's study on settlement and migration of German tribes based on place-names became one of the roots of further development, though the causal linking of tribes and place-names was disproved in the 1920s (Arnold, 1875). Meitzen presented a paper on the German house in its 'volkish' forms at the first meeting of German geographers (Meitzen, 1882). In 1895 he published his fundamental study on settlements and agriculture of European tribes, combining agricultural history and rural settlement forms to detect the specifically 'national' elements as defined by mind and mood (*Gemütslage*) and by legal traditions.

These and other bases were combined with an orientation towards *Landschaft* as an integrative phenomenological issue and with a regional concentration on Germany, including the areas of 'eastern colonization'. This is the reason why the research area became particularly susceptible to influences from the volkish and *Lebensraum*-side, including chauvinistic concepts of German supremacy (Filipp, 1979; Schultz, 1980). This does not mean that the fundamental studies of O. Schlüter on the morphology of cultural landscape, on formal settlement types and the process of formal development (1903), or that of R. Gradmann in Southern Germany (1913) were outspokenly nationalistic or even chauvinistic. Nor are they to be blamed for the fact that the rapid development of rural settlement geography occurred earlier in Germany than in other parts of Europe and that many aspects, terms and concepts became linked to the specific situation of the central part of Europe. There was a latent trend to apply a centrifugal perspective to marginal areas from a German point of view. There was a contradiction between the insistence on a single German tradition or history and a strong regional differentiation.

In the period we are dealing with in this chapter, rural settlement geography and its potential for nationalist tendencies had not been used for legitimation of the institutional and functional expansion of geography. It was in the 1920s and thirties that it became functional in a political context, linked to settlement schemes of inner colonization and resettlement, to *Blut und Boden* and nationalistic volkish strategies. Nevertheless, from its beginnings German rural settlement geography contributed to the strengthening of nationalistic feelings, or at least to a reassuring recognition of German specifity. This is partly due to the fact that there was little interest in and knowledge of other countries or cultural areas, as far as settlement types and their interaction with cultural, legal and economic development was concerned. As in *Volkskunde* and other areas of cultural research, this development of a subdiscipline 'from within' a nationalistic frame with strong regional limitation to one area in its formative period produced profound difficulties for transfer, application and recognition of concepts, terms and perspectives in other cultural areas. In these aspects, rural settlement geography marks the other extreme to oceanography and maritime geography.

COLONIAL RESEARCH AND COLONIAL GEOGRAPHY

The study of colonial geography, its roots, its ideological background and its role in the development of geography as a science in the last quarter of the nineteenth century has long been neglected in Germany. Compared to comprehensive studies on colonial politics and on the roots of German imperialism presented by historians and social scientists (Wehler, 1969; Smith, 1986), there are still only a few geographers dedicated to the topic. Schulte-Althoff (1971) published the first and for many years the only systematic study on German geography in the imperialist era; Zimmer (1978) studied the influence of colonialism on geographical education in German schools; Krings (1984) reminded us of the Congo Conference in 1884–5 and its effects; Heske (1987) published a critical review of the political implications in German colonial geography; and Bassin (1987a, 1987b) provided fresh insights into the nationalist and imperialist roots and contents in Ratzel's writings. If there is something common to these studies it is the critical concern with the rapid change towards nationalistic themes and trends after 1871, with its far-reaching roots and its function in strengthening the role of geography and its recognition as a science.

To understand the fundamental change in the years following 1871, it should be kept in mind that geographic exploration and research overseas intensified considerably in the middle of the century. As in other European countries, there was a growing interest by the educated public in expeditions and the exploration of exotic areas, matched by a series of new publications and the activities of geographical societies. Personalities such as A. Petermann, who had been active at the Royal Geographical Society in London from 1845 to 1854 and launched his *Petermanns Geographische Mitteilungen* in 1855, became promoters of intensified research, particularly in Africa. International co-operation and interchange was strong at this time. Many German explorers were integrated in expeditions organized by other countries. These geographers were primarily interested in scientific research; they did not respond to the colonial interests and projects discussed in public in the 1850s and more intensively in the sixties as a reflection of massive emigration and growing importance of overseas trade. There were exceptions such as Bastian (Central Africa), von der Decken (Eastern Africa) and von Richthofen (Far East), who favoured colonial acquisition by Germany.

The intensification of nationalist tendencies and the hopes and dreams for a strong national German state, particularly among the educated *Bürgertum*, were time and again disappointed. The resulting build-up of nationalistic dreams, with its many variations, exploded after 1871. The problem was that neither the system of government with a Reichstag struggling with specific problems of combining pluralism and effectivity, dissent and compromise, nor the political actions of outstanding figures such as Bismarck or Withelm II, were able to respond to this build-up. Within nationalist, colonialist and pan-German circles there was constant impatience, intensive observation of the national and colonial politics of other European countries, and open radical criticism of political action and inaction from above. The intensively published crossing of Africa by Stanley (1874–7) and the Brussels Conference initiated by King Leopold II in 1876 in order to intensify exploration in central Africa prepared that 'scramble for Africa' which became so important for the relation between the European powers in the 1880s. Research overseas, particularly in Africa, acquired a nationalist profile. This does not mean that the majority of German geographers participating favoured the rapid acquisition of colonies, but there was a tendency to judge research as a contribution to national glory and praise.

The rapid expansion of German geography after 1871 had produced a great number of geographers occupying the newly installed chairs who were considered *Stubengelehrte*, meaning they had no experience in empirical research at home, much less overseas. The nationalistic overtones and the reference to national *Bildungswert* were stronger than the scientific advances based on research, at least in colonial and overseas areas. Many of the outstanding geographers of the period were active in newly founded associations and societies. When in 1873 the *Deutsche Gesellschaft zur Erforschung Zentralafrikas* (generally referred to as *Deutsche Afrika-Gesellschaft*) was founded by Bastian, von Richthofen became vice-chairman. The change of emphasis at the beginning of the 1880s from exploration to openly colonialist functions and propaganda is visible in the appeal to found the *Deutsche Kolonialverein* (DKV), published in 1882 and signed by F. Ratzel, Th. Fischer, A. Kirchhoff and G. Gerland. In the DKV, as in institutions such as *Allgemeiner Deutscher Schulverband* (its transformation into the *Verein für das Deutschtum im Ausland* has been mentioned above), activities and criticism of official politics were defined by the difference between the specific political aims and the overall political performance.

Smith (1986, p. 256) has analysed the dualistic nature of German imperialist ideology, pointing out fundamental differences and persistent antagonism between *Weltpolitik* and *Lebensraum* ideology. As stated elsewhere (Sandner, 1989), I disagree with this rigid separation. What matters in this context is that colonial politics, as well as colonial research, were influenced by the relative weight of different aspects in imperialist ideology. One of these aspects was based on the experience of massive German emigration overseas (1871–85, totalling about 1.5 million), on concepts of securing settlement areas and on the stabilization of 'Germandom abroad', leading to the *Lebensraum* ideology stressed by pan-German circles. Another aspect was the combination of economic interaction, trade, control of raw materials and expansion of market areas with consequences for the management of colonial posessions via concessions to private enterprise. Wilhelmian expansionist *Weltpolitik* went beyond this aspect of the 'colonial question'.

Since these different motivations and interests could not be transformed into effective political action there was a great amount of criticism and antagonism to official politics and its representatives, even within nationalist and imperialist circles. At the same time there was an immediate reflection of the changes in the political sphere and the administration, the most important dates being 1884 (acquisition of German colonies in Africa), 1890 (dismissal of Bismarck, new style of Wilhelmian *Weltpolitik,* effects of the 'scramble for Africa') and 1903–5 (preparation of reforms in colonial politics and management which became effective in 1906). In order to show how geography responded to these changes and to the overall political sphere of the time, two personalities may serve as examples: Theobald Fischer (1846–1910) and Siegfried Passarge (1886–1957).

> The more Theobald Fischer reduced his scientific work on the Maghreb, the more he transformed his scientific insight to serve imperialism. He saw the German Reich as a future world power in trade, which should gain a foothold in Morocco. According to him, the 'Atlas-foreland' between the High Atlas and the Atlantic coast was suitable for agriculture and contained important mineral resources. Germany should acquire settlement areas in an independent state of Morocco and transform Mogador into a naval base. He tried constantly to attract interest in Morocco by lecturing trips and by articles in journals and newspapers. When all this propaganda was not effective, he transferred his attentions to the *Alldeutscher Verband,* which he had founded together with others and which was orientated to gain new *Lebensraum* including in oversea areas for the German state. His articles in the *Alldeutsche Blätter* produced a 'Morocco storm', but the politics of the government in 1903 caused him the greatest disappointment. (Engelmann, 1983, p. 98)

The last remark in this quotation is somewhat incorrect. This first 'Morocco Crisis' of 1905, which was produced by expansionist French interests in northern Africa and German counteraction, including the landing of troops in Tangier, led to a diplomatic defeat for Germany in the Conference of Algeciras (1906). The second 'Morocco Crisis' of 1911, with French occupation and a German gunboat present at Agadir for intimidation, ended up with Morocco becoming a French protectorate and Germany gaining some territory in the border region of Cameroon and French Congo. Both cases produced an outcry of nationalism about the 'incapacity' of German diplomacy and a move towards breaking up the *Entente* of European powers which blocked German expansion.

Fischer (sometimes referred to as 'Morocco Fischer') used his political activities in the 'colonial question' to extend and secure the infrastructure of his chair at Marburg University.

> Presumably correct in his estimation of his long-lasting contributions to research in Morocco and thus to German colonial interests in this region, possibly also after verbal arrangements prior to his voyage to Morocco in the spring of 1906, Theobald Fischer presented a new petition to the Prussian minister of education on September 11, 1905. (Leib, 1988, p. 91)

Now at last he got what he had been requesting for so many years: an assistant, a darkroom, a projector, and an increase in budget for buying books. In 1909 the Marburg chair ranked sixth out of the 23 geographical chairs at German universities as far as infrastructure was concerned (Regel, 1909).

The case of Siegfried Passarge may be used to point to the combination and interaction of different aspects in scientific and political performance. As described in his extensive autobiography, Passarge was influenced in his youth by his parental home and particularly by his father, a judge in East Prussia. There were early nationalist, racist and anti-socialist feelings and a youthful enthusiasm for following the exploits of African explorers, whose ranks he dreamed of joining. His schooldays ended in 1884, when many Germans hailed the acquisition of German colonies in Africa. During his student years at the universities of Berlin, Halle and Freiburg, dedicated to geology, geography and medicine (1886–92), Passarge was influenced by Haeckel, von Richthofen and Pechuel-Lösche.

After finishing his studies, Passarge was anxious to begin scientific research and 'to serve the Fatherland in the colonies' (Passarge, 1951, p. 130). Pechuel-Lösche provided the contact with the private *Kamerun-Kommittee*, founded in 1893 in Berlin by Vohsen and others. Its aim was to 'rescue' the hinterland of Cameroon from French expansion and to expand the colony, using an agreement with the British Royal Niger Company which backed these intentions. Vohsen, who had been director of the German East Africa Society and served as a consul in Sierra Leone, was a Jew, probably the only one with whom the profoundly anti-Semitic Passarge ever had relations of any depth. 'I spared his feelings, but I made no secret of my fanatical patriotic cast of mind' (Passarge, 1951, p. 364). In 1893 Passarge began his Cameroon expedition, which marked the beginning of his 'colonial period'. He was also active in South Africa (1896–9, serving the British West Chartered Ltd) and Venezuela (1901–2, serving the private El Caura Syndicate which intended to colonize the properties of the former president Crespo).

The African activities and experiences between 1893 and 1901 became decisive for Passarge. When as an octogenarian he wrote his autobiography, he dedicated 50 per cent of the volume to this period, but only a meagre 4 per cent to the period 1908–38, that is to his activities at Breslau and Hamburg universities. The formative elements of the 1893–1901 period can be classified under three aspects which interacted and changed in weight later.

The first refers to the effects of that specific combination of scientific aims and social and economic interests, including his career, and the primarily political functions which characterize the enterprises in which he was involved before getting a chair. In his autobiography he describes that the scientific output of the Cameroon expedition was rather meagre and that its primarily political aims were wasted by the incapacity of the government:

> I have long felt fury and shame about the incapacity, or let us say the harmlessness, of German state politics since 1890, that is since the dismissal of Bismarck. ... this period of corrupt parliamentarism ... with a Reichstag dominated by interests hostile to the state .... (Passarge, 1951, pp. 166, 325)

This criticism was nourished by constant comparison with the success of British colonial companies and by measuring the development of the German colonies against a list of requirements deemed essential in colonialist circles (which included members of the administration). For Passarge, these essentials included the conquest and pacification of the native tribes, the foundation of strong military and administrative stations, the rapid construction of railways, and the intensive development of infrastructure in the hinterlands. It favoured intensification of native farming and extraction of minerals, but rejected contracts to chartered companies as well as the large-scale settlement of German farmers. The outspoken admiration for the success of British colonial politics, particularly in India, combined with radical opposition to the activities of Rhodes, the British Southwest Africa Company and the De Beer Company in Southern Africa, allowed colonialist circles to judge their own government less by what it did than by its failures. In his colonial activities, which between 1895 and 1901 included 20 articles published in *Deutsche Kolonialzeitung*, 17 other publications on Africa, and public lectures, Passarge did not use geography for legitimation of his conclusions nor his political activities for legitimating geography, he was straightforwardly political. As with anti-Semitism, he bluntly objected to any suggestion that he was political or had political interests at all, acting 'as a strictly objective scientist', claiming that 'science and politics are opposites like water and fire'.

The second aspect refers to the development of his latent racism into a radical form, in which he combined his medical knowledge with Social Darwinism. His fight against the 'disastrous effects of Christian missionaries' and 'the soft treatment of negroes', whom he declared unfitted for cultural development due to a specific collective character, resulted in condemnation by the Reichstag. His experiences during the expeditions in Africa of 1893–1901 were crucial for his later writings on ethnography and for the development of his racist and deterministic 'law of development of the character of peoples' (Sandner, 1989).

The third aspect refers to his field observations and his fundamental empirical research in physical geography, particularly in Southern Africa, which produced a great amount of important insight. The Cameroon expedition was primarily a military and political enterprise. Von Richthofen was right when he stated in 1898 that Germany's disadvantages in the colonies, as compared with Great Britain and France, were due to a lack of exploration and research in the hinterland areas (Engelmann, 1983, p. 163). Passarge participated in the opening-up of these hinterlands, but German research really started 15 years later, with the studies of Hassert, Thorbecke and others. In Southern Africa things were different. Passarge was at the forefront of research, using his geological training in solid observations, which finally came together in his concept of *Landschaftskunde*.

The interaction of these aspects was influenced by changes in the political sphere and in his personal career. After describing his activities in colonial politics up to 1905 Passarge stated:

> But also later on I had no interest at all in colonial foreign politics, although occupied in the Hamburg Colonial Institute [the forerunner of

Hamburg University, which was founded in 1919]. For now there was a period of quiet development. It became clear that the disquiet which began around 1890, culminated in 1901, and finished in 1906, was caused by two personalities [Cecil Rhodes and Scharlach, who, according to Passarge, worked in favour of the De Beers Company]. (Passarge, 1951, pp. 382–3)

From 1903 Passarge was busy publishing the scientific results of his work, particularly on South Africa, which led to his *Habilitation* in 1903 and the assignments to the chairs at Breslau University (1905) and the Hamburg Colonial Institute (1908). But everything changed with the First World War:

Nothing has shattered me in all my life as much as the breakdown of Germany [in 1918]. There was only one way to restore the spiritual equilibrium: by doing scientific work and penetrating into the psychological problems which these ill-fated times have brought to the foreground. The problem of the law of development of the character of peoples and the problems of Jewry and its effects on the peoples of the world. (Passarge, 1951, p. 453)

Passarge's consequent radicalization and his specific combination of racism, anti-Semitism and nationalism after 1918 will not be dealt with here (Sandner, 1987, 1989; Fischer and Sandner, 1988). What matters is that here again there was an immediate response to the spirit of time. Passarge belonged to that second generation of geographers occupying chairs after 1871, who were trained as geographers and acted as productive researchers at the academic level. His outstanding contributions on geomorphology and *Landschaftskunde* persist. His interest in geographical education in schools was low, as were his direct contributions to this sector, including the teaching of colonial geography.

Unlike others of his generation he was directly involved in the political sphere, acting and reacting. His contributions in this sector during his 'colonial period' until 1905, as well as in the radicalizing 1920s, were primarily based on his political and ideological *Weltanschauung*, and did not aim to transfer geographical methods and insight to politics. They did not contribute to the legitimation and stabilization of geography as a science; they even had negative effects in the context of his anti-Semitic and anti-socialist fights in the twenties which caused a rejection not only by newspapers but included geography in general.

CONCLUSIONS

The problem of dealing with 'national identity', particularly in Germany, with its belated formation of a nation-state and its discontinuities, is mainly related to the definition of the subject. In national identity, collective historical consciousness and collective memory of past experience amalgamate with value judgements on the present and projections into the future. But what is the

meaning of 'collective' in this context and can there really be a 'deficit of
identity' or a 'loss of identity' as specified in so many studies on recent
German history? There is a great amount of literature dealing with these
problems, relating 'national consciousness' to 'nation-state', 'Fatherland' to
'nation', concentrating on the specific German 'identity problem' or on the
question 'what is "German" nowadays?' (Teppe, 1976; Pross, 1982; Weidenfeld,
1983; Weigelt, 1984; Alter, 1986).

But still there is that fundamental problem of defining the subject, that is
of relating the generalization included in 'the Germans' to a differentiation of
groups, segments, sectors and ideologies in a complex society. It is the same
problem we are confronted with when trying to describe the development
of 'German geography', which was never exclusively 'German' nor exclusively
'geography'.

The intention of this chapter has been to point to some specific relations
between overall nationalist tendencies and the development of geography as a
science becoming institutionalized after 1871. Discomfort is produced by the
lack of differentiation, which can be surmounted only by a thorough analysis
of what happened year after year or by comprehensive biographies.

After I had completed the draft of this chapter, H.-D. Schultz published
his voluminous study on geography as an educational area during the period
1870–1914 (Schultz, 1989). This is the first comprehensive study on the
relations between educational politics, the aims and progress of expanding
geographical education at different school levels, and the interaction between
university geography and geographical teaching in schools, each aspect including
conceptional change and adaptations.

Summarizing Schultz's study and also this chapter, it can be stated that
the period from 1870–1 to 1910–14 was formative for German nationalism
and imperialism, and at the same time it was formative for German geography,
including the relations between the universities and teaching in schools.
This period also covers the rapid expansion of nationalism and patriotism
during the 1890s, and the tension between, on the one hand, volkish traditions, culture and *heimatliche Landschaft* regarded as pre-industrial bases for
nationalist feelings, and on the other hand, a forward-looking orientation on
expanding world markets, economic modernization and national power. Ratzel's
Deutschland reflects the first and his *Anthropogeographie* the latter. The period
also included the transition from the concepts of the Bismarck era, defining
the German role in a generally European context, to the Wilhelmian era with
Social Darwinism and militant nationalism forming part of colonial and cultural
imperialism. Here, too, geography as a science – including methodological
discussions and the scientific advances – was related far more directly to the
fashions and adaptations of geography in schools than was ever admitted.

REFERENCES

Alter, P. 1986: 'Nationalbewußtsein und Nationalstaat der Deutschen'. *Aus Politik und
Zeitgeschichte*, B 1/86, 17–30.

Arnold, W. 1875: *Ansiedelungen und Wanderungen deutscher Stämme zumeist nach hessischen Ortsnamen*. Marburg: Elwart.

Bassin, M. 1987a: 'Race contra space: the conflict between German Geopolitik and National Socialism'. *Political Geography Quarterly*, 6, 115–34.

—— 1987b: 'Imperialism and the nation state in Friedrich Ratzel's political geography'. *Progress in Human Geography*, 11 (4), 473–95.

Bogulawski, G. von and Krümmel, O. 1884 and 1887: *Handbuch der Ozeanographie*, 2 vols. Stuttgart: Engelhorn.

Düwell, K. 1976: *Deutschlands Auswärtige Kulturpolitik 1918–1932. Grundlinien und Dokumente*. Cologne, Vienna.

Engelmann, G. 1983: *Die Hochschulgeographie in Preussen 1810–1914*. Erdkundliches Wissen 64, Wiesbaden: Franz-Steiner.

Filipp, K. 1979: 'Ziele und Wege der deutschen genetischen Siedlungsgeographie. Eine wissenschaftsdidaktische Betrachtung'. In *Recherches de géographie rurale* (Hommage à F. Dussart). Liège, pp. 3–20.

—— 1987: *Kritische Didaktik der Geographie. Prolegomena zur Emanzipation einer Disziplin*. Materialien zur sozialwissenschaftlichen Forschung 1. Frankfurt: Haag & Herchen.

Fischer, H. 1901: 'Einige Bemerkungen zur Gründung der Zentralkommission für Erdkundlichen Unterricht'. *Geographischer Anzeiger*, 2, 163–5.

—— and Sandner, G. 1991: 'Die Geschichte des Geographischen Seminars der Hamburgischen Universität im Dritten Reich'. In E. Krause, L. Huber, H. Fischer (eds), *Hochschulalltag im Dritten Reich. Die Hamburger Universität 1933–1945*. Berlin: Dietrich Reimer, pp. 1197–222.

Gradmann, R. 1913: *Das ländliche Siedlungswesen des Königreichs Württemberg*. Stuttgart: Engelhorn.

Hard, G. 1983: 'Die Dixziplin der Weißwäscher. Über Grenze und Funktion des Opportunismus in der Geographie'. In P. Sedlacek (ed.), *Zur Situation der Deutschen Geographie zehn Jahre nach Kiel*, 2nd edn. Osnabrück: Fachbereich 2, pp. 11–44.

Heske, H. 1987: 'Der Traum von Afrika. Zur politischen Wissenschaftsgeschichte der Kolonialgeographie'. *Ökozid*, 3, 204–22.

Hettner, A. 1907: *Grundzüge der Länderkunde*. I: Europa. Leipzig: Otto Spanner.

—— 1919: *Die Einheit der Geographie in Wissenschaft und Unterricht*. Geographische Abende im Zentralinstitut für Erziehung und Unterricht 1.

—— 1929: *Die Geographie. Ihre Geschichte, ihr Wesen und ihre Methoden*. Breslau: F. Hirt.

Kirchhoff, A. 1882a: 'Einleitung zu den Verhandlungen über Schulgeographie'. In *Verhandlungen des 1. Deutschen Geographentages zu Berlin 1881*. Berlin: D. Reimer, pp. 91–103.

—— 1882b: *Schulgeographie*. Halle a.d. Saale: Buchhandlung des Waisenhauses.

Krings, Th. 1984: 'Einhundert Jahre Berliner Kongo-Konferenz von 1884–1885: Ein Rückblick af die Hintergründe, Ergebnisse und Folgen'. *Die Erde*, 115, 295–304.

Leib, J. 1988: 'Zur Geschichte der Geographie und des Geographischen Instituts in Marburg'. In *Jahrbuch 1987* (ed. Marburger Geographische Gesellschaft), Marburg: Marburger Geographische Gesellschaft, pp. 86–111.

Mahan, A.T. 1890: *The Influence of Sea Power upon History, 1660–1783*. Boston, Mass.: Little Brown.

Meitzen, A. 1882: 'Das deutsche Haus in seinen volkstümlichen Formen'. In *Verhandlungen des 1. Deutschen Geographentages zu Berlin 1881*, Berlin: D. Reimer, pp. 58–88.

—— 1895: *Siedelung und Agrarwesen der Westgermanen und Ostgermanen, der Kelten, Römer, Finnen und Slaven*, 3 vols. Berlin: Besser.

Paffen, K. and Kortum, G. 1984: *Die Geographie des Meres. Disziplingeschichtliche Entwicklung seit 1650 und heutiger methodischer Stand*, Kieler Geographische Schriften 60. Kiel: Geographisches Institut.

Passarge, S. 1951: 'Aus achtziger Jahren. Eine Selbstbiographie'. Hamburg (unedited typescript, 522 pp.).

Penck, A. 1907: *Das Museum für Meereskunde zu Berlin*. Sammlung Meereskunde 1, Berlin.

—— 1912: 'Das Museum und Institut für Meereskunde in Berlin'. *Mitteilungen der Geographischen Gesellschaft Wien*, 413–33.

Pross, H. 1982: *Was ist heute deutsch? Wertorientierungen zur Bewußtseinslage der Nation.* Reinbek bei Hamburg: Rowohlt.

Ratzel, F. 1898: *Deutschland. Einführung in die Heimatkunde.* Berlin, Leipzig: W. de Gruyter.

—— 1900: *Das Meer als Quelle der Völkergröße.* München and Leipzig: Oldenbourg.

Regel, F. 1909: 'Die Geographischen Institute der deutschen Universitäten'. *Geographischer Anzeiger*, 10, 150–8.

Richthofen, F. von 1883: *Aufgaben und Methoden der heutigen Geographie.* Leipzig: Veit.

Sandner, G. 1987: 'Die Verbindung naturdeterministischer, sozialdarwinistischer und antisemitischer Ansätze in der naturwissenschfatlichen Geographie Siegfried Passarges im Vorfeld des Nationalsozialismus. *Nachrichtenblatt der Deutschen Gesellschaft für Geschichte der Medizin, Naturwissenschaften und Technik*, 37, 70–1.

—— 1988: 'Recent advances in the history of German geography, 1918–1945. A progress report for the Federal Republic of Germany'. *Geographische Zeitschrift*, 76, 120–33.

—— 1989: 'The "Germania triumphans" syndrome and Passarge's "Erdkundliche Weltanschauung": roots and effects of German political geography beyond geopolitics'. *Political Geography Quarterly*, 8, 341–51.

Schlüter, O. 1903: Die Siedlungen im nordöstlichen Thüringen. Berlin and Jena: H. Costenoble.

Schulte-Althoff, F.-J. 1971: *Studien zur politischen Wissenschaftsgeschichte der deutschen Geographie im Zeitalter des Imperialismus*, Bochumer Geographische Arbeiten 9. Paderborn: F. Schöningh.

Schultz, H.-D. 1980: *Die deutschsprachige Geographie von 1800 bis 1970. Ein Beitrag zur Geschichte ihrer Methodologie*, Anhandlungen des Geographischen Instituts – Anthropogeographie, 29. Berlin: Geographisches Institut der Freien Universität.

—— 1989: *Die Geographie als Bildungsfach im Kaiserreich.* Osnabrücker Studien zur Geography, 10. Osnabrück: Fachgebiet Geographie.

Smith, W.D. 1986: *The Ideological Origins of Nazi Imperialism.* Oxford: Oxford University Press.

Steger, H.-A. 1987: 'Mitteleuropäische Horizonte'. In H.-A. Steger and R. Morell (eds), *Ein Gespenst geht um …: Mitteleuropa*, München: Eberhard Verlag, pp. 15–30.

Teppe, K. 1976: 'Das deutsche Identitäsproblem. Eine historisch-politische Provokation'. *Aus Politik und Zeitgeschichte*, B 20–21/76, 29–39.

Wardenga, U. 1987: 'Probleme der Länderkunde? Bemerkungen zum Verhältnis von Forschung und Lehre in Alfred Hettners Konzept der Geographie'. *Geographische Zeitschrift*, 75, 195–207.

Wehler, H.-U. 1969: *Bismarck und der Imperialismus.* Cologne: Kiepenheuer & Witsch.

—— 1988: *Entsorgung der deutschen Vergangenheit? Ein polemischer Essay zum 'Historikerstreit'.* Munich: C.H. Beck.

Weidenfeld, W. 1983: *Die Identität der Deutschen*. Munich and Vienna: Hansa.

Weidenfeller, G. 1981: 'Der VDA zwischen "Volkstumskampf" und Kulturimperialismus'. In *Interne Faktoren auswärtiger Kulturpolitik im 19. und 20. Jahrhundert*, Stuttgart: Institut für Auslandsbeziehungen, pp. 17–26.

Weigelt, K. 1984: *Heimat und Nation. Zur Geschichte und Identität der Deutschen*. Mainz: v. Hase & Koehler.

Zimmer, P. 1978: 'Der koloniale Gedanke im Geographischen Unterricht des Deutschen Reiches – Ein Beitrag zur Geschichte des Erdkundeunterrichts'. Trier (unpublished thesis, 154 pp.).

5

Berlin or Bonn? National Identity and the Question of the German Capital

Mechtild Rössler

INTRODUCTION

> Berlin. The single most powerful geographic event of the past year was the collapse of the last century's political order in an act of landscape change: the breaking of a concrete wall ... We may already envision the reconstitution of Berlin as capital city of Germany, political, cultural and symbolic center of Europe's superpower.[1]

The history of Germany is a history of geographical changes in its territory, its frontiers and borders, and the location of the capital. Especially in the twentieth century with the two world wars, the question of the German territory stood at the centre of political developments. During the period of National Socialism Berlin became the 'centre of the German world empire' as *Reichshauptstadt* and became the symbolic place for the rise, fall and destruction of the Third Reich. The division of Berlin and Germany into four parts after the Second World War led to the separation between East and West with the so-called 'Iron Curtain'. This division also determined West Germany's orientation towards the West, its integration into a European federation, against other possibilities and the decision to make Bonn capital of the Federal Republic of Germany. The German Democratic Republic kept East Berlin as its capital. When the Berlin Wall fell in 1989, at first nobody asked about relocating the German capital, but with reunification in 1990 a heated debate on 'Berlin or Bonn' began (see figure 5.1). It was partially ended in 1990 with the decision to make Berlin the capital of a unified Germany in the unification treaty, but for the German people and the parliament the debate was not finished. With the decision on 20 June 1991 to relocate the government and legislature as well, the 'capital question' between 'Beethoven's Bonn' or 'Bismarck's Berlin' was answered. But the problems have not yet been solved.

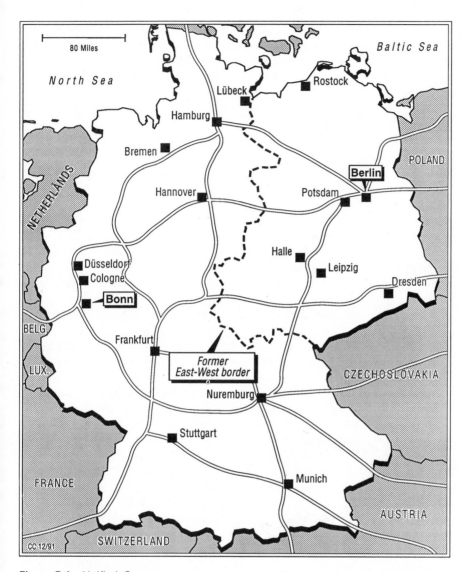

Figure 5.1 Unified Germany

This chapter concentrates on German territorial history and the symbolic places and symbolic aspects in the public discourse.

ON GERMAN TERRITORIALITY AND ITS HISTORY

> *Deutschland? aber wo liegt es? Ich weiß das Land nicht zu finden. Wo das gelehrte beginnt hört das politische auf* (Germany? Where is it? I don't know how to find the country where the scholarly Germany begins, the political one ends).
> Johann Wolfgang von Goethe, 1796[2]

The *Heilige Römische Reich Deutscher Nation* had no fixed capital at the centre; it shifted with the residence of the Kaiser. Thus, the *Nationalversammlung*, the national reunion assembly, held in Paulskirche in Frankfurt in 1848, could not decide on a permanent German capital, although some of the delegates proposed Frankfurt because of its central location. From that time onwards, it was a long way to Berlin, to Bonn, and back again to Berlin.

In the context of the German east colonization in the thirteenth century, the town Berlin-Cölln was founded and became a central market town. In the fifteenth century it was the residence of the Brandenburgs, and the population grew with the influence of Prussia, the founding of the Academy of Science in 1700 and the University in 1810. It became the first capital of Prussia.

Owing to the process of urbanization and industrialization, the city had a population of nearly one million when it became *Reichshauptstadt* (the capital of the German Reich) with the founding of the German Reich in 1871. The German empire was established as a political unit by Kaiser Wilhelm I. In comparison to the development of other European nation states this step came very late:

> Why France and even Poland should have attained political unity and some kind of a centralized political system before the middle ages were over, while Germany at the same time was disintegrating into a medley of little states too small and vulnerable for complete independence, too autocratic to form a federation, too selfish to visualize a Germany bigger than themselves, was perhaps the greatest tragedy of modern European history. In the long run it proved impossible to hold back the aspirations of German nationalism. The German Empire appeared late on the political map of Europe.[3]

The *Reichshauptstadt* Berlin was the cultural, political and industrial centre of central Europe at that time, something which was reflected in the changing face of the city. The history of Berlin represents the history of the Germans, the nation as well as the people. It was the capital for 74 years, from 1871 to 1945, during which time the German Empire, the Weimar Republic and the Third Reich rose and fell. In 1933 the German geographer Friedrich Leyden wrote a book on Berlin entitled *Geographie einer Weltstadt* (Geography of a World Metropolis).[4]

Hitler, who saw Berlin as the centre of Europe and the world, wanted to create a city to signify visually the power of the Nazis. Berlin was to be built as 'Germania' for the Thousand Year Reich, and he employed the architect Julius Lippert in March 1933 to create a truly 'German city'.[5] His successor was Albert Speer who developed the concept of the 'architecture of world power' in his first discussions with Hitler:

> all the architectural proportions of Berlin would be shattered by two buildings that Hitler envisaged on this new avenue. On the northern side, near the Reichstag, he wanted a huge meeting hall, a domed structure

into which St Peter's Cathedral in Rome would have fitted several times over ...[6]

He created a north–south avenue leading towards a triumphal arch. With its monuments, Hall concluded recently, 'Nazi Berlin would have been the ultimate City Beautiful.'[7]

The first big buildings to be constructed during the Third Reich were the *Erweiterungsbau* of the *Reichsbank* in 1934 by Heinrich Wolff, and the *Reichsluftfahrtministerium* by Emil Sagebiel in 1935–6, which later became the socialist *Haus der Ministerien* of the GDR. Sagebiel was also the architect of Berlin-Tempelhof airport built in 1935–6 as the 'aircross of Europe'.[8] During the so-called peace years of National Socialism another big project was realized: the *Olympia-Stadion* for the Olympic Games in 1936 (architect Werner March), which was a presentation and a showcase for the Third Reich.[9]

Architecture in the Third Reich had several functions, above all to stabilize the system while demonstrating power and economic strength. After 1937 programmes called *Neugestaltungsprogramme*[10] were initiated to rebuild Berlin, Hamburg, Munich and the *Gauhauptstädte*, the capitals of the Gaue. At the same time a new agency, the *Generalbauinspektor* of the Reich's capital Berlin, was set up with Albert Speer at its head. In any case, Berlin had top priority: it was the heart of Nazi Germany and it was to be built as the world centre of political and national power (see figure 5.2). It was to be Germania, the female symbol of the German Reich. They planned to finish the new construction of Berlin with the north–south axis and all the party and state buildings by 1950 and to rename Berlin 'Germania' at the opening of a world exhibition.[11] Hitler wrote after the occupation of France and Paris in 1940:

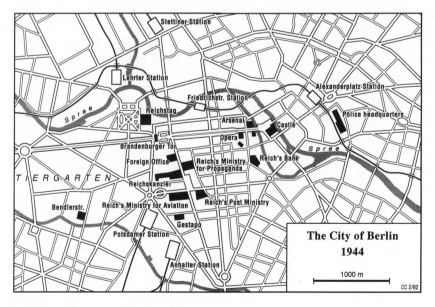

Figure 5.2 The city of Berlin, 1944

Berlin muß in kürzester Zeit durch seine bauliche Neugestaltung den ihm durch die Größe unseres Sieges zukommenden Ausdruck als Hauptstadt eines starken Reiches erhalten. In der Verwirklichung dieser numehr wichtigsten Bauaufgabe des Reiches sehe ich den bedeutendsten Beitrag zur endgültigen Sicherstellung unseres Sieges (Within a short period of time, Berlin must convey through its new architectural design that it is the capital of a powerful Reich, a justification of the greatness of our victory. I see the realization of this currently important architectural task for the Reich as the most significant contribution towards finally securing our victory). [12]

But the dreams and plans of Speer's 'Germania' were only partly realized and Berlin was destroyed in the Second World War and the Nazi empire ended. It was divided by the allies into four parts.

'BERLIN: A BROKEN CAPITAL' [13]

In the long run, however, the greatest difficulties came from the division of Berlin into sectors of occupation for Britain, France, the U.S.A. and the U.S.S.R. In theory, Berlin, like Germany itself, was to be governed as a unit. In practice, in Berlin as in Germany as a whole, the Russian-occupied area and the 'western' areas increasingly went different ways. The dividing lines were finally hardened in 1948, when West and East Germany adopted separate currencies. [14]

With the final division of Germany into the German Federal Republic – the *Bundesrepublik Deutschland* (the unified Western zones) – and the German Democratic Republic – the *Deutsche Demokratische Republik* (the former Soviet zone) – Berlin was divided into East and West too. East Berlin became the capital of the GDR, but West Berlin ceased to be the capital of the FRG. With the creation of the *Bundesrepublik Deutschland* in 1949, Bonn was chosen as the capital of West Germany. West Berlin was isolated due to the building of the Berlin Wall and was cut off from its surrounding area, the *Umland*. Also, Berlin lost its distinction as one of the cultural centres of Europe. Even the last architectural competition for a 'German capital' in 1958 did not preserve the historical centre of the old metropolis; it called for new construction. [15] Albert Speer planned a circular Autobahn in the 1930s as the boundary between the city and the *Umland*. The ring-road was built, but Berlin has since grown beyond it.

In the first years after the war, the idea of unification was still present, for example in traffic planning. However, East and West developed in totally different ways and had no mutual town planning or cultural exchange. In the 1950s and 1960s, two separate centres developed: in the western part around the Zoo and the so-called Ku'damm (*Kurfürstendamm*), and in the eastern part at the *Alexanderplatz*. The former centre of the city was divided by the Wall. Each part of Berlin in a way represented and symbolized the different political systems.

In principle, East Berlin was not legally a part of the GDR, but because it was in the former Soviet zone it became the capital of socialist Germany. It was rebuilt to display the thinking of the new socialist state and for the needs of this state. One can see the socialist planning in the large streets used for political parades, the 'people's palace' and the socialist housing complexes.

West Berlin had a different political status: with the new *Verfassung* (constitution),[16] which was accepted by the three Western allies, it had its own *Abgeordnetenhaus* (House of Representatives) and the laws of the *Bundesrepublik Deutschland* were only partly effective.

Since the Berlin blockade of 1948–9 and the air bridge created by the American *Rosinen-Bomber*, West Berlin has become the symbol for liberty and freedom. Berlin was a reminder of the history of division, the history of the ups and downs of the German nation.

BONN: THE NEW ORIENTATION TOWARDS THE WEST

After the destruction and division of Berlin and Germany, the Western zone had to decide again on a capital, and this proved not at all easy. Like the proposal by some delegates to the national reunion assembly in Paulskirche, Frankfurt was again under consideration but did not succeed. Bonn was not very popular and Frankfurt, geographically at the centre of West Germany, was strong financially and had been the 'secret capital'[17] after the war. Nevertheless, the final vote was 33 to 29 in favour of Bonn.[18]

The *Bundesrepublik Deutschland* had decided to take Bonn as the temporary 'provisional capital' in 1949 for several reasons. Bonn seemed appropriate for an orientation towards the West, towards the construction of a new Europe. Also, it was a decision favoured by the father figure of West Germany, chancellor Konrad Adenauer, the former mayor of Cologne, a town not far from Bonn.

Bonn, which was never mentioned as capital in the *Grundgesetz*, became the FRG's administrative centre, owing to the presence of the ministries, the *Ländervertretungen*, the embassies and consulates of foreign countries, and leading political societies and economic organizations. West Germany received a new identity with the *Wiederaufbau*, economic wealth, the European Community and parliament. West and East became two totally different states and societies with different cultural and political backgrounds and, most importantly, different identities, which can be seen partly in differences in language between East and West.

However, everything changed with the events of 1989. After the Berlin Wall fell and the 1990 reunification treaty was signed, Berlin was designated as the capital of unified Germany, but the location of the legislature and the parliament was left open. With this decision, though, the debate began in East and West.

'IS GERMANY REALLY UNITED?'

The Economist asked this question in early 1991,[19] while the *New York Times* pinpointed the fundamental issue of the debate on the German capital:

At the heart of the debate is the question of what Germany represents, a question that has confounded generations of Germans as well as foreigners. If Germany is a nation that prizes efficiency and order above all things, its capital should be Bonn, a quiet little town where there is never much to do besides conduct Government business. If, however, it chooses to immerse itself in the challenges of modern Europe, it could find no better place to do so than in swirling, turbulent Berlin.[20]

Positions in the German debate cut across all political and social boundaries: President von Weizäcker voted for Berlin for historical reasons. In contrast, the historian Fritz Fischer warned against moving the capital towards Prussia in his article 'Return to Prussia?'[21] He argued that the former capital of Prussia was centrally located in the state, but for unified Germany with its 1990 borders it would shift the orientation towards the East, only 50 kilometres from the Polish border. He called it a betrayal, a *Verrat*, of the founding principles of the Federal Republic German to move the capital to Berlin because the new Germany is based on the democratic traditions of the former *Bundesrepublik*, merely expanded by the five new *Länder*. The shift to Berlin would recall Prussia's militaristic history, Wilhelmian world politics, and the continuity of the German Reich. Berlin is far from the West, while Bonn represents the West European connection. Fischer's arguments were based on the respective histories of Berlin and Bonn but, as with all other combatants, he uses pictures and images, a symbolic structure which underlies and influences the public discourse.

'IMAGES OF BONN AND BERLIN: THE SYMBOLISM OF TERRITORY'

The debate on Berlin/Bonn goes deeper into history and the symbolism of territory than one realized during the 1989 events: 'Some of those who favour Berlin say they want Germany to accept a new historical role. They say that because Berlin ... represents the future of both Germany and Europe.' Not only the Germans had fixed ideas about the different places: *The Economist* raised the 'capital question' of 'Beethoven's Bonn' or 'Bismarck's Berlin'.

Some ethnologists analysed the symbolic structure of the process of political transformation[22] that included the symbols of the flags used by people for different purposes at demonstrations. They considered the significance of the German colours *Schwarzrotgold* (black, red, gold) and the European flag with the twelve stars on a blue ground to mean the integration of a unified Germany into a united Europe. Often the GDR flag was used at the so-called Monday demonstrations in 1989, with the Stalinist symbols removed. The 'national aspect'[23] remained at the centre, but switched from 'we are the people' to 'we are one people', clearly indicating the unification of the two 'half-nations'.[24] The German flag became the symbol of the 'new, old nation' and was displayed at the huge unification ceremony in Berlin in front of the *Reichstag* on 3 October 1990. Berlin was the symbolic place for a symbolic ceremony to bring two identities together as one nation.

However, a sense of being a nation (*Nationalgefühl*) cannot come about with the destruction of the Wall or with a ceremony and a unification treaty. There is no secure German identity and the identity crisis, the estrangement between East and West, is evident.[25] The result of the historical division is 'a city with no periphery but two centers, each a showcase for rival ideologies'.[26]

Berlin itself illustrates the problem: the mixture of Prussia's monuments at the centre of the city with the GDR façades such as Honecker's people's palace (*Volkspalast*),[27] or the domain of avant-garde yuppies with areas taken over by squatters in Kreuzberg in the West. The history of the two half-nations is visible in the architecture which has led to endless discussions between urbanists, town planners and politicians about how to create the new capital from the old divided one. The original centre is 'empty' now,[28] but will soon be filled with new symbols of capitalism and Western economy.[29] Daimler-Benz, the Mercedes company, for example, has bought 60,000 square metres near the *Potsdamer Platz*; the Sony company has followed. Berlin will be the 'construction site'[30] of Germany.

Wenn man aber denkt, die Einheit Deutschlands bestehe darin, daß das sehr große Reich eine einzige große Residenz habe, und daß diese eine große Residenz, wie zum Wohl der Entwicklung einzelner großer Talente, so auch zum Wohl der Masse des Volkes gereiche, so ist man im Irrtum (If one thinks, though, that Germany's unity as a lay empire consists in its having a single large capital, one which is conducive to benefiting the development of highly talented individuals as well as the masses, then one is mistaken). (Johann Wolfgang von Goethe to Eckermann, 23 October 1828)[31]

SOLUTIONS AND CONCLUSIONS: 'MY CAPITAL OR YOURS?'[32]

The unification treaty signed by West and East Germany on 3 October 1990 named Berlin as the capital, but the decision concerning the seat of government and parliament was left open. The discussion began with financial issues: the move to Berlin would cost between 50 and 60 billion German marks, the independent study of the Prognos Institute in Basel stated.[33] 'After all, the Germans seemed to want to shift their capital 600 kilometres (375 miles) east from the "federal village" of Bonn on the Rhine to Berlin, the former capital of Bismarck's Prussia and Hitler's Reich',[34] *The Economist* wrote five days before the final decision. On 20 June 1991 the votes were counted in the Bundestag. Similar to the situation after the Second World War when Bonn was chosen, Berlin won by a slight majority (338 to 320). Berlin, the old capital, became the new one. The reunification of Berlin corresponded to the unification of Germany, and once again Berlin became the symbolic location of the German identity. In December 1991 the decision was taken to move some of the ministries to Berlin and to leave eight in Bonn. So the story of Berlin and Bonn is endless and the discussions will go on.

Berlin bears the marks of victories and decline, of destruction and reconstruction, of the division and reunification of the German nation. The 'Mythos Berlin'[35] represents the German identity with all its contradictions. Berlin should pave the way for a German nation with different identities, one which will incorporate the history of the two Germanies and the history of other nationalities living in the country.

Racial, class and national dichotomies as well as the growing racism in Germany indicate the problems of national identity and the problems with it, too. Racism demonstrates the fundamental uneasiness concerning the unification of the two Germanies into one nation: it is an attempt to construct an internal identity that excludes outsiders.

Berlin will be one of the political and cultural centres of Europe, the pivot between North and South as well as between East and West. It could be a centre for new developments in a multicultural society, the exchange of varied experiences and a critical reflection of national symbolism and identity.

NOTES

1 Denis Cosgrove, '... Then we take Berlin: cultural geography, 1989–90'. *Progress in Human Geography*, 14(4) (Dec. 1990), 560–8.

2 Quoted in Thilo Lang (ed.), *Oh Deutschland, wie bist Du zerrissen! Lesebuch der deutschen Teilung*. Munich: Droemersche Verlagsanstalt, Th. Knaur Nachf, 1988, p. 73.

3 Norman J. G. Pounds, *Divided Germany and Berlin*. Princeton, N.J.: Van Nostrand, 1962, p. 36.

4 Leyden worked for the Foreign Office, but had difficulties after 1933 because he was Jewish. He was forced to go into exile in The Netherlands, but he was sent to a concentration camp in 1940, where he was killed. See Mechtild Rössler, *Wissenschaft und Lebensraum. Geographische Ostforschung im Nationalsozialismus*. Berlin: Dietrich Reimer, 1990.

5 Meyers Lexikon. Leipzig: Bibliographisches Institut, 1936, p. 1215.

6 Albert Speer, *Inside the Third Reich: memoirs*. New York: Macmillan, 1970, p. 74.

7 Peter Hall, *Cities of Tomorrow: an intellectual history of urban planning and design in the twentieth century*. Oxford: Basil Blackwell, 1988, p. 200.

8 Wolfgang Schäche, 'Architektur und Stadtplanung während des Nationalsozialismus am Beispiel Berlin'. In Hans J. Reichhardt and Wolfgang Schäche, *Von Berlin nach Germania. Über die Zerstörungen der Reichshauptstadt durch Albert Speers Neugestaltungsplanungen*, Berlin: Transit Verlag, 1985, pp. 15–17.

9 See also the film 'Olympia' by Leni Riefenstahl, which deals with the counterpart of Nazi architecture: the architecture and shape of Aryan bodies.

10 It was based on the *Gesetz zur Neugestaltung deutscher Städte*, 4 October 1937.

11 Reichhardt and Schäche, *Von Berlin nach Germania*, p. 25.

12 Adolf Hitler, *Hauptquartier*, 25 June 1940. Quoted in Schäche, 'Architektur und Stadtplanung', p. 32, n. 26.

13 Rudolf Walter Leonhardt, *This Germany: the story since the Third Reich*. Harmondsworth, Middx: Penguin, 1966, p. 85.

14 T.H. Elkins, *Germany*. London: Christophers, 1960, p. 257.

15 Wolf Jobst Siedler, 'Berlin – Spiegel deutscher Geschichte'. In *Deutschland, Europa und die Welt*, Gütersloh: Bertelsmann, 1986, pp. 30–4, esp. p. 33.

16 From 1 September 1950.
17 *Der Spiegel*, no. 17, 22 April 1991, 47.
18 33 for Bonn, 29 for Frankfurt.
19 *The Economist*, 23 February 1991, 45.
20 Stephen Kinzer, 'Berlin and Bonn partisans square off as vote for Germany's capital nears'. *New York Times*, 17 June 1991.
21 Fritz Fischer, 'Rückkehr nach Preußen? Die Bundesrepublik sollte auch künftig von Bonn aus regiert werden'. *Die Zeit*, 14, 29 March 1991.
22 G. Korff, 'Rote Flaggen und Bananen – Politischer Symbolismus des Vereinigungsprozesses'. *Schweizerisches Archiv für Volkskunde*, 86(3–4) (1990), 130–60.
23 Ibid., 134.
24 K.H. Bohrer, 'Und die Erinnerung der beiden Halbnationen'. *Merkur-Deutsche Zeitschrift für Europäisches Denken*, 44 (March 1990).
25 Ibid., p. 185.
26 Karl E. Meyer, 'Germany's once and future capital'. *New York Times*, 25 June 1991.
27 Jürgen Leinemann, 'Ein Herz und eine Mitte'. *Der Spiegel*, no. 17, 22 April 1991, 50–77, esp. 55.
28 Ibid., 68.
29 Bohrer, 'Und die Erinnerung der beiden Halbnation', 145.
30 Die 'Baustelle Deutschlands', after Leinemann, 'Ein Herz und eine Mitte', 77.
31 Reprinted in Lang (ed.), *Oh Deutschland*, p. 153.
32 *The Economist*, 16 March 1991.
33 See Wolfgang Hoffmann, 'Die Kosten des Umzugs'. *Die Zeit*, 22 February 1991.
34 *The Economist*, 15 June 1991, 48.
35 Leinemann, 'Ein Herz und eine Mitte', 77.

REFERENCES

Berger, Alfred 1970: *Berlin, 1945–1969: a postwar history of Berlin, excerpts from authoritative sources*. Munich: Gersbach.
Berlin, an architectural history, guest-edited by Doug Clelland, 1983. New York: Martin's Press. Series title: *Architectural Design Profile 50*.
Bohrer, K.H. 1990: 'Und die Erinnerung der beiden Halbnationen'. *Merkur-Deutsche Zeitschrift für Europäisches Denken*, 44, 183–8.
Brandt, Willy 1957: *Von Bonn nach Berlin; eine Dokumentation zur Hauptstadtfrage*. Berlin-Grunewald: Arani.
'The capital question' 1991. *The Economist*, 15 June.
Carriere, Bruno 1987: 'Berlin, une histoire sans fin'. *La Vie du rail*, no. 2106.
Cosgrove, Denis 1990: '… Then we take Berlin: cultural geography, 1989–90'. *Progress in Human Geography*, 14(4), 560–8.
Deutsche Gesellschaft für Auswärtige Politik. Forschungsinstitut. Dokumente zur Berlin-Frage, 1944–1966. Mit einem Vorwort des Regierenden Bürgermeisters von Berlin. Hrsg. vom Forschungsinstitut der Deutschen Gesellschaft für Auswärtige Politik e. V., Bonn, 1967. Munich: Oldenbourg.
Elkins, T.H. 1960: *Germany*. London: Christophers.
Erickson, John 1983: *The Road to Berlin: continuing the history of Stalin's war with Germany*. Boulder, Col.: Westview Press.
Fischer, Fritz 1991: 'Rückkehr nach Preußen? Die Bundesrepublik sollte auch künftig von Bonn aus regiert werden'. *Die Zeit*, 14, 29 March.
Fragen an die deutsche Geschichte. English: *Questions on German history: ideas, forces,*

decisions from 1800 to the present: historical exhibition in the Berlin Reichstag: catalogue c. 1984. [organized by the German Bundestag]. English edn of the 9th updated German edn Bonn: German Bundestag, Press and Information Centre, Publications Section.

Gedmin, Jeffrey (guest ed.) 1990: 'German unification'. *World Affairs*, 152(4).

Grass, Gunther 1990: *Two States, One Nation?* San Diego: H.B. Jovanovich.

Hall, Peter 1988: *Cities of Tomorrow: an intellectual history of urban planning and design in the twentieth century.* Oxford: Basil Blackwell.

Hanock, Donald and Welsh Helga (eds) 1991: *German Unification: process and outcome.* Boulder, Col.: Westview Press.

Hoffman George W. (ed.) 1989: *Europe in the 1990s: a geographic analysis.* New York: John Wiley.

'Is Germany really united?' 1991. *The Economist*, 23 February, 45.

'Kaff oder Metropole?' 1991. *Der Spiegel*, no.17, 22 April, 34–45.

Kinzer, Stephen 1991: 'Berlin and Bonn partisans square off as vote for Germany's capital nears'. *New York Times*, June 17.

Korff, G. 1990: 'Rote Flaggen und Bananen – Politischer Symbolismus des Vereinigungsprozesses'. *Schweizerisches Archiv für Volkskunde*, 86(3–4), 130–60.

Lang, Thilo (ed.) 1988: *Oh Deutschland, wie bist Du zerrissen! Lesebuch der deutschen Teilung.* Munich: Droemersche Verlagsanstalt, Th. Knaur Nachf.

Laux, Hans Dieter 1991: 'Berlin oder Bonn? Geographische Aspekte einer Parlementsentscheidung'. *Geographische Rundschau*, 43, 740–3.

Leinemann, Jürgen 1991: 'Ein Herz und eine Mitte'. *Der Spiegel*, no. 17, 22 April, 50–77.

Leonhardt, Rudolf Walter 1966: *This Germany: the story since the Third Reich.* Harmondsworth, Middx: Penguin.

Leyden, Friedrich 1933: *Groß-Berlin. Geographie einer Weltstadt.* Breslau: Hirt Verlag.

Marsh, David 1989: *The New Germany at the Crossroads.* London: Century.

Matthews, Christopher 1991: 'After the wall: image'. *San Francisco Examiner*, 28 April, 20–3.

Mattox, Gale A. and Shingleton, Bradley (eds), *Germany Old and New.* Boulder, Col.: Westview Press.

Meyer, Karl E. 1991: 'Germany's once and future capital'. *New York Times*, 25 June.

Mikes, George 1974: *How to Unite Nations.* London: Coronet Books.

Pommerin, Reiner 1989: *Von Berlin nach Bonn: die Alliierten, die Deutschen und die Hauptstadtfrage nach 1945.* Cologne: Bohlau.

Pounds, Norman J. G. 1962: *Divided Germany and Berlin.* Princeton, N.J.: Van Nostrand.

Reichhardt, Hans J. and Schäche Wolfgang 1985: *Von Berlin nach Germania. Über die Zerstörungen der Reichshauptstadt durch Albert Speers Neugestaltungsplanungen.* Berlin: Transit Verlag.

Rössler, Mechtild 1990: *Wissenschaft und Lebensraum. Geographische Ostforschung im Nationalsozialismus.* Berlin: Dietrich Reimer.

Siedler, Wolf Jobst, 1986: 'Berlin – Spiegel deutscher Geschichte'. In *Deutschland, Europa und die Welt.* Gütersloh: Bertelsmann, pp. 30–4.

Speer, Albert 1970: *Inside the Third Reich: memoirs*, trs. from the German by Richard and Clara Winston. New York: Macmillan.

Tent, James F. 1988: *The Free University of Berlin: a political history.* Bloomington: Indiana University Press.

Verheyen, Dirk 1991: *The German Question: a cultural, historical, and geopolitical exploration.* Boulder, Col.: Westview Press.

'Wie Bonn Bundeshauptstadt wurde' 1991. *Der Spiegel*, no. 17, 22 April, 45–55.
Windsor, Philip 1963: *City on Leave: a history of Berlin, 1945–1962*. New York: Praeger.

6

Nationalism and Geography in Modern Japan, 1880s to 1920s

Keiichi Takeuchi

Nationalism in the broader sense is understood as feelings of support for political institutions and of attachment to culture and regions. But similar feelings having been found even in primitive societies. The term 'nationalism' is generally used in a narrower sense to refer to political movements seeking or already exercising modern state power and justifying such actions on the basis of certain tenets: (a) the nation must have an explicit and peculiar character; (b) the interests and values of the nation take priority over the interests and values of all other territorial units; and (c) the nation must be, as far as possible, politically independent; this usually requires that the nation must at least have attained political sovereignty. Nationalism in this modern sense generally rests on the premises that the nation consists of all the inhabitants within the territorial frame of that state or region and that all these inhabitants have as a common denominator a kind of national identity or an identification with the national territory and the national culture. In the contemporary world the concept of the nation-state often borders on the fictitious in that within many nation-states formed in the nineteenth century there are now to be found a large number of nationalist movements or what might be termed 'mini-nationalism' on the part of ethnic minorities.

In Japan, prior to the middle of the nineteenth century, the cultural and territorial identification with the nation, that is Nippon, existed only among the ruling classes, the *samurai* and vassals of the imperial court. It was true that, in the face of the menace of the Western powers in the 1840s, nationalist sentiments gained ground among the general populace. But in the situation of the time where Japan was divided into numerous feudal clans (*han*) not all the people had the feeling of being one with the nation as a whole. Only after the Meiji Restoration in 1868 did a sense of national identity rapidly form, under the pressure of the institutional and ideological policies of enforcement instigated by the Meiji government. The institutional aspect was to be found in a political unification brought about by means of fiscal imposition and national conscription. In the ideological sphere, many intellectuals declared that Japanese culture was both unique and homogeneous and also pointed out the necessity for the modernization of the country, citing advanced Western countries as examples of successful modernization.[1]

After the establishment of the compulsory education system in 1872, geography came to be made much of as workable material for the encouragement of nationalism and especially as an effective means of driving the Japanese people to further modernization efforts by contrasting the situation in prosperous Western countries with the abject conditions in Asian countries that had fallen under Western domination. It should be noted here that, with regard to the relationship between geography and nationalism, even at this initial stage certain characteristics of Japanese nationalism were already in evidence, due to Japan's physical isolation from other countries, and the related seclusionist policy of the shogunate government for more than 200 years up to the middle of the nineteenth century. Statements to this effect can be found in geographical writings and school textbooks of geography in the 1860s and 1870s. It is true that, where geographical conditions and historical background were concerned, the majority of the Japanese people had never had experience of direct contact with foreign peoples. (Britain is an insular country as is Japan, but the British people have always had a great many contacts with the Continent.) The diffusion of geographical knowledge of other countries certainly inspired in the Japanese people a desire for contact with foreign peoples, and moreover served to awaken a consciousness of the uniqueness of Japanese culture and of the necessity to build the country into a strong and rich state.

As mentioned above, pre-Meiji Japan was divided into numerous feudal clans; in parallel with this, another situation existed which had arisen in the second half of the eleventh century at commencement of rule by a central shogunate government, successors to which for the most part remained in power throughout the following centuries. These historical circumstances greatly contributed to the formation of the cultural and economic unity of Japan. In modern Japan, since the Meiji Restoration, no regionalist or separatist movement has arisen that takes the form of nationalism in the state or micronationalism. Certainly, there were, and still are, ethnic minorities, such as the Ainus in Hokkaido and the Ryukyu people of the south-western islands. But the Ainus were too few in number to set up a micro-nationalist or regionalist movement in the face of massive colonization by the Japanese; and the Ryukyu people, who had been subject to the double sovereignty of Japan and China, became annexed to Japan under the political and military pressures exerted upon them at the beginning of the Meiji period. There was also the fact that the Ryukyu people were anyway too similar in culture to the Japanese people to be able to make a clear distinction between them and the Japanese.

Along with the diffusion of geography as school material around the end of the 1880s, there appeared a certain number of publications which systematically introduced and applied the Western geography methods of the day. The readers of these publications were generally intellectuals, but it was the schoolteachers of geography who made extensive use of them. In contrast with the authors of geographical writings of the early Meiji period, who had been more or less self-made scholars obtaining their geographical knowledge of other countries from run-of-the-mill school textbooks compiled in Western countries, the authors of the above publications had usually received their higher education in Japan and carried out further systematic studies in geography either in Japan or abroad. Moreover, they applied the geographical methods pertaining to the man–environment paradigms they had developed to the interpretation of the geographical characteristics of Japan. None the less, while the geographical works they published exercised considerable influence, they were not academic geographers in the strict sense of the term. The establishment of chairs of geography at higher normal schools took place at the end of the 1880s and the chairs of geography at imperial universities were created in 1907 in Kyoto and in 1811 in Tokyo, by which time the authors mentioned above were already established in other careers.

Three important authors of geographical books appearing at the end of the last century, Kanzo Uchimura (1861–1930), Inazo Nitobe (1862–1993) and Shigetaka Shiga (1863–1927),[2] were graduates of the Sapporo Agricultural College in Hokkaido. Founded in 1876, this College constituted one of the projects instigated by the Japanese government for the colonization of Hokkaido. For the work of colonization, the Japanese government enlisted the aid of a number of foreign experts, who subsequently arrived in Japan. Among them was William Smith Clark (1826–86), who came from the United States in order to fill the post of principal of the College. He had previously studied natural sciences at both Amherst College and Göttingen University, and at the time he received the invitation to Japan he was president of the Massachusetts College of Agriculture. He exerted a good deal of influence at the Sapporo Agricultural College as both a fervent Protestant and an enthusiastic observer of nature. In point of fact, both Uchimura and Nitobe were baptized when they were students at the Sapporo Agricultural College; and it was the influence of Clark and other foreign teachers that induced many graduates in the early days of the College to develop an interest in natural history via training in agronomy.

According to his later writings, in his middle school-days Kanzo Uchimura was already deeply interested in geography and history and when, after graduating from the Sapporo Agricultural College, he gained the opportunity to pursue studies in fishery at Amherst College in the United States, he spent much of his time there studying geography. Later, in fact, in 1923, he wrote that he had already had the intention of becoming a geographer when he enrolled in the Sapporo Agricultural College at the age of 17. And in his diary entry for 19 June 1919 we read that his writings on geography were the fruit of a two-year study of geography and history at Amherst College. Subsequently, after his return to Japan in 1888, he published writings on geography together with writings on Christianity; and in 1892 he produced 'Japan: its mission', written in English and published in the English-language journal *Yokohama Mail*. In this paper, he analysed the vocation of the Japanese nation on the basis of geopolitical considerations. According to him, much of the Japanese archipelago was open to the Pacific, having as it did a number of favourable port sites; but where traffic with continental Asia was concerned, only Kyushu had good harbours. He insisted, too, that because of the mountainous nature of much of the natural conditions prevailing in Japan, it was necessary for the Japanese to intensify agricultural usage, in order to enrich the country. In 1894, he published *Chirigakuko* (Considerations on Geography), his third book, following two books on Christianity. The first part of this book consists of a systematic presentation of general geography, and the second part consists of a regional geography of the continents of the earth. The fundamental theme of this book is an analysis of man–environment relationships, and in fact the second edition of this book was published in 1897 under the revised title *Chijinron* (Discussions on Earth and Man).

In his geographical writings, Uchimura several times made reference to *Earth and Man: lectures on comparative physical geography, in its relation to the history of mankind* by Arnold Guyot (1849) as an important work of its kind. He also referred to *Physical Geography* by the same author as well as to *Geographical Studies* by Carl Ritter translated by the Reverend W.L. Gage, Peschel's *Vergleichende Erdkunde*, and *The Earth as Modified by Human Agencies* by George P. Marsh as being of significance for him. When he made use of the terms 'vocation' or 'mission' he meant them in the Christian sense; but, while a firm advocate of Protestantism, insisted on a properly Japanese interpretation of Christianity and cast a somewhat jaundiced eye on the activities of foreign missionaries in Japan, whether Catholic or Protestant. He was nationalist in the strict sense in his interpretation of what the 'vocation' of the Japanese nation consisted of, and in his English-language article published in 1894 justifying the Sino-Japanese war. But his nationalism differed considerably from the nationalism encouraged by the Japanese government of that time; hence he was compelled to resign from a lectureship at the First Higher School of Tokyo because, due to his Christian convictions which made it unthinkable for him to consider the emperor on a par with God, he refused to stand up at a reading of the Imperial Rescript. After 1902 he opposed the Anglo-Japanese Alliance Treaty because it called for Japanese involvement in British international politics and this he could not condone; subsequently he also opposed the Russo-Japanese war in 1904.

The career of Shigetaka Shiga differed greatly from that of Uchimura. Even before his enrolment at the Sapporo Agricultural College he had harboured a considerable antipathy towards Christianity, and once in college he strongly resisted the pressures brought to bear upon him to convert to Christianity. His studies at the Sapporo Agricultural College awakened in him an interest in natural history, especially geography. In 1866 he seized the opportunity to board a trainee warship operated by the Japanese navy which proceeded to visit the Southern Pacific islands and Oceania. He took along with him Charles Darwin's *Journal of Researches into the Geology and Natural History of the Various Countries Visited by H.M.S. Beagle, 1832–1836* (1839) which, as far as he was concerned, constituted an important guidebook. His account of this voyage, the *Nanyo-jiji* (Affairs of the Southern Seas), is a mixture of scientific reports on the islands and countries visited and geopolitical advocacy of

Japanese advances into the Southern Pacific area. Later, he taught geography at the Tokyo Semmon Gakko (the present Waseda University); the transcripts of his lectures were published for the first time in 1889 under the title *Chirigaku kogi* (Lectures on Geography).

Shiga did not specify the works to which he referred but, judging from his writings, we may suppose that he was relatively well informed with regard to the geographical writings of his day. For example, at the very beginning of his 'Lectures on Geography' he defined geography as 'the science of observing all phenomena relating to the earth' and stated that 'man could not exist without the earth, but nothing happens that does not involve man. We need to recognize the fact that, among the hundreds of disciplines extant, the study of geography is the most urgent and necessary.' His interpretations of earth-surface phenomena were formed in an environmental context, but at no time was he a determinist. In the same book, after pointing out several geographical characteristics of Japan derived from her insular position, he averred that 'man is not influenced only by his relationship with the land or by geographical factors alone'. He insisted on the importance of the moral character of the people for the formation of a nation, citing as an example the achievements in the form of land reclamation work by the Dutch people. Of paramount concern to him at all times was the Japanese nation, in connection with which fact he stated that 'it is vital for as many Japanese as possible to learn the geographical position of Japan in the world. In order to achieve this task, the study of geography is very necessary.' It was along these lines that he wrote the *Nihon fukeiron* (The Japanese Landscape), published in 1894, in which he extolled its beauty and distinctive features, comparing it with the landscapes of the West and of China, and basing his postulations on his extensive knowledge of physical geography. This work was widely read and contributed to the encouragement of a nationalist sentiment among the people; for Shiga's work induced in people what can only be described as a peculiarly Japanese mode of feeling, consisting of the awareness of a national identity based on pride in the beauty of the Japanese landscape.

Recent studies of Shiga's writings have revealed, however, that his work relied heavily on English writings on alpinism and natural scenery, notably the *Handbook for Travellers in Japan* (London and Yokohama, 1891), a joint work by Basil Hall Chamberlain (1850–1935) and Willen Benjamin Mason (1854–1923).[3] In point of fact, Shiga's viewpoint with regard to scenic beauty as hypothesized in 'The Japanese Landscape' greatly differed from the traditional Japanese viewpoint that tended towards the miniaturistic and the contrived. In Shiga's view, a beautiful landscape involved nature in its wild state with mountains and the precipitousness of mountains as important components, after the fashion of Alpine scenes. The popularity of Shiga's book indicated that the aesthetic sensibility of the Japanese people had become transformed, by that time, through the process of modernization, and that Shiga attained success in his efforts to rouse the nationalistic instincts of the Japanese people by his reappraisal of scenic beauty; a reappraisal based, moreover, not on traditional Japanese mores but on the aesthetic values of the Western world. Generally, most studies underline the nationalist aspect of Shiga's thought.

He was a member of parliament from 1903–4 and supported the war against Russia, joining the besieging army in Port Arthur in order to observe the battle there. But after 1910, when he travelled round the world, his attitude towards world affairs and what he saw as the tasks confronting the Japanese nation in the international sphere gradually underwent a change. He became, in fact, somewhat pessimistic regarding Japan's development in the world sphere, what with the growing boycott of Japanese immigrants in America and elsewhere and the increasing racial discrimination that he experienced at first hand in South Africa and other places. His last work, *Shirarezaru kuniguni* (Unfamiliar Countries) appeared in 1925; in it he reiterated the need for an open-minded attitude towards international affairs, an opinion that was diametrically opposed to the increasingly chauvinistic attitude of the Japan of the 1920s.

Inazo Nitobe, who after his graduation from the Sapporo Agricultural College studied agronomy in the United States and Germany, never actually used the term 'geography' in any of the titles of his works; but in his *Nogyo Honron* (Principles of Agronomy) published in 1908, he referred to a number of geographical books such as Marsh's *Earth as Modified by Human Action* and F. Ratzel's *Anthropogeographie*. Chapter 6 of this latter book presents village forms classified according to their origin and functions. Relying mainly on A. Neitzen's methods, the work of classification in this particular chapter constituted the first systematic description of settlement geography in Japan. He discussed the influence of physical conditions on agriculture on the basis of the development of agronomical technology. While he developed a somewhat physiocratic view, he was careful to point out that the importance of agriculture to the national economy differed according to country and the stage of economic development.

He was, moreover, a nationalist in the sense that he constantly strove for a better understanding abroad of the Japanese and of Japanese culture by giving lectures abroad and writing several books in English on Japanese culture. Also a Christian and a friend of Kanzo Uchimura, he however exercised a broader influence than that of the latter, in the capacity of a professor who advocated moralistic and idealistic principles. His pacifist convictions led him to serve a term as Deputy Secretary General of the League of Nations during 1920–6; but he felt compelled to resign from this post at the onset of Japan's now explicit acts of armed aggression in China.

Furthermore, in the history of Japanese geography, Nitobe fills an important niche because of the influence he exerted on certain Japanese geographers who, in one way or another, played a part in the early development of geography in Japan. Tsunesaburo Makiguchi (1871–1944), author of *Jinsei Chirigaku* (Geography of Human Lives), published in 1903 and widely read by schoolteachers, particularly those who aspired to obtain a middle-school teacher's licence, later participated in the regular study meetings on Japanese folklore held at Nitobe's home from 1910. In Makiguchi's second book, *Kyoju no togochushin toshite no kyodoka kenkyu* (Homeland Studies as Focuses of Schoolteaching), published in 1912, it is possible to discover the influence of Nitobe, especially in the chapter dealing with observations on the village.

Another member of the study group held at Nitobe's home was Michitoshi Odauchi (1875–1954), who published a large number of pioneering studies on settlement geography in the 1920s and 1930s. A part-time lecturer who read geography at several universities, Odauchi was admitted as a member of the Association of Japanese Geographers, which had been founded in 1925 and constituted what was, up till the time of the Second World War, a highly exclusive circle of academic geographers. In spite of his acceptance into the Association, however, Odauchi was considered an 'outsider' by orthodox academic geographers.

After the Second World War Odauchi was already in old age, and a reappraisal of his work began to develop among a large number of geographers belonging to the younger generation of scholars in academic geography. In 1946 Odauchi addressed a short critical remark to academic geography circles stating that what he regretted most was that geographical academia had disarticulated itself from the tradition of agronomy. He did not specify exactly what he meant by the term 'tradition of agronomy'; but, judging from his previous writings, we may safely suppose that he was referring to the thought expressed in the geographical writings of the former students of the Sapporo Agricultural College, particularly Kanzo Uchimura, Shigetaka Shiga and Inazo Nitobe. There is no doubt that Japan's geographical academia has failed to avail itself of or develop a large part of the intellectual heritage bequeathed to it by the pre-academic geographers of a few generations ago. To an extent, the 'outsider' tendencies evinced by Odauchi found a parallel in Shiga, upon whose death in 1927 Naomasa Yamasaki (1879–1929) composed an obituary for the *Chirigaku hyoron* (Geographical Review of Japan), the organ of the Association of Japanese Geographers, in which he commented that 'The Japanese Landscape' was without doubt effective in implanting knowledge of and interest in geography in the mind of the general reader in Japan, at a time when people knew nothing about the subject. In short, for this authoritative academic geographer, who was also professor of geography at that stronghold of academia, the Imperial University of Tokyo, the role of Shiga was not so much that of scholar but rather popularizer of geographical and geological knowledge among the Japanese people.

Further serious consideration should be given to the geographical thought of the period between the 1880s and the 1920s, which was broadly formed under the influence of agronomical disciplines and given expression in social practice. While the practitioners themselves were Meiji-period nationalists, they none the less came to adopt a critical stance with regard to the official policies of the age, since the brand of nationalism they professed differed from the dominant ideological trends comprising the nationalism of the imperial Japan of the twentieth century.[4]

NOTES

1 For further details see K. Takeuchi 'How Japan learned about the outside world: the views of other countries incorporated in Japanese school textbooks, 1868–1986', *Hitotsubashi Journal of Social Studies*, 19 (1987), 1–13.

2 An excellent biological and bibliographical study of Shiga is presented by S. Minamoto, in his 'Shigetaka Shiga, 1863–1927', *Geographers: Bibliographical Studies*, 8 (1984), 95–105. Otherwise the topics dealt with in this paper have already been discussed in more detail in K. Takeuchi 'Landscape, language and nationalism in Meiji Japan', *Hitotsubashi Journal of Social Studies*, 20(1) (1988).

3 Minamoto, 'Shigetaka Shiga'.

4 I discuss this topic in K. Takeuchi, 'The influence of Japanese imperial tradition and Western imperialism on modern Japanese geography'. In Neil Smith and Anne Godlewska (eds), *Geography and Empire*, forthcoming.

7

Russian Geographers and the 'National Mission' in the Far East

Mark Bassin

In Asia we too are Europeans

What is left to be done in Asia? Only to place the second son of the Sclavic Czar on the throne of Genghis. It would be a vast step in the progress of the Mongol race and of civilization and worthy of the great advance and the great epochs of the 19th century, and the only means by which nearly half of the inhabitants of the earth can be Christianized and brought within the pale of commerce and modern civilization .... For without Russian interposition, the Mongol race must go down in internecine religious wars, pestilence, and famine, pressed as they now are on all sides by the irresistible force of Christian powers .... The blood of Japhet has triumphed over that of Sham. The curse of Noah is about to be accomplished, the prophecy fulfilled, and Asia Christianized.

Perry McDonough Collins, 1860

Russland ist durch seine Stellung und Bedeutung vor allen Ländern dazu berufen, der Erforscher des asiatischen Kontinents zu werden.

Gustav Radde, 1861

INTRODUCTION

The critical importance of geographical exploration in the process of imperial expansion in the nineteenth century is well appreciated and requires no particular comment. In virtually every country that sought to participate actively in the contest for territories and colonies in the non-European world, the pressing need for topographical surveys, charts of coastlines and potential port facilities, evaluation of natural resources, reports on indigenous peoples and a spectrum of other information ensured that geographers, and the organizations that sponsored them, would have an illustrious and universally acknowledged role to play in the national endeavour.[1] What is perhaps not so well appreciated is that the efforts of explorers were not limited to obtaining what might be called 'practical' geographical information, such as those items just mentioned. Imperialism was after all not only a pragmatic policy and programme of political action. At the same time, it was a mentality, a *Weltbild*, and a system of values that fitted within the larger ideological constellation of the colonial protagonist's nationalism. The far-off and little-known lands that were to be appropriated and colonized had to be explored not only in terms of their mineral wealth and transportation routes, for these *terrae incognitae* had to be assimilated *conceptually* as well and somehow rendered meaningful within the same general ideological framework that produced the imperial impulse. In the present chapter, I shall consider how Russian geographers confronted this task and undertook to 'discover' – for themselves and for the broader Russian public – hitherto unknown territories in the far south-eastern corner of Siberia such that they could be fitted within the vision of a national mission of civilization and enlightenment in Asia.[2] In this process, we shall be able to appreciate the specific contribution these geographers made to Russia's evolving national identity and sense of self-purpose in the nineteenth century.

NATIONALISM AND RUSSIA'S MISSION TO THE EAST

Hans Kohn and others have demonstrated the extent to which the emergence of nationalism in nineteenth century Europe was a reaction to a major crisis of national self-identity in the country in question.[3] Russia was no exception, and the 1830s and 1840s witnessed the emergence there of a nationalist doctrine centred on the problematic and increasingly painful problem of Russia's relationship to the societies and cultures of Western Europe. For a century after the promulgation of the Petrine reforms in the early eighteenth century, Russia's educated elite had remained diligently faithful to Peter's attempts to 'Europeanize' Russia and bring it as it were into the fraternity of European nations. This endeavour, the nationalists of the 1830s declared, had not only proved a grand failure but moreover had been fundamentally misguided from the outset. Europe would never accept Russia as one of its own and indeed should not do so, for Russia was a very different sort of nation. Russians had worried far too long about Europe and slavishly tried to imitate it, they insisted, and in so doing had failed to recognize, had indeed even denigrated, their own unique culture and civilization. The time had come for Russia to break away, to concern itself solely with its own special needs and pursue an independent path toward its future development. Exactly what this independent path looked like or precisely where it would lead were far from clear, however, and the nationalists consequently cast around rather desperately for any means or activity that might further promote the country's welfare and help bring it closer to realizing a glorious destiny. What they sought was nothing less than a national mission, a grand undertaking that could establish their national virtue and the fact that they indeed possessed a capacity for creative national achievement independent of the West.[4] It became apparent almost immediately that at least one alluring mission might lead Russia to the East, to Asia.

Imperial powers have historically taken satisfaction in the beneficial 'civilizing' effects assumedly exercised by their dominion over foreign peoples and regions, and the Russian nationalists seized enthusiastically upon this option. As the bearers of enlightenment and Christianity to the stagnant and backward peoples of Asia, they reasoned, they could actually compete with other European countries and conclusively demonstrate that they stood at the same high and worthy level of development. Indeed, in their own backyard they had available a broad geographical arena upon which to work, an arena which moreover was territorially contiguous and thus apparently much more easily and naturally accessible to them than to the rest of Europe. From this standpoint, the Russian nationalists could even argue that the task of civilizing Asia belonged first and foremost to Russia. Of all of Europe, demanded the Orientalist V.V. Grigor'ev in 1840,

> Who is closer to Asia than us? ... Yes, if the science and civic life of Europe must speak to Asia through the mouth of one of its peoples, then of course it will be us. Is it not clear that Providence preserved [Asia] as if intentionally from all foreign influence, so that we Russians would find it ... more capable of, and inclined to accept those gifts that we [alone] bring to it![5]

These sentiments came to a explosive head in the early 1850s, when collective European belligerence erupted against the Russians in the form of the Crimean War. The message for them was clear: Europe had definitively declared its hostility against Russia and the intention not to admit the country into its ranks. More than ever, therefore, Russia faced the imperative to turn away from the West to new directions that would offer Russia the opportunity for constructive and creative national effort, and more than ever it was clear that this new path would lead to the East. Aleksei Khomiakov captured this spirit in a letter to the nationalist historian Mikhail Pogodin at the outbreak of the war 'All of our attention has been directed toward European affairs, [while] our true interests summoned us to intensified activity in the East, which could have become ours very easily.'[6] And after the war had taken its desperate turn for Russia, Pogodin himself elaborated upon these sentiments. With near-maniacal jubiliation, born quite obviously out of the deepest despair, he sketched a dizzying programme of activity:

> Leaving Europe in peace and in expectation of more favorable circum-stances, we should turn all of our attention to Asia, which we have practically ignored, although it is predestined [for our domination] ....
> Let Europeans live as they know how and spread out in their territories as they like; for us, half of Asia lies waiting – China, Japan, Tibet, Bukhara, Khiva, Persia – if we want to expand our territorial holdings. And perhaps we ought to, in order to spread the European element throughout Asia and to let Japhet tower above his brothers .... Lay new

roads into Asia or seek out old ones, establish communications along at least the traces of those lines indicated by Alexander the Great and Napoleon, set up caravan routes, girdle Asiatic Russia with railroads, send steamships up its rivers and across its lakes, and connect it with European Russia .... Asia, Europe, influence upon the entire world! What a magnanimous (*velikodushnyi*) future awaits Russia! And is there anything here that is impossible?[7]

There was a subtle but notable ambiguity in Pogodin's logic here, for at the same time that Russia turned *away* from the West toward Asia, it was apparently doing so as a representative of this very West, in order to, as he put it, 'spread the European element'. In the sections that follow, we shall see how Russian geographers responded to the spirit of Pogodin's febrile exhortations and how they also reflected the ambiguity of his position.

THE CHANGING SITUATION IN THE FAR EAST

In the second quarter of the nineteenth century, events in the Far East and on the Pacific took a turn that within a few short decades would completely transform the political and social order of these parts of the world.[8] For centuries Western powers had been making attempts at economic penetration into the populous markets of imperial China, and for centuries these attempts had been effectively resisted by a Chinese government that was manifestly uninterested in allowing foreign access into the country's interior. In the late 1830s Chinese resistance was finally challenged by Great Britain in the form of the Opium Wars. With the Treaty of Nanking, which ended the conflict in 1842, as well as the so-called 'unequal treaties' concluded the following year, the Chinese were constrained to make a variety of major concessions that would strengthen the British presence in China and allow them to expand radically their opportunities for commercial activity. Other Western powers quickly followed suit, and throughout the rest of the decade, similar concessions were sought and obtained by the French, Belgians, Americans and others. With this, the Middle Kingdom was exposed for the 'paper dragon' or helpless giant that it had become in the face of the military and economic dynamism of Europe. In the mid-1850s, under different circumstances and without a military engagement, Japan followed China's lead and signed a treaty with the American Commodore Matthew Perry, similarly granting access to interior markets. Thus the exotic societies of East Asia were 'opened' to Western interests, and the prospect this offered for virtually all Western observers was positively intoxicating.

The Russians, no less than anyone else, followed the events unfolding far away in the Pacific, and no less than anyone else their imaginations were captivated by them. The sentiment expressed by Alexander Herzen that the Pacific seemed truly destined to become 'the Mediterranean of the future' was shared by much of Russian public opinion,[9] and the determination was strong that Russia should not forgo participation in the dynamic arena of activity that was taking shape in the Far East. Indeed, there was an issue that served directly to link the country geographically as well as politically with this arena. The Amur river valley, a frontier region on the borders of south-eastern Siberia and northern Manchuria, had been occupied by Russian Cossacks for some five decades in the seventeenth century, before they were forced to evacuate by a superior Chinese military force in 1689.[10] Since that time, Siberian merchants and officials had regularly noted the desirability of acquiring at least navigational rights along this river, but the events of the early decades of the nineteenth century served to recast the significance of the Amur valley and refocus the attention of the country upon it in a major way.[11]

By regaining possession of these territories, which the Russians believed belonged to them by historical rights and had been effectively stolen by the Chinese, they would have a convenient riparian artery running from deep in the interior of Siberia to the Pacific, along which the country could participate and share in the benefits from the blossoming commercial activity in this part of the world. The Amur region itself was reputed to be generously endowed with natural resources and precious metals and – quite unlike the rest of desolate and barren Eastern Siberia – to contain rich agricultural lands. Possession of the Amur valley, moreover, would insure that Russia could retain Manchuria and northern China securely within its political and economic sphere of influence, and prevent these regions being staked out and carved up by other European powers. Yet while all these considerations were undeniably weighty, they were none the less most meaningful to those few Russians who knew something about these territories and Russia's position in the Far East. There was, however, another dimension to the developments on the Pacific that had a potential intrinsic appeal for much broader circles of nationalist Russian opinion. The advance into the Amur valley was plainly an advance into Asia, and as such could provide the country with a desperately sought-after opportunity to pursue its national mission of bringing progress and civilization to the grateful heathen masses. Before such an appeal could exercise its full effect, however, these territories had to be 'discovered', and this was largely to be the work of Russian geographers and explorers.

NATIONALISM, GEOGRAPHY, AND THE REDISCOVERY OF THE AMUR

Geographers figured enthusiastically in the ranks of the nationalists, and for them the prospect of a national mission in the East carried an entirely special significance. It was they, after all, who more than any were responsible for providing the basic information which Russian activities in the remote and little-known reaches of Asia would require. The founding of the Russian Geographical Society in 1845 was a clear indication that geographers were not only aware of this challenge but indeed quite anxious to respond to it. The impetus behind the establishment of the Society was intensely nationalist, and at the inaugural ceremony its president made clear the intention that it would devote its scholarly attention exclusively to the study of the *otechestvo* or Russian fatherland and to the enhancement of the country's welfare and future independent development.[12] That the effort to promote Russian interests in Asia would be a major concern of the Society was apparent from the outset, in the fact that the very first large-scale project undertaken was a translation of Carl Ritter's multivolume *Erdkunde von Asien* for use in Russia. Indicatively, this was in fact only a selective translation of those volumes dealing with regions that were geographically contiguous to the Russian empire and thus most specifically meaningful in terms of the imminent imperial project.[13] The contributions of the Society towards the furthering of Russia's mission to the East were by no means limited to such bibliographic undertakings, however, and geographical exploration to Asiatic parts of the Empire and beyond formed one of its major areas of activity throughout the nineteenth century.[14]

Nowhere was this special role of geographers in Russian expansion more pronounced or critical than in the Amur valley, for in the early 1840s the region was a veritable *terra incognita* for the Russians. Its natural-geographical characteristics were entirely unknown, as were the extent and even the precise course of the river itself. Two unresolved issues in particular were important as the question of Russian claims on the river began to be discussed. By the letter of the treaty signed with China in 1689, in which Russia had relinquished all claims to the river, a line of stone markers was to be erected delimiting the official boundary between the two countries. It was not known in St Petersburg exactly where these markers were, or even if they had ever been set up. The point was a significant one, for the further the boundary line ran to the south, the more enhanced were potential Russian claims to the territories of the Amur valley. The second area of uncertainty, rather more important, was the widespread belief that the mouth of the Amur was so shallow and filled with sand bars that it rendered the river effectively unnavigable for ocean-going vessels and consequently useless to the Russians.[15]

What made these two issues so particularly important was the circumstance that, throughout the 1840s and early 1850s, the Russian government was not interested in promoting the Russian presence in the area. Indeed, far from endorsing the forward policy urged by the nationalists, the hyper-conservative ministers of Tsar Nicholas I (1825–55) were instead concerned with precisely the opposite: to maintain quite strictly the *status quo* in the Far East. At the time, Russia conducted a trans-Mongolian caravan trade with China that was moderately lucrative, and the efforts of the government were directed first and foremost at insuring that the income derived from customs duties collected at the border depot at Kiakhta remained intact. In view of this concern, it was held to be categorically inadmissable to alienate the Chinese and threaten their co-operation in the border trade by levelling demands for such unilateral territorial concessions as the Amur valley.[16] Given this policy orientation, the unresolved questions about the geography of the region were extremely useful for the government, for they offered a convenient and as it were neutral excuse for resisting all proposals for Russian activities in the Amur valley.

Needless to say, this official opposition was enormously frustrating for Russia's nationalist forces. The inevitable conclusion was that if Russia's national interests were to be pursued, and an arena for its constructive activity in the East secured, they would have to act on their own initiative and with their own resources, independently from official governmental intentions and if necessary even surreptitiously. The extent to which they were prepared in this way to contravene official policy was an indication of the depth of their commitment to their own vision of Russia's future. And in so far as the first challenge on the Amur was a geographical one – in other words, to confront the uncertainties just mentioned and offer some more positive and detailed geographical information about the south-east corner of Siberia – geographers and explorers took the lead in resurrecting the Amur issue and advancing the cause of expansion in the Far East. As we shall see, they were animated in this endeavour by the conviction that in advancing Russia's position on the Pacific they were securing an arena in which the country could exercise its new-found national mission as the Civilizer of the East.

The first explorer to visit the Amur in the period under consideration was Alexander von Middendorf. The original idea for his expedition had been conceived in the 1820s by the naturalist Karl von Baer, who was interested in investigating the relationship of organic life to climatic conditions in various environments. With the sponsorship of the Academy of Science, three sites were chosen for study – the Taimyr peninsula in north-central Siberia, Yakutia, and the Shantar Islands off the southern coast of the Sea of Okhotsk[17] – and the execution of the expedition assigned to von Baer's young colleague Middendorf. Middendorf set out in 1842 (the year in which Britain concluded the Treaty of Nanking with China), spent some two years on the Taimyr peninsula, travelled to Yakutsk in early 1844, and proceeded in the autumn of the same year to the Okhotsk Sea. Upon completion of his observations on the Shantar Islands, Middendorf's instructions directed him to return to Yakutsk and thence to St Petersburg.[18] At this point, however, his expedition took an entirely unanticipated turn, for the explorer was intent to press further south and survey the mouth of the Amur and the river itself.

In venturing to the Amur, Middendorf's primary goal was to seek out the border posts that had ostensibly been erected in 1689.[19] While still in Yakutsk, he sent a communication to the Academy of Science explaining his intention to alter his itinerary and requesting permission to travel to the mouth of the Amur and proceed upriver, returning to Russia via Irkutsk rather than Yakutsk. As may be expected in view of the official governmental policy, the Academy of Sciences rejected his request out of hand and a letter was dispatched from St Petersburg forbidding him to carry out his plan. Middendorf, however, no doubt anticipating the rejection of his request, left Yakutsk for the Okhotsk Sea before the reply from the capital could reach him, and 'paid no attention' as he proudly wrote in his subsequent account of his journey, 'whether [his Amur trip] corresponded to the intentions of the central administration or not'.[20] He reached the mouth of the Amur in December 1844, carried out his investigations as he travelled upstream, and arrived back in St Petersburg some four months later. On the basis of his observations, Middendorf was able to demonstrate conclusively what advocates of the 'Amur issue' had been claiming, namely, that the Russian–Chinese border of 1689 was in fact located considerably further to the south than the government suggested.[21]

It was an unmistakable testament to the intense appeal exercised by the Amur issue even at this very early date that a young naturalist would, by flaunting what he understood perfectly well to be the express wishes not only of his sponsoring organization but of the Russian government as well, risk his career for the sake of it. In his account of the expedition published some years later, Middendorf discussed the Amur and the issues that had compelled him to act as he had. He spoke of the practical economic benefits to Eastern Siberia and Russia overall that the acquisition of this region – reputedly rich in agricultural and other natural resources – would bring, but it was clear that he attributed a national significance to this acquisition that went well beyond these material considerations. Referring to the prospect of the incipient American occupation of California, he pointed out that the Americans and Russians were simultaneously emerging on the Pacific as neighbours, and further noted that both of these actions brought to a successful completion a world-historical process of colonial expansion by European peoples – the Anglo-Saxons to the west and the Slavs to the east – that had begun centuries earlier.[22] The point that Middendorf wanted to emphasize most, however, was not the conclusion of one historical period but rather the dawning of another: a period in which Russia's national energies would be focused upon the civilizing of those regions over which they had extended their influence. In this manner, he incorporated the acquisition of the Amur valley into the vision of a Russian mission in the East.

In regard to the unsettled territories of the Russian Far East, Middendorf saw the advance of civilization in two different respects. On the one hand, he depicted it as the struggle of man against the uncontrolled elemental forces of Nature (or 'savagery' (*varvarstvo*), as he put it), in other words, the attempt to civilize a wild and inhospitable realm and bring it under rational human control for purposes that would be beneficial for human society. That such a process would eventually take place in this region was quite inevitable, he noted, for in so far as the Amur provided the only access into remote Siberia from the Pacific, Western naval traffic 'and with it [Western] civilization were sooner or later bound to penetrate' up it.[23] Nevertheless, it was the Russians who, approaching overland from the west, were the ones to take matters resolutely into their own hands and actually set about accomplishing this, and their achievements reflected magnificently upon them. Some 15 years after his expedition, after the region had been formally annexed from the Chinese empire and the Russian occupation had begun, Middendorf could write with enormous satisfaction of the beneficial effects these developments had had. 'The Amur region has been aroused, as it were, from an enchanted sleep that lasted thousands of years .... Man has declared war ... on nature: he is combing the oceans for whales, he is annihilating [*istrebliat*] primeval forests with fire and iron, he is building homes and farms.'[24] By taming nature and establishing at least the rudiments of their own social and economic order, the Russians were demonstrating their capacity for constructive achievement and their ability to contribute to the progressive development of Western civilization.

The other demonstration of the advance of civilization through Russian agency was in regard to the societies of backward Asia. As we shall see, the geographers who were to follow Middendorf's tracks to the Amur region expressed themselves in this regard more emphatically and at far greater length than he did himself. Nevertheless, he made it quite clear that he as well recognized the significance of this aspect of Russian activity in the Far East. The Russian territorial advance to the Pacific coast brought it into direct and intimate contact with China, and in this position Middendorf perceived what was at once a novel responsibility and an opportunity for the Russians. As they were at work settling and developing Siberia, he speculated, 'the Middle Kingdom will undergo a renaissance, [and] its tightly packed population is bound to spread out'. Middendorf apparently was suggesting that Chinese from the south could be used to occupy desolate parts of Siberia. Beyond this, however, he spoke in more general terms about Russia's future tasks in regard to Asian society as a whole: 'The resurrection of ossified and petrified Asia is a task, the solution of which will be up to our future generations. And then ...?'[25]

Although the information brought back by Middendorf regarding the location of the boundary markers played a significant role in stimulating interest in the Amur region, a considerably thornier issue remained to be resolved. This was the question as to precisely how navigable the river was. The government accepted the opinion of numerous Russian naval officers that Sakhalin was in all likelihood not an island but a peninsula that connected to the mainland by an isthmus north of the Amur estuary. Such a geographical arrangement would make direct access to the Amur from the Sea of Okhotsk and the North Pacific impossible, thereby cutting the river off from the principal arenas of Russian settlement and activity in the Far East – the Okhotsk coast, Kamchatka, and Russian America – and rendering it effectively useless for the Russians. Even if Sakhalin were in fact an island, the argument continued, the mouth of the Amur was nevertheless so shallow and filled with sand bars that it would still make entry or exit by ocean-going vessels impossible. These were substantial objections by any measure, and the advocates of annexation appreciated that until the navigability of the Amur estuary and its accessibility from the north could be proven, there would be no reasonable hope for their cause.

It was with just this intention that two young members of the recently founded Russian Geographical Society, Alexander Balasoglo and Genadii Nevel'skoi, began in 1847 to make plans for a naval expedition to the Far East. Balasoglo, a geographer who worked in the archive of the Foreign Ministry, was one of the most knowledgeable people about the Far East in the capital.[26] He had taught for several years in the Naval Academy, where the cadet Nevel'skoi was one of his students. Balasoglo was driven by a passionate, almost fanatical, nationalism which pronounced Russia to be the equal of all of Europe.

It is time for Russia to understand its future, its calling in humanity. ...
Russia is precisely another Europe, a Europe between Europe and Asia,
between Africa and America – a marvellous, unknown and new country.
In it, and only in it, are concentrated all the threads of world history.
... The Slavic soul is the chosen vessel for the melding of all peoples
into humanity.[27]

As part of this magnificent destiny, Balasoglo fervently insisted that Russia would play a role in the East and, indeed, that Russia possessed a very special prerogative in this regard:

> The East belongs to Russia unchangeably, naturally, historically, voluntarily. ... It was bought with the blood of Russia already in the prehistoric quarrels of the Slavs with the Finns and the Turks, it has been suffered for in Asia through the Mongol yoke, it was welded to Russia by the Cossacks and has been earned from Europe by protecting it from the Mongols. [28]

We do not have comparable statements by Nevel'skoi, but on the basis of his close friendship with Balasoglo and his co-membership in the Russian Geographical Society it may be assumed that he fully shared in his colleague's sentiments. [29] Both of them saw the prospect of a Russian advance in the Amur valley as a necessary step toward Russia's self-assertion in Asia and the full realization of a burgeoning national destiny. Moreover, both were resolutely determined to make their own contribution, as geographers and explorers, to this dynamic process.

By 1849 the plans were set and the expedition organized. At the last minute, Balasoglo's participation in the project was interrupted by his arrest in connection with his activities in the Petrashevskii circle in St Petersburg. Distressed by this development but undaunted in his resolve to further Russia's destiny in the Far East, Nevel'skoi proceeded without him on the sloop *Baikal*, which had been built under his commission the year before in Helsinki. Nevel'skoi's official mission was to bring supplies from Kronstadt to the port of Petropavlovsk on the southern tip of Kamchatka, after which he had permission to proceed south to investigate the southern Okhotsk coast. His explorations in the summer and autumn of 1849 confirmed precisely what he expected: there was no isthmus connecting Sakhalin to the mainland, and the mouth of the Amur moreover was navigable by ocean-going vessels of any size. Nevel'skoi was sent back to the Okhotsk Sea in 1851 to determine a suitable location along the southern coast for a port. He was expressly forbidden to venture south to the Amur. By this point, however, his enthusiasm for advancing Russia's position in the Far East was not to be repressed, and like Middendorf he simply violated his orders. He not only sailed into the Amur estuary but actually established an outpost there, the first Russian settlement in the region in a century and a half. [30] With this, the definitive occupation of the Amur valley by the Russians had begun.

GEOGRAPHICAL 'DISCOVERY' AS CULTURAL CONSTRUCTION

The letters and diaries that have survived from Nevel'skoi's mission enable us to follow the thoughts of the participants in a way that was not possible with Middendorf. As we do so, it quickly becomes apparent that the process of 'exploration' or 'discovery' in which they were engaged was complex and multifaceted; indeed, it might be said that not one but at least two quite discrete processes were taking place. The first, and more immediately apparent, of these we have just summarized and corresponds to what is traditionally understood by the notion of discovery: navigating and describing unknown waters, surveying coastlines, establishing navigational conditions and so on. The second process was more subtle and corresponded not so much to the need for 'objective' geographical information as to the needs of the new nationalist consciousness. For the intrepid Russians aboard the *Baikal* were not merely filling in a blank region on the map, they were at the same time locating an arena that clearly stood in need of the very civilization and enlightenment that the Russians were desperately seeking to provide. From this standpoint, the entire project of 'exploration' and 'discovery' took on a significantly enhanced meaning, for it transcended a concern with naive geographical reality and became an essentially psychological exercise in cultural perception, interpretation and construction.[31]

This enterprise commenced as soon as Nevel'skoi disembarked from his ship in the early days of August 1851. He related that he immediately encountered a group of Giliaks, a tribe indigenous to the lower Amur, who were being lorded over by an arrogant official from Manchuria. A brief exchange served to convince Nevel'skoi that the presence of the Manchu was resented and that the natives entertained a 'concealed hostility' to their 'oppressors'. Nevel'skoi reported with satisfaction how he was able to excite the awe and barely concealed glee of the assembled Giliaks by drawing his revolver and threatening the offender, and their reaction convinced him that if it came to a confrontation, the natives were already sure to be on his – the Russians' – side. As Nevel'skoi became more familiar with the natives of the Amur, he learned that the Manchus were not their only problem. For some years, British and American ships had been sailing up the Tartar Straits. These 'white men', the natives complained, terrorized them, plundering their provisions and furs and committing wanton atrocities. 'The Giliaks, not knowing whom to turn to for aid and protection, and having not a single means of defending themselves, do not know how to repulse and punish the intruders.' Nevel'skoi recognized his opportunity to announce and justify Russia's new presence in the region, and did not fail to seize the occasion. Affirming that the Russians have always considered the river valley to be a part of their empire and its inhabitants to be imperial subjects, he proclaimed his arrival as their saviour and he made the following proclamation: 'In order to protect you poor indigenous peoples who are his subjects from the offences of foreigners, the Great Tsar [*Pili-Pili-Djanguine*, or very very rich old man] has decided to erect military posts in the Bay of Hope/Nadezhda and on the mouth of the Amur, a decision which I, as the emissary of the Great Tsar, solemnly declare to you.'[32] Thus, in the space of a single morning, an exploratory party began to implement one of the most precious principles of Russia's national mission of salvation.

Even more than Nevel'skoi himself, however, the 'discovery' of the Amur as an arena for the exercising of Russia's mission was pursued by his wife Catherine who, in keeping with Russian military custom, faithfully accompanied her spouse on all his missions. While he was occupied with naval operations and the construction of the new settlements, she turned her full attention to the indigenous groups of the region. In this way she was effectively the first Russian to have extended contact with them, and in her reactions we can see outlines of an attitude taking shape that would become characteristic of others who followed. Her initial evaluation, of course, was uncompromisingly negative, for the groups she encountered seemed to her to represent in every way the lowest level of civilization imaginable. In a letter to her sister, written a few days after her husband's speech, she depicted these first impressions:

> We often see the ugly faces of the Giliaks lurking about us. Despite their characteristic cowardice, the expression on their faces is ferocious and cunning. Their clothing consists of a shirt of dog-skin, their shoes are made out of sealskin, and they wear their black hair, rough as a horse's mane, braided into pigtails: of which the men have many but the women only two which they fasten together with a cord. They have enormous hats, the form of which is vaguely reminiscent of Tyrolean caps, crafted largely out of tree bark and decorated along the brim with awful designs. It is all atrociously ugly.[33]

In the weeks that followed, Catherine Nevel'skoi elaborated upon these impressions in a flow of letters. Her daughter assembled this correspondence after Catherine's death and, making full use of her mother's vivid imagery, summarized the picture they depicted in the following terms:

> Overcoming her disgust, [Catherine] feeds the filthy, stinking, bloodthirsty Giliaks in their hovels (for it is quite impossible to call them homes), and these savages, warmed by her goodheartedness, trustingly tell her everything she wants to know. She doesn't shun their ugly unwashed wives and children, but rather, stifling her fear and repugnance, combs their hair, sews their clothes and teaches them how to sew. ... With them watching, she herself digs in the mud and plants potatoes, demonstrating in front of their very eyes that, contrary to their primitive fear, God does not punish the tiller of the soil with death, but rather rewards him for his labour.[34]

Here we have an early and unmistakable indication of the fact that with the Amur territories the Russians were acquiring far more than a link to the Pacific and the resources of a river valley. Beyond this, the region represented a vast panorama of precious ethnographic material, which could be effortlessly identified as savage and uncivilized and upon which the Russian nationalists accordingly had full opportunity to exercise their mission as civilizers and enlighteners. Indeed, the groups that Madame Nevel'skoi encountered must have been even more alluring from this standpoint than anything available in China or Japan, for stagnant and decrepit as the latter societies may have been, the unknown tribes of the Amur were unquestionably more so. In the case of the Amur, moreover, there was at least the chance that they could be Russia's and Russia's alone, something which could never be the case with the larger countries of East Asia.

The high point, it would seem, of her activities among the Giliaks was her missionary work, in bringing them the word of the Gospel. 'She herself learns their barbaric language so that she can speak with them, in order to instill in them the concepts of God, of goodness, of love for that which is close to them, [of concern] about their future life, and – with sensible arguments and on the strength of her own conviction and her charm – in order to convert them to Christianity.' By her own description, this message proved irresistible to the savages, who were so 'captivated and filled with timid admiration for our pious Sunday rituals and other religious festivals' that they came on their own volition *en masse* and asked for sacred ablution. There was little wonder that the natives would begin to deify her and, by extention, to recognize all their Russian masters as benefactors and saviours. 'Observing her, living close to her, these pitiful half-people, half-animals began to see her as a divine being, and to see the Russians as defenders and friends, honest, firm, and good-hearted.'[35] With this comforting conclusion, the Russian nationalists could successfully capture for themselves in the desolate wilderness of the far-eastern *taiga*, half-a-globe away from home, precisely that role and corresponding self-image they so desperately sought.

The exploratory activities we have been examining produced the desired results. The resistance of the government was gradually overcome, and throughout the decade of the 1850s the Russian occupation of the Amur valley proceeded apace. In 1854 a flotilla under the command of the governor-general of Eastern Siberia, Nikolai Murav'ev, made a descent down the river from its headwaters to its mouth, the first such Russian venture in nearly two centuries. Although its immediate goal was to help prepare for a possible naval engagement with England in the Pacific, it made clear at the same time that the Russians were going to advance claims of their own in the region. Four years later, in the wake of Russia's calamitous defeat in the Crimean War, the demise of Nicholas I and the ascent of his son Alexander II, Murav'ev negotiated the Treaty of Aigun with the Chinese establishing Russian jurisdiction over the left bank of the Amur and the river itself. These acquisitions were formalized and expanded in the Treaty of Peking, signed in 1860, by which Russia acquired the Ussuri river (a southern tributary of the Amur) and the territories between it and the Pacific coast as well. With this, Russia established its position as a modern Pacific power, and in recognition of his services toward this end the tsar elevated the governor-general to the nobility as Graf Murav'ev-Amurskii: Count of the Amur.

The activities of explorers on the Amur expanded along with the Russian presence there. While the most critical 'geographical' dilemma had been resolved by Nevel'skoi, a great deal of reconaissance and surveying work remained to be done in connection with the settlement, economic development and military fortification of the region. At the same time, a different task was becoming increasingly important: the need to introduce the country's newly-acquired *terra incognita* in the remote far-eastern reaches of Siberia to the Russian public west of the Urals. This task fell naturally to those with the most intimate knowledge of the region, and thus by the mid-1850s a trickle of letters, reports, articles and eventually books by these explorers began to appear in European Russia. In these materials, and especially in those designated for the popular periodical press, the psychological process of 'discovery' that we have observed on a private level with the Nevel'skois was re-enacted publicly, for a broad audience. Along with engaging descriptions of the physical geography of the region, its exotic flora and fauna, the Russian advances in the Pacific were imbued with the eminently virtuous auras of constructive activity, salvation and enlightenment, and in this way rendered meaningful in terms of a larger nationalist vision. Indeed, there can be little doubt that for most European Russians, it was precisely in this latter regard that these advances and territorial acquisitions portended the greatest significance.

Gustav Radde, for example, a naturalist dispatched by the Geographical Society in St Petersburg to study the biotic life of the Khingan mountains around the middle Amur, wrote a brief article in the popular illustrated weekly *Illiustratsiia* (The Illustration). In it, he endorsed Middendorf's optimistic anticipation that with the arrival of the Russians, the primeval and savage wilderness would yield to the forces and intentions of civilization. 'The riches of all realms of nature here are innumerable', he confirmed, but suggested that ossified China had done nothing to make them available for exploitation, such that one could do little more than admire them. It was not until the 'onset of [real] human activity takes place' – until, that is, the region was settled and developed by an advanced nation such as the Russians – that 'the age-old silence of the forests will disappear and the period of civilization will commence'. Pointing to the settlements of Cossacks being established at that moment along the Amur, he concluded for his readers with satisfaction that this moment had already arrived.[36]

Yet more graphic was the prospect of Russia as the benefactor and patron of the 'primitive' tribes of the region; and accounts of geographers picked up on Nevel'skoi's depiction of the Russians as the saviours of these helpless peoples from the pernicious domination of the evil Manchurians. Mikhail Veniukov, who was subsequently to become a well-known explorer and popularizer of Russia's advance into Asia, began his career during this period in the Far East, where he was the first Russian to explore the headwaters of the Ussuri river. His primary task was to prepare topographical surveys, but in his account of his mission, published in the journal of the Russian Geographical Society, he devoted considerable discussion to the native peoples that the Russians encountered in the region. Conversations with the Goldi peoples living along the Ussuri, he recounted, 'convinced even me that they bless their fate for the fact that the Russians have appeared on the Ussuri, [for the latter] are able to rule subject primitive peoples without ruining their life, and [moreover] have been long awaited there as saviours [*izbaviteli*] from the cruel yoke of the Manchurians'.[37] At the same time, he was explicit in his distress at the abysmally low level of cultural development of Russia's new charges, and referred to this fact in explaining why this rich river valley, despite its abundant and rich resources, should be so sparsely populated:

Here is manifested in all its force that unalterable law which determines that the successes of humanity even in the propagation of the race are in direct correspondence with the mass of blessings that are supplied by civilization. The hunters and gatherers who inhabit all of East Asia are limited in their demands by their ignorance, wander in the vast forests among the wild mountains, exposed to all the destructive influences of Nature. Finally, unable to withstand the cruel contact with organized tribes, these peoples will forever be unable to grow and multiply. ... Entire Goldi families die out under the influences of the more powerful Manchurians ....[38]

While Veniukov's topographical description of the Ussuri valley might have presented only a mixed interest for many readers, no one could have remained unmoved by his comments on the pathetic Goldi. The implications were eminently clear: the Russians had no choice but to adopt these children of nature and assume responsibility for their continued well-being. This would be accomplished through developing in this area 'all types of activity with which European civil life is rich', that is, Russian settlement and economic and social development.[39] Radde, for his part, depicted how this very process was taking place: 'The savage Giliak is submitting to the strictness of [European] laws, and the poor Goldi and the Amur Tungus – who have done nothing but suffer oppression up to this point – are joyful at the protection granted them by the Russian Cossacks.'[40] The satisfaction that a Russian in the country's European capitals might have derived from such depictions can be easily imagined, for they served as virtual eyewitness accounts of the easy success Russia was enjoying in its endeavours to exercise its national mission in Asia.

A particularly vivid example of how explorers' depictions of the Amur were able to render the region meaningful within the framework of Russian nationalist concerns can be seen in an account written by Nikolai Przheval'skii. Przheval'skii was later to gain world renown with his four expeditions into Central Asia, but his first scientific foray was to the Ussuri valley in 1866–7. His task was to survey how Russian settlements that had been established in the region some years earlier were faring but, like Veniukov, he devoted almost as much attention to the indigenous inhabitants. Przheval'skii's report appeared in the widely read *Vestnik Evropy* (The European Messenger). He obviously shared Veniukov's cultural bias and dim view of 'simple' societies, and he wrote in the following graphic manner about an Orochei tribesman he happened to encounter:

What a small difference there is between this person and his dog! Living like a beast in its lair ... he forgets all human strivings and, like an animal, cares only about filling his stomach. He eats meat or fish, half-cooked on coals, and then goes hunting, or sleeps until hunger compels him to get up, start a fire in his stinking smoky hovel and once again feed himself.

This is how he spends his entire life: for him today is no different from yesterday or tomorrow. Not feelings, desires, joys, hopes – in a word, nothing spiritual or human – exist for him.[41]

In spirit, Przheval'skii's reactions are identical to those of Catherine Nevel'skoi, and in both cases the same psychological process is taking place. The encounter with such utter degeneracy and human decay disgusts, but at the same time the prospect of such abjectly needy ethnographic material immediately works to stir a sense of responsibility, indeed an imperative to do everything possible to help the savages better their wretched lot. In this way, Przheval'skii's account once again has the effect of securing a ready arena upon which an ambitious national mission of salvation could be exercised. This is precisely what we have observed, to repeat, with Madame Nevel'skoi; the difference, of course, is that in the pages of one of Russia's most popular journals Przheval'skii transforms what for the admiral's wife was a personal experience into one which the entire reading public can effectively share.

Przheval'skii does not leave this imperative to the imagination of his readers; rather, he speaks quite candidly about it. He describes the desire of Koreans living adjacent to Russia's new southern border in the Far East to settle on Russian territory, and he recommends that they be allowed to do so. They should not, however, be settled on the border region near their old homeland, but should be moved further to the north, to Lake Khanka or even to the Amur. The point was to relocate these Koreans to places where Russian settlements already existed, so that they could live

> among our peasants, with whom they could become better acquainted and from whom they could assimilate something new. Then, gradually, the Russian language and Russian habits, together with the Orthodox religion, would begin to penetrate to them, and perhaps with time a heretofore unknown miracle would take place: the regeneration in a new life for these groups who came from tribes as stubborn and immobile as the other peoples of the Asiatic East.[42]

It was precisely because they could offer such a palpable prospect of this sort of *chudo* or miracle that Russian geographers and explorers were able to contribute to and stimuate the nationalist preoccupation with a national mission. Przheval'skii made the accomplishment of this miracle into an uncomplicated affair indeed, involving no more than the occupation and settlement of the region. The superior Russian ways would then presumably diffuse among the native peoples as if by osmosis.

RUSSIA'S ROAD TO EUROPE LEADS TO THE EAST

Up to this point, Russia's turn to the East and the assertion of a mission in Asia have been depicted essentially as an emphatic, defiant and unambiguous turn away from Europe. Yet as we shall consider now, it was in fact much more complex. For despite all the sincere anti-European passion that the nationalists espoused, it nevertheless proved impossible even for them to overcome the preoccupation with European values and expurgate the desire to establish the country's identity as a European nation that had been an intrinsic part of Russia's national psychology for at least a century. The result, already noted above in regard to Pogodin's tirade, was a fundamental ambiguity, indeed a veritable ideological schizophrenia, that produced within Russia's nationalist doctrines something resembling the split face of Janus. This condition remained submerged and invisible as long as the nationalists cast their eyes to the West and were able simply to affirm Russia's national exclusivity and superiority, but it emerged in full force the moment the geographical arena shifted to Asia. For in Asia the Russians advanced, as we have seen, as the bearers of civilization and enlightenment, and there was no question whatsoever that the civilization and enlightenment they bore was ultimately that of the Christian West, in other words Europe.

From this state of affairs, a conclusion was to be drawn that was not necessarily harmonious with other elements of the nationalist doctrine but was inescapable nevertheless; namely, that by shifting its attention to Asia, Russia might actually be able accomplish what had for over a century proved to be manifestly impossible in Europe itself. In the East, in other words, Russia could finally demonstrate just how successfully it had been able to assimilate the social and cultural patterns and standards of the West. This sobering insight was affirmed with characteristic temerity by no less a Russian nationalist and resolute anti-European than Fyodor Dostoevsky himself, who in the final pages of his *Diary of a Writer* observed: 'In Europe we [Russians] were hangers-on and slaves, but in Asia even we are masters. In Europe we were Tatars, but in Asia we too are Europeans.'[43] From this standpoint, it is clear that Russia's turn to the East and the assumption there of a national mission of salvation was, despite all nationalist rhetoric, not necessarily a turn away from Europe at all, but rather simply another means for confronting the West and trying to deal with Russia's own sense of national inadequacy in this regard.

Some of this rather contorted logic can be seen in an article by the geographer Peter Semenov. Semenov was an ardent nationalist, who from an early date had advocated the need for Russia to concern itself with its own future in the East. He dominated the work of the Geographical Society for over half a century, and in this capacity was one of its staunchest patrons and advocates of the geographical study of Asia. Semenov was particularly stimulated by the annexation of the Amur and, although he did not visit the region himself, he was an authority on it and contributed to the geographical literature devoted to it. His article, appearing in 1855 in the journal of the Geographical Society, was ostensibly devoted to a physical-geographical description of the region, but like his colleagues he took the opportunity to explain the significance of Russia's advances in a more general nationalist framework. Over the past 30 years, Semenov stated, Russia had made great advances in the study of Asia, conducting explorations around the Caspian sea, in Central Asia, Mongolia and of course in the Far East. 'By all of these routes', he concluded, 'Russia moves forward for ... the general interest of humanity: the civilizing of Asia.'

As Semenov developed his exposition, however, it became clear that Asia itself was not the only, indeed not the primary, object in this civilizing mission, for the cardinal point was the manner in which it reflected upon Russia's relationship to Europe. Semenov repeated the attempt of the orientalist

Grigor'ev to capture the mission of civilizing Asia for Russia alone, with the claim that Russia occupied a special geographical and historical position between East and West and therefore was more naturally suited than the rest of Europe to bring civilization to the former. Like Grigor'ev, Semenov also alluded to divine intentions in this regard, suggesting that Russia was in effect God's chosen instrument to effect this grand mission:

> Chosen by God as an intermediary between East and West, having received its Christianity in the capital of an Eastern Empire [Byzantium] and spent its adolescence as the European hostage of an Asiatic tribe ... Russia is equally related to Europe and Asia and belongs equally to both parts of the world. For this reason it is more capable than other nations of fulfilling the role which its geographical position and history have designated for it.

It was ultimately much more than geographical location or historical pre-conditioning, however, that indicated Russia's enhanced suitability *vis à vis* the West for the task of civilizing Asia. Beyond this, Semenov felt that the actual record of accomplishment thus far demonstrated it as well. While the Russians had always acted in a manner that was consistently beneficial to the local Asiatic population and put their interests first, the European nations were interested in nothing more than callous subjugation and merciless exploitation. At every moment, he asserted, Europeans 'who have attained the highest stage of contemporary civilization, oppress hundreds of millions of peoples of a different, less developed race [and] exploit these Asiatics for their own mercenary ends like inert matter. [The Europeans] convert them like stupid animals into the blind instrument of their material interests, and do not given them even a single ray of their own enlightenment.' Thus, European activity in Asia was of a purely predatory nature, uninspired by the paternal desire to help the ignorant Asiatic savages or half-savages that animated Russian activities there. Indeed, the nationalist Semenov strove most emphatically to disassociate his own country from the brutal and bloody legacy left by the Europeans in the non-European world.

> The Russians do not annihilate – either directly, like the Spanish at the time of the discovery of America, or indirectly, like the British in North America and Australia – the half-wild tribes of Central Asia [and the Far East]. Rather, they gradually assimilate them to their civilization, to their social life and their nationality.

'For this reason', Semenov continued triumphantly, 'each new step of Russia into Asia is another peaceful and sure victory of human genius over the wild, still unbridled forces of Nature, of civilization over barbarism ....'

The illogic underlying Semenov's argument is remarkable. He was attempting nothing less than to maintain the standard Russian nationalist stance *against* Europe in a geographical and social context that necessarily transformed Russia itself *into a representative of Europeanism*. The standard claim for Russian exclusivity really lost its meaning at this point and had to be replaced by a vaguer claim, namely that Russia is simply better than other European powers at carrying out its mission. This, however, amounted in turn to nothing more than the assertion that Russia could beat the West at its own game. In Asia, Russia in effect not only became European but indeed could claim to be *better* than Europe at propagating its own – the West's – principles. Yet despite this apparent hyper-confidence, the same resentments and insecurities that were part and parcel of Russia's nationalist ideology are apparent in Semenov's presentation as well, and he felt it necessary to end his essay with the following defiant thought:

> Just let the children of this West say now that we still stand on a low level of civilization! If this low level is already producing such marvellous fruits for the interests of humanity in general, then we are fully justified in expecting even more from a higher [level of civilization], which will quickly develop in view of the rapid pace characterizing the history of our development.[44]

CONCLUSION: GEOGRAPHERS AND RUSSIAN NATIONAL IDENTITY

One of the most fundamental dilemmas hexing Russia's national identity has come from the fact that the country's intermediary geographical and cultural position between East and West obstructs what many if not most Russians since the time of Peter the Great at least have striven to accomplish, namely, to identify their country fully and unambiguously with Europe. The vision of a national mission to the East in the nineteenth century must be seen as an attempt to deal with and perhaps even resolve this paradoxical situation. It was hoped and believed that success in and recognition by Europe of Russia as the bearer of European enlightenment and civilization to the East could effectively establish an unmistakable affinity with the West. In this chapter I have tried to demonstrate the function that geographers and explorers filled in this messianic enterprise. Their contribution went significantly beyond the familiar and well-appreciated mission of supplying purely physical-geographical information – although such information was certainly supplied and was of critical importance – and included aspects of the articulation of the imperial ideology itself. Indeed, to the extent that some palpable vision of the distant territories and regions in question was a part of this ideology, it may be said that its articulation actually depended largely upon the contribution of geographers. From this standpoint it is clear just how explorers to the Far East such as Nevel'skoi, Middendorf, Veniukov and others were participating in the overall effort to define and at the same time to normalize a 'national identity'. We may note in conclusion that they were not especially successful in this endeavour, for the stubborn dilemma frustrating Russia's national identity persisted and is as apparent today as ever.[45]

ACKNOWLEDGEMENTS

I am most grateful to David Hooson for his committed support of my early research on Russian exploration in the Far East, and in particular for his encouragement of my participation in the present volume.

NOTES

1 See J. D. Overton, 'A theory of exploration', *Journal of Historical Geography*, 7 (1981), 57–70; L. H. Brockway, *Science and Colonial Expansion: the role of the British Royal Botanic Gardens*, New York: Academic Press, 1979; M. Bassin, 'The Russian Geographical Society, the "Amur Epoch", and the Great Siberian Expedition, 1855–1863', *Annals of the Association of American Geographers*, 73 (1983), 240–56.

2 For a fuller treatment of the themes dealt with in this essay, see M. Bassin, *The Vision of a Siberian Mississippi: nationalist ideology, imperial expansion, and Russia's annexation of the Amur river valley, 1840–1865*, Cambridge: Cambridge University Press, forthcoming.

3 Cf. H. Kohn, *Prophets and Peoples: studies in 19th century nationalism*, New York: Collier, 1961; E. Lemberg, *Nationalismus*, 2 vols, Stuttgart: Rowohlt, 1964; B. C. Shafer, *Nationalism: myth and reality*, New York: Harcourt, Brace, and World, 1955.

4 H. Kohn, 'Messianism' in his *Revolutions and Dictatorships: essays in contemporary history*, Cambridge, Mass.: Harvard University Press, 1939; J. Brun-Zejmis, 'Messianic consciousness as an expression of national inferiority: Chaadaev and some samizdat writings of the 1970s', *Slavic Review*, 50 (1991), 646–58.

5 V. V. Grigor'ev, *Ob otnoshenii Rossii k Vostoku*, Odessa: n.p., 1840, p. 8.

6 Cited in N. P. Barsukov, (ed.), *Zhizn' i trudy M.N. Pogodina*, 22 vols, St Petersburg: Tip. M.M. Stasiulevicha, 1888–1906, vol. XIII, p. 16.

7 M. P. Pogodin, 'O russkoi politike na budushchee vremia' in *Istoriko-politicheskie pis'ma i zapiski vprodolzhenii Krymskoi voiny 1853–1856*, Moscow: V.M. Frish, 1874, pp. 231–44, esp. pp. 242–4.

8 Cf. F. Wakeman Jr, *The Fall of Imperial China*, New York: Free Press, 1975; H. McAleavy, *The Modern History of China*, London: Weidenfeld & Nicolson, 1967.

9 A. I. Gertsen, 'Kreshchenaia sobstvennost'' [1853], in *Sobranie Sochinenii v tridtsati tomakh*, Moscow: AN SSSR, 1954–65, vol. XII, pp. 94–117, esp. p. 110.

10 M. Bassin, 'Expansion and colonialism on the eastern frontier: views of Siberia and the Far East in pre-Petrine Russia', *Journal of Historical Geography*, 14 (1988), 3–21, esp. 12–15.

11 On the Amur issue in the nineteenth century, see P. I. Kabanov, *Amurskii Vopros*. Blagoveshchensk: Amurskoe Knizhnoe Iz-vo, 1959; R.K.I. Quested, *The Expansion of Russia in East Asia, 1857–1850*, Kuala Lumpur: University of Malaya Press, 1968; M. Bassin, 'A Russian Mississippi? A political-geographical inquiry into the vision of Russia on the Pacific 1840–1865', unpublished PhD thesis, University of California–Berkeley, 1983.

12 P. P. Semenov-Tian-Shanskii, *Istoriia poluvekovoi deiatel'nosti imperatorskogo russkogo geograficheskogo obshchestva 1845–1895*, 3 vols, St Petersburg: Bezobrazov, 1896, vol. III, pp. 1317–18.

13 K. Ritter, *Zemlevedenie Azii*, trans. P. P. Semenov et al., 6 vols, St Petersburg: n.p., 1856–95.

14 Bassin, 'The Russian Geographical Society', 248–53.
15 A. I. Alekseev, *Amurskaia Ekspeditsiia 1849–1855 gg*, Moscow: 'Mysl'', 1974, pp. 6–9.
16 Bassin, 'A Russian Mississippi?', 139–142ff.
17 V. B. Sochava, 'Stranitsy iz proshlogo russkoi geografii (zhizn' i deiatel'nost' A.f. Middendorfa)', *Sibirskii geograficheskii sbornik*, 2 (1963), 215–36, esp. 226; P.B. Iurgenson, *Nevedomymi tropami Sibiri*, Moscow: 'Mysl'', 1964, pp. 9–10.
18 A. Middendorf, *Puteshestvie na sever i vostok Sibiri*, 2 vols, St Petersburg: Imp. Akademiia Nauk, 1860–77, vol. I, p. 14; V.F. Gnevusheva, *Materialy dlia istorii ekspeditsii Akademii Nauk v XVIII–XIX vekakh*, Trudy Arkhiva AN SSSR, vol. 4. Moscow and Leningrad: AN SSSR, 1940, pp. 202–3.
19 Middendorf, *Puteshestvie*, vol. I, pp. 158–9.
20 Ibid., pp. 137 [quote], 113; N.G. Sukhova, 'Sibirskaia ekspeditsiia A.F. Middendorfa', *Vestnik Leningradskogo Universiteta* (Seriia geologicheskaia i geograficheskaia), 6:1 (1961), 144–50, esp. 146.
21 Middendorf, *Puteshestvie*, vol. I, pp. 159, 166–7.
22 Ibid., p. 188.
23 Ibid., pp. 137, 188.
24 Ibid., p. 27.
25 Ibid., p. 188.
26 On Balasoglo, see S.S. Tkhorzhevskii, *Iskatel' istiny. Dokumental'naia povest'*, Leningrad: Sovetskii pisatel', 1974.
27 [A.P. Balasoglo], 'Proekt uchrezhdeniia knizhnogo sklada s bibliotekoi i tipoglafiei', in V. Desnitskii (ed.), *Delo Petrashevtsev*, 3 vols, Moscow: AN SSSR, 1937–51, vol. II, pp. 16–47, esp. pp. 41, 43.
28 Ibid., p. 44.
29 On Nevel'skoi, see A.I. Alekseev, *Genadii Ivanovich Nevel'skoi, 1813–1876*, Moscow: 'Nauka', 1984; I. Vinokurov and F. Florich, *Podvig Admirala Nevel'skogo*, Moscow: Gos. Iz-vo kul'tumo-prosvetitel'noi literatury, 1951.
30 G.I. Nevel'skoi, *Podgvigi russkikh morskikh oritserov na krainem vostoke Rossii 1849–1855 gg. Pri-amurskii i pri-ussuriiskii krai*, St Petersburg: L.S. Suvorin, 1897, pp. 58ff; Bassin, 'A Russian Mississippi?', 149–53.
31 Paul Carter has recently examined this process in his splendid study of the 'discovery' of Australia: *The Road to Botany Bay: an exploration of landscape and history*, Chicago: University of Chicago Press, 1987.
32 V. Vend, *L'Amiral Nevelskoy et la conquête définitive du fleuve Amour*, Paris: Librarie de la *Nouvelle Russe*, 1894, pp. 76–7.
33 Cited in ibid., p. 147; for a slightly different version, see Nevel'skoi, *Podvigi russkikh morskikh ofitserov*, p. 436.
34 Nevel'skoi, *Podvigi russkikh morskikh ofitserov*, pp. 418–19.
35 Ibid., p. 419; Vend, *L'Amiral Nevelskoy*, pp. 164–5.
36 [G. Radde], 'Pri-Amurskaia oblast'', *Illiustratsiia*, 3:57 (1859), 101–2, esp. 101.
37 M.I. Veniukov, 'Obozrenie reki Ussuri i zemel' k vostoku ot nee do moria', *Vestnik Imperatorskogo russkogo geograficheskogo obshchestva*, 25:ii (1859), 185–242, esp. 190–1 (quote), 200, 204.
38 Ibid., 232–3.
39 Ibid., 228.
40 [Radde], 'Pri-Amurskaia oblast'', 102.
41 N.M. Przheval'skii, 'Ussuriiskii krai. Novaia territoriia Rossii', *Vestnik Evropy* (1870), vol. 5: 236–67; vol. 6: 543–83, esp. 569.
42 Ibid., 576.

43 F.M. Dostoevskii, *Dnevnik pisatelia*, 3 vols, Paris: YMCA Press, n.d., vol. III, p. 609.
44 P.P. Semenov, 'Obozrenie Amura v fiziko-geograficheskom otnoshenii', *Vestnik Imperatorskogo russkogo geograficheskogo obshchestva*, 15:vi (1855), 227–54.
45 Cf. M. Bassin, 'Russia between Europe and Asia: the ideological construction of geographical space', *Slavic Review*, 50 (1991), 1–17.

8

Ex-Soviet Identities and the Return of Geography

David Hooson

The break-up of the Soviet Union in the last few years has also been a breakdown, to use a psychiatric analogy, in which it seems that a complete collapse is an essential precondition for the creation of new identities from the ruins of the former superstate. For the whole lifetime of all but a very few people in the world, we have become accustomed to viewing the Soviet Union as a huge monolithic, obtrusive presence on the world map, more than twice as large as any other country. Add to this the magnified perception accorded by the widely used Mercator projection, plus notions of an impregnable 'heartland' and the lack of knowledge in the outside world about what was happening in this apparently secretive, sealed-off *terra incognita*, known to be bristling with nuclear weapons, and you have all the ingredients for a nightmarish obsession, which could be fostered by the Pentagon and shared by many in the Western world. It was therefore perceived as *sui generis*, defined by an overarching ideology and a highly centralized institutional structure, with a large population perceived as a melting-pot producing 'Soviet' men and women. The recent sudden collapse of the apparently all-powerful, established 'communist' institutions, the evaporation of the ruling ideology and the breakdown and pauperization of this vast country have forced us to ask: what is left?

It is at this point that geography comes back into its own. The end of ideology and imposed centralization has meant the return of geography, just as it signals the return, rather than 'the end', of history. We are required to redraw our mental maps of this enormous slice of the earth's surface, and this means rediscovering the regions which have a profound meaning for the peoples who have inhabited them – often for a very long time – and whose significance is expressed in a strong sense of identity which includes a recognized and loved natural environment. This is part of their life blood and their collective soul and they will, given half a chance, defend it against despoliation. It is these regions, and identities, which we need to discover and reckon with as the disintegration of the former Soviet Union proceeds apace.

WHY THE SOVIET UNION SURVIVED SO LONG

It is necessary to recall, in these nihilistic and anarchic days, that many people in the world viewed the Soviet Union not as a global threat but as the last best hope for mankind. Many within the Soviet Union believed this too, even in the face of superhuman difficulties, disappointments and disasters. There were indeed some relatively bright periods when it seemed, both inside and outside the Soviet Union, as if it might not only survive but perhaps even inherit the earth. One such was the few years around 1960, during which I happened to visit the country for the first time. Compared to the period when Stalin was alive, and also to the 1970s and 1980s, this was a time of achievement, hope and even justifiable pride, in spite of some awful episodes like the crushing of the Hungarian uprising, the building of the Berlin wall and the Cuban missile crisis.

There was a spirit of reform in the air, coupled with an ebullient self-confidence, a ferment of new ideas and publication and a relatively open attitude towards the outside world. Both *glasnost* and real *perestroika* were, in many ways, more in evidence than in Gorbachev's time, and economic growth was proceeding more quickly than it had been (or would be later) and compared well with America at the time. Large new deposits of oil and gas were discovered in the Volga and Siberia. It was able to feed itself, without imports.

Moreover, after commonly being written off after the devastation of the Second World War, it was suddenly *the* major new factor in the world. Superpower status was quickly accorded it after the stunning symbolic announcement represented by the orbiting Sputnik. In the decade after the war it had apparently established a vast empire, from Berlin to Hanoi – with a third of the world's people ruled from Moscow. Further, this was the time when most of the former colonies of the European powers were gaining their independence and looking around for suitable models and allies. Many, not predisposed to accept the lead of their former oppressors, turned to the Soviet Union as the up-and-coming power and exemplar. As a resident of Washington at this time, I was able to observe the electric effect all this had on the United States and benefited personally from the sudden, almost frenetic, infusion of resources to restructure and boost American education, science and 'Soviet Studies' in particular. Although the Sino-Soviet split was occurring, the monolithic image persisted in US military and political thinking right through the Vietnam War.

In retrospect, these (Khrushchev) years were a high point from which it has been downhill all the way ever since for the life of the average Soviet citizen and for the prestige of the country in the world. It is difficult to believe now that in those days the Soviet 'threat' to 'catch up' with the United States by the end of the century was given credence by quite well-informed people in the West. But by 1984, when I was in China, I found that pity and contempt, rather than fear or admiration, were the dominant sentiments expressed towards the Soviet Union. In particular, the fact that the Russians were not able to feed themselves was felt to be a disgrace. The Soviet Union had by then become surrounded by hostile communist states and within Russia by hostile non-Russians, many of whom, such as the Georgians or Estonians, were also envied by Russians for their perceived higher living standards. Bogged down in Afghanistan and drained by a string of dependent client states from Ethiopia to Vietnam and Cuba, the Soviet Union, highly armed though it was, had certainly lost the shine it had in the world 20 years earlier.

One of the lost opportunities following the fall of Khrushchev, and the one most relevant to the topic of this chapter, was the reversal of his plan to decentralize economic activities (the *Sovnarkhoz* project) to the regions across the country, with a corresponding reduction of the powers of the central ministries in Moscow and their large, stultifying, out-of-touch bureaucracies.

THE MULTINATIONAL EMPIRE

The foregoing historical sketch of the changing fortunes of the Soviet Union is inserted to emphasize the importance of the global and national context and atmosphere to the ability and desire of the separate cultural and regional identities to assert their distinctiveness and press for more control over their own lives.

Looked at today, it seems as if the fatal mistake of the Bolsheviks in setting up the USSR in 1922 was to structure the new federal state explicitly on ethnic lines. After the Civil War and the breaking away of Poland, Finland and the Baltic States from the empire, the new country had a much larger percentage of ethnic Russians (two-thirds) than in 1914. In an effort to stop further defection by non-Russian peoples, many of whom (including Ukrainians and Georgians, for instance) had already declared their independence, the 'ethnic' constituent republics were formed, masterminded and often deliberately gerrymandered by Stalin. Some nationalities and new 'republics' were virtually invented or created by arbitrary amalgamation, especially in Central Asia; some of these 'divide-and-rule' actions, notably the Nagorno-Karabakh Armenians assigned to Azerbaijan, are now coming home to haunt the affected regions.

Although the right of secession was written into the Soviet constitution, it was understood by all to be a sham and the federal structure of republics did not stop the Soviet Union from becoming more and more centralized with respect to real power of any sort, despite the formal protestations of devolution. This pent-up frustration goes far to explaining the alacrity with which independence movements mushroomed when Gorbachev initiated his *glasnost* policy in 1987. But in so doing they have come up against the arbitrary nature of the established republic boundaries and have seen internal secession movements grow, as in Georgia which has several nationalities who want to contract out from that republic. Another rather bizarre aspect of the formal secession option enshrined in the Soviet constitution was that a full republic could only be set up bordering a foreign country, because otherwise secession would make an unacceptable hole in the middle of the country! Thus the Turkmens and Tadzhiks were allowed full republic status, whereas the Volga Tatars and the Bashkirs, with substantially larger populations, were denied.

However now the chickens are coming home to roost with a vengeance. In hindsight the setting up of an explicitly ethnic federal structure for the USSR actually set it up, in another sense, for its 'legal' disintegration 70 years later.

THE ROOTS OF DISCONTENT

In retrospect, the catastrophe at Chernobyl in April 1986 appears to have been the catalyst for the cataclysms of subsequent years in the Soviet Union. Amongst other things, it ushered in an unprecedented flood of information and the official license to disseminate and discuss it – the *glasnost* phenomenon which has entered the languages of the world. But while Gorbachev's hope clearly was to channel this new energy into the turbines of economic *perestroika*, it has turned out to be largely directed towards nationalist and ecological concerns: hitherto these had either been ignored or smothered with platitudes about the fraternity of nations or the necessarily benign stewardship of natural resources and environment under socialism. All this, together with the general breakdown of the economy with its concomitant demoralization, pervasive pessimism and chronic anxiety, has created a naturally strong incentive for the republics to break free of a 'union' that seemed to be dragging them down while offering few long-term benefits. This is the context in which the tensions among ecological, ethnic and economic factors have been played out and are still actively unfolding in the various regions of the former Soviet Union. Three decades ago, when the Soviet Union was riding high in economic growth and world prestige, when ecological damage was nowhere near as bad as it is now (the Aral Sea was its normal size then, for instance), whatever ethnic stirrings and discontents there were (and there were many) could not get off the ground. Now, with the disappearance of the Union (the ideology, the police-state atmosphere and the partocracy – though of course they have often just changed hats) the interplay between the economic, ethnic and ecological factors has come centre stage.

THE AFTERMATH OF CHERNOBYL

The profound impact of the Chernobyl disaster, from the local to the international scale, cannot be overemphasized. Locally, at the head of the Kiev reservoir on the Dnieper river, the model city Pripyat remains deserted alongside many other towns and villages in neighbouring parts of the independent states of Ukraine and Belarus. The long-term health toll looks increasingly ominous as time goes on, and the contamination of soil, vegetation and water is still a clear and present danger, with all its psychological fallout of uncertainty about food, health and life itself.

Downstream of Chernobyl is the best farmland of Ukraine, while about half the population of the former Soviet Union is located within about 500 miles of the sinister reactors. It is hard to exaggerate the effect the accident had in re-igniting anti-Moscow sentiments in Ukraine and Belarus and in accelerating the assertion of sovereignty rights and eventually independence. The overwhelming vote for independence in Ukraine in late 1991 appears to have been the greatest surprise for the Russians. It was certainly a bitter disappointment for Solzhenitsyn who, in a rare pronouncement in Moscow (*Komsomolskaya Pravda*, 18 September 1990) called for a new Slavic state, comprising Russia, Ukraine and Belarus (plus northern Kazakhstan!) and for the jettisoning of all the non-Slavic republics. Considering also the common cultural history of these Slavs, off and on for over a millennium, quite apart from the crucial economic position Ukraine has occupied in the Russian and Soviet empires since the nineteenth century, this makes some sense. Also, considering issues such as the control of the Crimea and the Black Sea Fleet, quite apart from the Chernobyl fallout, this fraternal Slavic relationship could be one of the most sensitive of the whole panoply of ex-Soviet identity problems.

The ethnic factor in Ukraine is not, however, in itself an intractable one since, although over one-fifth of its population is ethnic Russian, by far the largest group outside Russia, most of them voted for Ukrainian independence and, provided they are not unduly discriminated against and the Ukrainian economy does not collapse below the Russian, it need not explode. Even in the Crimea, Russian claims are complicated by the fact that large numbers of Tatars, whose 'autonomous' republic it was before 1944, have now returned, while in the case of the Trans-Dniester region of Moldova, where the Russians want to secede from the latter republic, Ukraine seems to be taking a neutral stance.

THE EX-SOVIET MIDDLE EAST

Serious tensions, often going back centuries, are resurfacing in the non-Russian republics of the Trans-Caucasus and Central Asia, which had been colonies of Russia since the nineteenth century; again the interplay between nationalist economic and ecological forces is evident. For instance, the Turkic-speaking Azerbaijanis complain that their oil at Baku has been exploited for the benefit of Russia for over a century, leaving them with a polluted land and a distorted economy. Their neighbouring long-time 'enemies', the Armenians, are choking on the fumes from their chemical factories and bemoaning the disappearance of their Lake Sevan due to hydroelectric and irrigation projects, as well as the destruction of vineyards during Gorbachev's anti-alcohol campaign. They even blame the Russians for the excessive damage and loss of life in their recent disastrous earthquake through shoddy construction work, while they are seething with indignation over the territory of Nagorno-Karabakh where their kinsmen were placed by Stalin under Azerbaijani jurisdiction in the 1920s, reportedly to placate the Turks. The on-going war over this territory between Armenians and Azerbaijanis is the worst so far in the former Soviet Union, accompanied by economic blockade as well as massacres.

Georgia, which is not involved in this dispute, has serious internal ethnic problems, with several secession movements simmering, the South Ossetians of the Caucasus being the most vociferous (they wish to join their kinsmen across the mountains), not to mention the serious internecine explosions among the Georgians themselves.

The Central Asians, mostly Turkic-speaking Moslems, have been under the Russians for little more than a century, a tiny fraction of their thousands of years of settlement in the regions of mountains and deserts. But they are the victims of what is probably the most serious single ecological catastrophe in the former Soviet Union: the shrinking of the Aral Sea, which is less than half its size in 1960 and is both symptom and cause of very serious health and economic problems over a wide area. The Central Asian people, who have been increasing at about four times the rate of the Slavs, blame their colonial experience at the hands of the Russians for distorting their economy by decreeing cotton monoculture in place of the formerly diversified economy. The loss of the precious water from the two major snow-fed rivers through wasteful irrigation methods and, in particular, the massive siphoning off of water from the Amu Darya into the Kara-Kum Canal, has led directly to the shrinking of the sea. The extinction of a flourishing fishing industry, the blowing of poisonous salts and pesticides on to farmland and an alarming increase in serious diseases, such as throat cancer, together with a very high infant mortality, have led to a major all-pervading crisis in the region, for which the Russians are being blamed. Long-time Russian residents of Uzbekistan and other republics, even though they were happily settled there, are escaping from the increasing hostility of the native people back to Russia, but there is little in the way of a welcome waiting for them there. The final straw for the Central Asians was the cancellation in 1986, the year of Chernobyl, of long-laid plans to divert southward water from the north-flowing Siberian rivers to 'save the Aral Sea'. This was seen as the final proof that the Russians, having raped their natural environment and economy, are callously leaving the Central Asian people to die. And yet all the Central Asian economies have become heavily dependent on Russia for their livelihood, which underlines once more the intimate connections between nationalism, economy and ecology.

RUSSIA ITSELF

Even within the vast so-called Russian Federation the same triangular tensions are legion. The opening salvo in the modern Soviet battle for environmental conservation was the move in the mid-1960s to stop the pollution by a pulp mill of Lake Baykal, which contains one-fifth of the world's fresh water. This has since become a rallying point for Russian Siberians in their desire to get more independence from Moscow, especially control over their environment and natural resources. Echoes of the American revolution: Russia cannot take Siberia for granted, and since most of 'Russia's' oil and gas (not to mention gold) is under Siberian land, Siberian identity is asserting itself. Even within Siberia, the environment and reindeer-herding economy of the Nentsy people, in the far north, has been ravaged (the word *ecogenocide* is now often used by Soviet scientists). Their homeland has been the site of the most important natural gas exploitation from which Russia earns much of its hard currency, but their protests have apparently stopped further drilling there. Gold and diamonds in Yakutia (Sakha) and oil in Tatarstan (with chemical pollution of its Volga river) are other cases of economic and ecological issues among two of the most prominent and assertive of the ethnic identities within the Russian Federation. But even among people as dominantly Slavic as those of the Russian Far East and the St Petersburg region, there are resentments against the former 'Centre' for excessive control in the past, and they want to establish their own free trading zones, with Japan and the Baltic, respectively, in mind.

The three Baltic states – the first to regain their independence after 50 years of Soviet occupation – are faced with serious ecological problems deriving from Soviet industrial projects (including a very large and dangerous nuclear plant in Lithuania). Especially in Latvia and Estonia, most of the large ethnic Russian populations who have come in with these projects seem to want to stay. Coupled with their continuing dependence on Russian oil and gas, their dilemmas are great and varied even after independence and they are looking to their Baltic neighbours – Germany, Sweden and Finland – just as the Central Asian republics are looking to Turkey, Iran and even Saudi Arabia.

This brief survey has only touched on some of the myriad problems which have emerged in the former Soviet Union. The question of identity is clearly the most insistent to have surfaced after the long freeze. But it is not enough to treat it as a purely ethnic or cultural question. What is involved here is a re-search for the real regions of cultures, economies *and* environment which mean something (or in some cases everything) to the peoples who inhabit them. The process of crystallization of these regions, beyond the bald and flawed 'Republic' boundaries of today, promises to be long and painful but inevitable and ultimately right.

9

'National Unity' and National Identities in the People's Republic of China

Lisa E. Husmann

With the end of the Cold War China remains one of the last of the multinational empires and the final stone wall of communist ideology. While the ultimate effects that recent world-wide political changes will have on China have yet to be seen, it is significant that the economic chaos following in the wake of disunity among formerly centralized socialist states is serving as fuel for the Chinese government's propaganda machine. Recent reports of world events in the *People's Daily*, for example, emphasize the hardships suffered by formerly united Soviet citizens to justify the use of force in suppression of the Tian'anmen demonstrations of June 1989. At a news conference in April 1991, Premier Li Peng commented on the matter: 'If we had not taken the measures that we were forced into taking on that occasion, China today might be bogged down in economic chaos and decline as well as political instability.' Li continued, saying that China had now achieved 'political stability and unity', but added, without citing specific examples, that 'there are still some factors which could lead to instability' (Holley, 1991).

One issue which could potentially disrupt stability and unity in the People's Republic is what is called in Chinese the '*(shaoshu) minzu wenti*', translated as 'the minority question' or alternatively 'the minority problem', as the Chinese term 'wenti' conveniently implies either or both of these notions. While China is often regarded abroad as a homogeneous state, some 8 per cent of its population is made up of members of 55 officially recognized minority nationalities (*shaoshu minzu*). Although this percentage may seem small compared to China's dominant nation (the Han), 8 per cent of 1.13 billion still represents 91 million people, a substantial number. Indeed, many of China's so-called 'minorities' have populations larger than sizeable independent states of today. At 9.8 million, the Man (Manchu) nationality, for example, has a population comparable to that of Belgium; the Zhuang nationality, with a population that exceeds 15 million, is similarly close to that of Saudi Arabia (see table 9.1).

Table 9.1   China's 10 largest national groups

National group	Population (1990)
Han	1,042,482,187
Zhuang	15,489,630
Manchu	9,821,180
Hui	8,602,978
Miao	7,398,035
Uighur	7,214,431
Yi	6,572,173
Tujia	5,704,223
Mongolian	4,806,849
Tibetan	4,593,330

Source: *Beijing Review* (December 1990)

The challenge posed by the minority problem in China today is two-fold: first, it involves convincing non-Han groups of the validity of what the central government (as well as most members of the Han nationality) consider to be China's 'natural boundaries'; secondly, it entails getting non-Han groups to turn the focus of their national identity away from the periphery and towards the 'centre' of China, homing in somewhere near Beijing, it is hoped. Strategies for tackling this challenge have taken many forms, including national language and education policies, expanded transportation networks and the creation of new 'cultural hubs', as well as government-sponsored (overwhelmingly Han) migrations from 'core' to 'periphery'. Exacerbating the minority problem is the fact that China is plagued by a contradictory minority policy that promotes both autonomy and assimilation. Consequently, minority policy in China's past and present history has oscillated 'between poles of pluralism and ethnocentric hegemonic repression', swayed by external as well as internal political developments (Gladney, 1991, p. 91).

MINORITY LANDS AND THE IMPORTANCE OF THE MINORITIES PROBLEM

A survey of China's linguistic geography reveals the importance of the distribution of China's national minorities across the 'Chinese motherland'. While the Han are concentrated mainly in the densely settled eastern lowlands, the minority groups are spread out along mainly western (including north-western and south-western) reaches of China, their domains comprising about half the total territory of China. This distribution – from an economic and strategic point of view – in part explains the special considerations minority groups receive from the Chinese government, which include rights to theoretical autonomy, special considerations for entrance into universities, and exemptions from China's one-child policy.

In terms of resources, China's minority areas contain large quantities of unworked mineral and petroleum deposits, the majority of China's forestland, as well as a large percentage of the livestock herds from which the country derives its meat and wool. In addition, the sparsely populated minority regions have supplied China with what might be best described as a 'population safety-valve' for absorbing emigrants from overpopulated Han areas in the east. Government-sponsored migrations for 'development of the border-regions' have led to a dramatic increase of Han in traditionally non-Han areas since 1949. According to Chinese figures, by 1982 out of the five officially designated 'autonomous regions' of China, only *Xizang* (Tibet) and *Xinjiang* (Eastern Turkestan) retained a non-Han majority.

Strategically, appeased minority groups in China have meant strengthened border defences against attack from neighbouring states. However, as the threat of Soviet invasion into China's north and west borders has effectively disappeared with the collapse of that empire, the need for placating minorities in these border regions is taking on a new and urgent dimension: namely, the potential threat of political liaisons between national identities which transcend China's state boundaries. Although these transborder national identities are today most salient among China's Mongolian and Turkic (including Uighur, Kazakh and Kirghiz) populations, whose kinsmen are fast making changes just across the border, it also applies to less heard-from groups, such as the Shan and the Miao, whose populations continue across the border into Burma and Thailand, and into Laos and Vietnam, respectively.

UNITY AND DIVERGENCE OF HAN NATIONAL IDENTITIES

From its earliest usage in the first millennium BC, the Chinese term for China, *Zhongguo* (literally, the 'Middle Kingdom'), has been used to distinguish the 'we' at the centre from the 'them', the barbarians, of the periphery. National identity for China's Han majority is still largely based on this Confucian concept of spatial and authoritative hierarchy, and the Han today look toward the 'core' of the Middle Kingdom – the Wei River Valley in China's Central Plains, the so-called 'cradle of Chinese civilization' – for their national identity (Chang, 1986, p. 192).

The Han Dynasty (206 BC–AD 220) was a major formative period in the Han culture, and continues today to be extolled in Chinese history books as one of the Golden Ages of China's past. Not only was the Han empire the most populous that had existed in human history, surpassing the contemporary Roman empire, but it was also marked by a great expansion in human knowledge. The manufacture of paper and porcelain were but two of its technological contributions. China's majority Han population looks to this era as the historical foundation of its national identity, a fact which is evident in the name as well as other cultural terminologies used by the Han nationality today. The spoken Chinese language, for example, is known as *Hanyu* (the 'language of the Han'), Chinese characters are called *Hanzi* (the 'letters of the Han'), and a gentleman is referred to as a *hao Hanzi* (a 'good son of Han'). The fall of the Han Dynasty saw the beginning of an extended period of disunity within the Middle Kingdom. Still, the political unity achieved under the Han Dynasty became increasingly symbolic of an ideal. As Moser (1985, p. 30) notes: 'In the Europe of the Middle Ages, the concept of a universal Roman empire held on long after the reality was gone. So it was in China; the Han concept that one emperor should rule over "all under Heaven" survived centuries of chaos.' In modern times, this long-standing ideal of Han political cohesion has been restructured according to the ideals of the state as 'national unity'. While the ideal of unity remains an enduring sentiment among many Han today, calls for 'national unity' are *not* eagerly accepted by many peoples in China whose heritage lies outside the Han tradition.

Among the many acts of standardization implemented during the first unification of the Middle Kingdom (in the third century BC), were those designed to unify the written language, then used differently in various parts of the emergent empire. The success of the campaign far outlasted the dynasty under which it was carried out: the Chinese script used throughout China today is a direct descendant of the early standardized form and is considered by the Han nationality to be tangible evidence of their cultural continuity through time. To this day the Chinese written language is often cited as a major element in Han cultural cohesiveness (see, for example, Wang, 1982, p. 57).

However, while language is often a crucial factor in national identity, it is by no means the whole story. In the case of the Han nationality, while unity survived as an ideal it was not as readily maintained in practice. Long periods of political disunity saw the development of subcultures within the Middle Kingdom, especially as Han migrations away from the Central Plains led to the converging of Han with the non-Han cultures of the 'periphery'. Physical geography has also played a role in the development of Han subcultures. The Chinese saying '*Nan chuan, Bei ma*' ('in the South the boat, in the North the horse') describes the situation of open terrain in much of northern China as compared to the south where the presence of lakes and rivers often isolated communities from their neighbours. Today, the differential development is revealed in the linguistic geography of Han China: whereas across northern China there exists a homogeneity of Mandarin dialects, southern China is characterized by a complexity of mutually unintelligible Han languages, including Shanghainese, Cantonese and Taiwanese, to name but a few.

Among Han subcultures there have also developed distinctive cuisines, customs, land settlement patterns, as well as established behavioural stereotypes:

> The alleged personality differences between the people of each province and/or dialect group are spoken about by Han Chinese as assuredly (and probably with about the same accuracy) as are the differences in 'national character' within Europe by Europeans. The people of Shaanxi are considered miserly, those from Hunan [including Mao Zedong] hot-tempered, the Cantonese sharp businessmen. These stereotypes strongly influence the expectations of Han Chinese today about the probable behavior of any new acquaintance from another province. (Moser, 1985, p. 6)

Thus, while Han identity in China has focused on unity and cultural continuity through time, within the Han subcultures that have developed under the influence of political divisions, non-Han contacts and differential landscapes, there also exists a strong sense of identification with as well as loyalty to one's native place (*yuanji*). Today, ties of clan, province, and dialect endure as strong forces distinguishing Han sub-identities.

'NATURAL BOUNDARIES' AND THE 'CHINESE' NATIONAL IDENTITY

The concept of 'natural boundaries' for the Middle Kingdom has historically been used as an argument against expansion into the periphery (see, for example, Yu, 1967, p. 67). Nevertheless, the political territory of the Middle Kingdom expanded and, as it did, so did the Han conception of their 'natural boundaries'. By 1795 China controlled Eastern Turkestan, Mongolia, Tibet and all of south China from the south-westernmost province of Yunnan to Taiwan. It was a high-water mark that established, at least as far as most Han are concerned, the traditional boundaries of their country. It is essentially these same borders which Nationalist President Chiang Kai-shek outlines in his famous treatise, *China's Destiny*:

> So far as China's physical configuration is concerned, her mountains and rivers form one integral system. If we take a bird's eye view of the country from the west to the east, we shall find in the north, the Tian Shan and Altai Mountain ranges starting from the Pamir Plateau and extending eastward; in the center, the Kunlun Range extending from the Pamir Plateau to the plains of southeastern China; and in the south the Himalaya Mountains extending from the Pamir Plateau to the mid-southern Peninsula. Between these mountain ranges there are the Amur River, Yellow River, Huai River, Yangtze River and Pearl River valleys. It is along these river valleys that the Zhonghua (Central Hua) nation has grown and developed. There is not a single piece of territory within these areas, therefore, which can be torn away or separated from China, and none of them can form an independent unit by itself. (Chiang, 1947, pp. 8–9)

The concept of 'natural boundaries' under the Nationalist government was accompanied by an extreme assimilationist policy which promoted the notion that 'the various stocks in China constitute not only one nation but also one race'. This interpretation of race and nation was used as justification for 'why the entire Zhonghua nation, so solid in its make-up, is destined to live gloriously or perish ignominiously as a whole' (Chiang, 1947, p. 13). Although this policy continues to be upheld by the Nationalist government of Taiwan (where only the 'High Mountain' (*Gaoshan*) Austronesian peoples are recognized as forming a separate nationality), in the People's Republic the policy has been criticized as perpetuating the reactionary concept of 'great-Han chauvinism' (*da-Han zhuyi*) (Gladney, 1991, p. 283).

Formulated during the 1930s, the nationality policy of the People's Republic was influenced by developments in the USSR, as well as by the Long March which gave early communist leaders (including Mao Zedong and Deng Xiaoping) an extended tour of non-Han lands and an appreciation of the strategic importance of non-Han political support. The pro-minority attitude promoted by the communist party surfaced in several forms, including changes of terminology regarding minority-related issues. The term 'amalgamation' (*ronghua*) for instance, was substituted where 'assimilation' (*tonghua*) had been used earlier. Descriptions of minorities as 'raw' (*sheng*) or 'cooked' (*shu*) – adjectives traditionally used to describe degrees of Sinification – were avoided, as were other demeaning terms including 'tribe' (*buluo*) and 'aborigines' (*fan*). In addition, graphic pejoratives within Chinese characters (such as graphic components carrying the meaning of 'dog' or 'insect') were replaced by less offensive elements.

Modelled on minority policies enacted in the Soviet Union, the original *Constitution of the Soviet Republic* (of China) was written in 1931 by the underground Chinese Communist party. As were the many acts of linguistic cosmetics performed by the party, the attitudes promoted within the constitution were similarly strategic in enlisting support among non-Han peoples disgruntled by Chiang Kai-shek's nationality policies. Most notably, Article 14 of this document promises not only privileges to China's minority groups, but also the right to secede:

> The Soviet government of China recognizes the right of self-determination of the minorities in China, their right to complete separation from China, and to the formation of an independent state for each minority. All Mongolians, Tibetans, Miao, Yao, Koreans, and others living on the territory of China shall enjoy the full right to self-determination, i.e. they may either join the Union of Chinese Soviets or secede from it and form their own state as they may prefer. (As cited by Schwarz, 1971, p. 49)

In subsequent versions of China's constitution, however, 'rights to self-determination' are replaced by an offer of limited regional 'autonomy'. As Gladney (1991, p. 88) points out:

> The transition in Chinese terminology from 'self-determination' (*zi jue* or 'self-rule' *zi zhu*) to 'autonomy' (*zi zhi*) is slight, but for the minorities it represented a major shift in policy. Unlike the Soviet Union, where requests for Baltic and Central Asian secession had to be taken seriously because that right was written in the constitution, any such request in China is regarded as criminal.

Today, a basic discrepancy between the national consciousness of the Han and many of the non-Han peoples of China is in the interpretation of just where China's 'natural boundaries' lie. According to the historiographic interpretation of the Han, regions such as Tibet, Inner Mongolia and Eastern Turkestan – while still considered peripheral to the Han core – are none the less integral parts of the Middle Kingdom. As such, non-Han inhabitants of these regions are just as much 'Chinese', (*Zhongguo Ren*, that is 'People of the Middle Kingdom') as the Han are. This interpretation, in which Chinese identity is defined along lines of political incorporation into the state (citizenship), is propagated in school textbooks across China – including those written in minority languages. One of the first sentences learned from a Uighur language textbook (published by the Central Nationalities Institute in Beijing) reads (in Uighur): '*Junggo bizning ulugh watan*' ('China is our magnificent fatherland') (Central Nationalities Institute, 1984, p. 53).

The extent to which non-Han peoples have come to accept their 'Chinese' identity often parallels the degree of their conviction that the regions they inhabit indeed fall within China's 'natural boundaries'. These sentiments vary, of course, from group to group as well as among individuals within any given minority. For example, while Tibetan resistance to Chinese political and cultural incorporation is strong in Lhasa, in neighbouring Qinghai Province (once part of Tibet) relations between the Han and Tibetan population are relatively amicable, and intermarriage between the two groups is not uncommon.

An increasingly common complaint among China's minority nationalities has been over the issue of resources. Significantly, resource-related conflicts have been on the increase even in the south-west, among China's 'assimilated' minority groups. A main source of recent tension in the south-west has been the establishment of state-owned rubber plantations in minority areas. Minority nationalities there (mainly the Yi and Dai peoples) complain that the plantations are taking their land, and fights between local populations and Han plantation workers have ensued.

Disagreements over resources are, of course, essentially the same as those concerning the validity of state-conceived 'natural boundaries'. The perspective of the Chinese government, that all the territory controlled by the People's Republic belongs to China's population as a whole, is naturally extended to the resources within those territories. In the words of a spokesman for China's State Nationalities Affairs Commission (here, regarding the development of oil fields in Eastern Turkestan): 'All of this wealth belongs to the people of all nationalities of our country. They constitute indispensable material bases on which to build our great motherland into a powerful modernized country' (Xue, 1982, p. 35).

CATEGORIES, CRITERIA, CONTRADICTIONS: IDENTIFYING CHINA'S MINORITY GROUPS

In the 1950s, systematic research on ethnic groupings in China began, spurred in large part by strengthened relations between China and the Soviet Union. (Not only was ethnographic research in China based on the model that had been developed and implemented in the USSR but, in addition, among the Soviet advisors who came to China were many well-trained ethnographers.) By 1955 over 400 groups had registered applications for separate national minority status (Fei, 1981, p. 60). The criteria used by the newly created State Commission for Nationality Affairs in identifying which groups qualified were those set down by Stalin (originally for use in the USSR). According to

Stalin's definition: 'A nation is a historically constituted, stable community of people, formed on the basis of common language, territory, economic life and psychological make-up manifested in a common culture' (as cited in Franklin, 1973, p. 60).

The first census of the People's Republic, conducted in 1953, registered 41 nationalities. In the census of 1964, taken after the minority nationality identity campaigns of the 1950s, 53 groups were registered. Although today there still remain several groups, such as the 'Boat People' of the south-east coast, who continue to seek minority status, the number of officially recognized minority nationalities has stabilized at 56 in both the 1982 and the 1990 censuses. Recognition of these 56 groups was supposedly based on Stalin's definition of a nation. However, a comparison of the groups that were chosen with the approximately 350 other applicant groups that were refused recognition reveals numerous contradictions in the application of these criteria. Fei Xiaodong, a leading ethnologist involved in the classifications of the 1950s, cites the officially recognized Jingpo nationality as an example of breach of the 'common language' criteria. Similarly, the people of the Miao nationality do not share a 'common territory', and many recognized groups, such as the Zhuang, do not share a 'common economic life' (Fei, 1981, pp. 62–77).

In his recent study of the Hui (the so-called 'Chinese Muslim') nationality of China, Dru Gladney concludes that the decisions for recognition of national groups in China were often based at least as much on political considerations as they were on cultural criteria. According to Gladney:

> Clearly, a cultural theory of ethnic identification, whether the Stalinist or cultunit approach, is inadequate to account for the Hui as a distinctive ethnic group ... The leaders of the Chinese Communist Party had other reasons for accepting some groups such as the Hui, and rejecting others, such as the Chuanqing Blacks or the Sherpas. (Gladney, 1991, p. 71)

The Chinese government's 'other reasons' for recognition of specific groups have most often been tied to political and economic considerations. In the case of the Hui, for example, Gladney points out that in recent years this group has played a significant role in China's efforts to maintain close political-economic ties with Middle Eastern countries: 'As a result of this contact, construction of the Xiamen International Airport was partially subsidized by the Kuwaiti government. Kuwait has also assisted in the building of a large hydroelectric dam project along the Min River outside of Fuzhou' (Gladney, 1991, p. 284).

However, while proving lucrative in the case of the Hui, the links that connect several of China's minority groups with a larger identity that extends beyond the political boundaries of the People's Republic has also given the Chinese government considerable grounds for concern. The presence of Mongols, Kazakhs or Shan across the border serve as a constant reminder of historically based national identities – identities that the Chinese government has long tried to get these groups to forget. Not only do these transborder cultural ties present major distractions among groups that are supposed to be redirecting their 'national' attentions toward Beijing and their new 'Chinese' identity, but they also provide historic alternatives to China's 'natural boundaries' as well as cultural and historic contradictions to the state-sponsored ideal of a 'united Chinese motherland'. With the rise of nationalism among their ex-Soviet counterparts across the border, the issue of national identity in China's Inner Mongolian and Eastern Turkestan regions has warranted increased governmental vigilance in recent times. Although less publicized than ethno-political issues occurring in Tibet, these two regions have come to the fore as ethnic hot spots, increasingly demanding of 'national' as well as international attention.

THE CASE OF INNER MONGOLIA

Mongolian national consciousness is rooted in Ghenghis Khan's unification of nomadic tribes of the Central Asian steppes, and their successful creation of a vast empire in the early thirteenth century. Although this empire quickly collapsed in the fourteenth century, Mongolians today – whether living within the political realm of China or the former Soviet Union – look to Ghenghis Khan as their national hero. In the former People's Republic of Mongolia, where reverberations from the collapsing Soviet empire led to the ousting of the Moscow-linked dictator in March 1990, the resurgence of Mongolian nationalism has taken several forms. Buddhist monasteries are being restored and reoccupied, the traditional Mongol script is replacing Cyrillic, Soviet-influenced place-names – including the name of the Mongolian capital Ulan Bator (meaning 'Red Hero') – are being changed to their pre-revolutionary names, and statues of Lenin and Stalin are being torn down and replaced by those of Ghenghis Khan.

While the exit of Soviet troops from Mongolia was a welcome development in the eyes of the Chinese government, the swift overthrow of communism and the rise of nationalism there has been an increasing source of concern for the effects it has been having across the border. The region of Inner Mongolia, incorporated into China as the Inner Mongolian Autonomous Region in 1947, is home to more ethnic Mongolians – over 3 million – than the Mongolian state itself (the former People's Republic of Mongolia, and before that 'Outer Mongolia'), where just over 2 million Mongolians reside.

In May 1991 a crackdown on two organizations formed by ethnic Mongolian intellectuals and party cadres was carried out by the Chinese government in Inner Mongolia. Documents ordering the crackdown (obtained and published in July 1991 by the Asia Watch group) describe what the Chinese officials perceived as insidious influences from across the border:

Between the end of 1989 and the beginning of 1990, the 'Mongolian Democratic League', the opposition party Organization of the Mongolian People's Republic, sent people disguised as traders to our region to establish contacts and stir up trouble, claiming that 'the time has come for the reunification of the 'three Mongolias' [Inner, Outer, and Buryat Mongolia] ... The two illegal organizations ostensibly used the discussion

of 'national culture' and 'national modernization' in public. But in fact
... their real aim was to oppose the leadership of the Communist Party,
the socialist system, to entice a national split and undermine the unification of the motherland. (Asia Watch, 1991, pp. 9–12)

The formation and suppression of the two groups marks the latest phase
in an apparently fast-developing pro-Mongolian ethnic identity movement in
the Inner Mongolian Autonomous Region. As early as 1981 Mongolian students
began protests against the growing Han presence in Inner Mongolia. (While
the Mongolian population of Inner Mongolia – at 3.3 million – is large
compared to the Mongolian population across the border, their numbers are
dwarfed by the population of Han in Inner Mongolia at 17.3 million.) More
than two dozen Mongolians were placed under arrest for their involvement
with the two organizations which, according to the Asia Watch report were
'dedicated to researching and promoting traditional Mongolian culture and
identity'. Regardless of the intent of these groups, the severity of the crackdown reflects the extent to which the Chinese government feels threatened by
such national movements, and demonstrates the criminality with which threats
to 'unification of the motherland' are equated.

THE CASE OF EASTERN TURKESTAN

In Western language sources, the region of north-western China is often
referred to as 'Chinese Central Asia' or 'Eastern Turkestan'. Since its incorporation into China as a province in 1884, the region was given the Han name
Xinjiang (literally, the 'New Frontier'), appropriately reflecting the wild-west
attitude many Han still feel toward the region. In 1955, under new nationality
policies introduced by the Communist Party, the region was redesignated as
the Xinjiang Uighur Autonomous Region. In addition to the titular Uighur
nationality, other groups in the region include the Kazakh, Kirghiz, Tajik,
Uzbek and Tatar nationalities.

'Turkestan', meaning 'land of the Turkic people', is typically conceived of as
including the regions of Kazakhstan, Uzbekistan, Turkmenistan, Kirghizstan,
as well as north-western China. As a linguistic region, however, 'Turkestan'
is much larger: speakers of Turkic languages include such geographically dis-
parate groups as the Azerbaijan, Tatar and Yakut nationalities, and 'Republican
Turkish' (the official language of Turkey) contributes a sizeable number of
speakers to this group. To the extent that the languages within the Turkic
group are mutually intelligible (which – with the exception of Yakut and
Chuvash – they are in varying degrees), linguistic geography helps to define
the outer limits of the 'Turkic' national identity.

In addition to language, religion too plays an important role in the cultural
unification of peoples within Turkestan (the term being used here in its
more narrow sense as including the 'Chinese' and formerly Soviet regions of
Central Asia). The spread of Islam to Turkestan dates back to the tenth
century and has been a major factor in creating common cultural links among

traditionally antagonistic settled and nomadic peoples in the region. Religious beliefs and culture also engender a degree of cohesion with other Muslims living beyond the realm of Turkestan. Islamic rites such as Ramadan and the pilgrimage to Mecca, for instance, foster cultural ties with neighbouring Islamic countries which are of increasing political importance. In 1984 over 1400 Muslims left China to go on the Hajj; by 1987 this number had increased to over 2000. As a by-product of these cultural-religious institutions, long-distance communication networks have been established. Eberhard (1982, p. 65), for example, notes that even at the height of the Second World War Muslims from China passed through Turkey and Middle Eastern countries on their way to Mecca, bringing news from China. Today this network is strengthened by contacts of a more permanent nature. According to Gladney (1991, p. 63), for instance: 'several Hui students (from China) are presently enrolled in Islamic and Arabic studies at the Al-Azhar University in Egypt'. In addition, many Turkic exiles (many post-1949) from Eastern Turkestan have established communities across these countries. Exiled Uighur nationals for instance, allegedly comprise a sizeable percentage of the Saudi Arabian air force.

While providing cultural links within the Muslim world, religion in Eastern Turkestan has, at the same time, created barriers against assimilation into both the Han and the state-defined 'Chinese' culture. Traditionally, Muslim groups in China have not become members of the Han communities in their neigh-bourhoods, but have instead established their own tightly organized communities according to Islamic beliefs and rules. Such a Muslim community (called *umma*) has a religious leader (*imam*) who provides religious education (including the teaching of Arabic for reading the Koran) and serves as a mediator when conflicts arise among community members. Moreover, Islam is monotheistic and thus requires separate temples from those of the Han, where multiple deities as well as ancestors have traditionally been worshiped. Even with the dismantling of religious and community structures in the decades following 1949, the issue of food taboos has still largely kept the two groups separated. In most Han cuisines, the meat is predominantly pork and the cooking fat is pork grease – indeed, pork was declared by Chairman Mao 'a national treasure'. Outside the home, the *halal* alternative for China's Muslim population is provided in separate pork-free establishments (often operated by members of the Hui nationality). Although usually available in the north-west, such establishments are more difficult to come by in many other regions of China. The major inconveniences thus faced (as anyone who has 'gone Moslem' in travelling through China will appreciate), is an added incentive for Muslims in China to take their travels in more culturally compatible countries. (The black market for foreign currency is very active in Xinjiang. Among other things, hard currency is necessary for travel outside of China – big business with would-be pilgrims to Islamic holy sites abroad.)

In addition to language and religion, physical traits also distinguish many Eastern Turkestan nationalities from China's Han majority. (Even among China's non-Turkic and more 'sinified' Moslem group, the Hui physical dis-tinctiveness from the Han is often a source of pride. Gladney (1991, p. 24), for example, notes that: 'Upon one's first arrival in a Hui village or home, the locals frequently bring out the individual with the largest nose, longest beard, fullest eyebrows, most extended earlobes, and say: "Look at this guy, he's a real *Hui!*"') A series of personal correspondences exchanged in the early 1950s between Chu Chia Hua (a minister in the Nationalist government) and Mohammed Emin Bugra (a Uighur exiled in Istanbul) candidly demonstrate some of the difficulties these dissimilarities in culture and appearance posed to the Chinese government in their endeavours to convince the Turkic population of their absolute affinity with the 'Chinese civilization'. In one such letter, written in June of 1952, Mr Chu begins:

> My Dear Mr Emin;
> As I have been interested in the well-being of your compatriots in the border Provinces for several decades and as I have enjoyed your friendship in all sincerity, I cannot allow myself to remain silent when I am confronted with opinions with which I personally cannot agree. No Chinese, whether he be in government service or not, has ever thought of Xinjiang in other terms than one of homogeneity and equality with the rest of the country. Could you possibly point out any individual who has considered it as lying beyond the pale of Chinese civilization? (Bugra, 1954, p. 18)

In a reply letter to Mr Chu, Mohammed Emin Bugra gives the following response:

> Honorable Minister;
> I believe that you exaggerate this. There exist in China true scholars and fair-minded persons who know that Turkestan holds a high place among the countries of Asia and that it possesses a completely different identity from China. The uneducated classes who know nothing of Turkestan hear the invented Chinese names as 'Xi-yu' (Western Regions) or 'Xinjiang' (New Frontier) and therefore believe it is Chinese and its inhabitants Chinese. However, upon coming in contact with a native of Turkestan and noting their difference in mien, language, and other characteristics, they form the opinion that they are of a completely different nationality. When we were in China we would meet with this sort of incident daily; in fact, when seeing us these classes would point us out to each other as 'Indu-ren' (Indians) or else 'Alabu-ren' (Arabs). I am positive that the true scholars will not conform to your claims. Few people who place political interests above everything else will give frank opinions, but they will not be able to conceal the facts. (Bugra, 1954, p. 32)

Beyond cultural factors, another major barrier to the incorporation of Eastern Turkestan into a greater China and 'Chinese' identity has been the great distance and inaccessibility of the region from China's capital. Until recently, the only route suitable for land travel between Xinjiang and the core area of China has been through the difficult Gansu Corridor. Moreover, the cultural centres of Eastern Turkestan have historically been located in the extreme west of the region. Ancient trading centres such as Kashgar and other cities associated with strategic passes and low-lying gaps through the Tian Shan mountains (including the Yili river valley and the Dzungarian Gates) afford relatively easy access from Soviet Central Asia and contribute to the region's geographical orientation toward the West. Even after Xinjiang's official incorporation as a province of China in 1884, non-Han inhabitants as well as local Chinese administrations set up in *Xinjiang* were more or less independent from the central authorities in the Chinese capital. For much of the nineteenth and well into the twentieth century, the major political influence was not from China but instead emanated from Russia. As late as 1945 the Yili region in north-western Xinjiang was the centre of a Soviet-backed separatist regime called the 'Eastern Turkestan Republic' nominally headed by an Uzbek and supported by other non-Han nationals (McMillen, 1979, pp. 15–26).

Since 1949, however, the Chinese government has sought to overcome the powerful geographic influence of the Xinjiang topography. This was accomplished by redirecting the internal communications system away from its historical centrifugal focus across the borders, toward Urumqi in the political centre of Xinjiang, and then by linking Urumqi to Beijing by railroad. Significantly, extension of the railway through the Yili valley (which had been delayed since the severing of Sino-Soviet relations in the 1960s) was finally completed in 1991. The new line links Yili (and by extension, Beijing) with neighbouring Kazakhstan.

The Chinese government has also sought to strengthen its control over the region by sending large numbers of cadres and other personnel of Han stock to Xinjiang. Han in-migration has succeeded in significantly changing the content of Xinjiang's population. According to one estimate, prior to 1949, 94 per cent of the region's population was Muslim and non-Han (Eberhard, 1982, p. 62). In the decades following the establishment of the People's Republic, however, the number of Han migrants to Xinjiang increased dramatically. Within the seven years from 1954 to the end of 1961, for example, total net immigration reached 1,523,200, the 'boom' period for migrants to Xinjiang. According to Yuan (1990, pp. 62–3), there were several reasons for this increase:

The establishment of the Production and Construction Corps in 1954, even now still composed of 90 per cent Han ... provided numerous permanent residential areas mainly for the Han. The 'volunteers' promoted and encouraged by the government, and 'self-drifters' attracted to the state-owned farms from densely populated and rural areas, rushed into Xinjiang. Many prisoners and exiled cadres who made political mistakes in the fifties were forced to go to Xinjiang to participate in 'labour reform' and 'labour education' ... Xinjiang was one of the main areas to receive 'the right-wingers' resulting from the 'anti-rightist' movement of 1957 and 1958.

Although Han in-migration has often meant an increase in new industries, agriculture, as well as educational and medical facilities, Han movement into non-Han lands is resented by many Turkic, Tibetan and other non-Han peoples as a government attempt at whole-scale cultural and political incorporation. However, while Han migration has indeed altered the content of Tibet's population, the case of Xinjiang is far more extreme. According to Chinese census figures, by the end of 1985 the population of Tibet stood at 1,960,000, of which only 110,000 were Han or of non-Tibetan origin (Yuan, 1990, p. 64). In contrast, Xinjiang in 1984 had a total population of 13,440,800, of which 5,346,300 were Han – a Han proportion of nearly 40 per cent. Moreover, recent reforms in China's rural policies now allow surplus rural manpower to move without permanent registration. As Yuan (1990, p. 64) points out: 'For six months to a few years, temporary migrants skilled in various trades work outside their hometowns; Xinjiang is one of the most popular regions [for these so-called "self-drifters"] since opportunities for work are good.' A recent report estimated that such temporary migration in Xinjiang has reached about 200,000 per year (Yuan, 1990, p. 64).

In the summer of 1990, the Khunjerab Pass (the main southern access to Xinjiang via Pakistan) was closed and travel in north-western China was restricted following a series of clashes between the People's Armed Police and non-Han groups in northern Xinjiang in April 1990. Although the initial clashes were reportedly set off by the decision of Chinese authorities to stop the privately financed construction of a mosque, the main motive for discontent in the region was an increasing feeling of resentment toward the sustained wave of Han migrants into Xinjiang (Peters, 1991, pp. 152–4). Government reports that label the disturbances as specifically 'Kirghiz' are difficult to confirm. It does, however, serve China's governmental purposes that the discontent be characterized as occurring strictly within the Kirghiz population, as opposed to a coalition among Turkic or Muslim groups in the region.

The dangers of religious or political collaborations among national groups within Xinjiang, or beyond with groups in the former Soviet Central Asian Republics, have been real enough to 'bring the policy of divide and rule to the forefront of Chinese governmental programs' in the region (Peters, 1991, p. 155). These efforts, of course, are not new, nor are they confined to the Chinese state. Language programmes in both China and the Soviet Union, for example, were devised to 'divide and rule' their Turkic populations. While in 1959 Arabic was replaced with Latin as the new script for Turkic-speaking peoples in China, from the late 1930s to the 1940s the Cyrillic script had been introduced for the same Turkic languages spoken in the USSR. As a result, although these groups still shared a common language, their written materials were no longer comprehensible beyond the border. This linguistic 'divide and rule' tactic, however, had only limited success. In China, while Latin was taught to schoolchildren throughout Xinjiang, at home many parents continued to educate their children using the traditional Arabic script. By 1984 Beijing, as a face-saving measure, instituted a revised Arabic-based writing system.

The more that 'divide and rule' policies succeed in Xinjiang and the more Han immigrants enter the region – inevitably diluting the Turkic and Muslim cultures there – the less hopeful the outlook for would-be Eastern Turkestan separatists. On the other hand, national identity and its perceived distinctiveness from 'Chinese' culture continues to be resilient among many groups of Eastern Turkestan. As recently as March 1992, for example, Tomur Dawamat (government chairman of Xinjiang) announced that a 'handful of secessionists have been colluding with outsiders in stepping-up sabotage and other activities with the goal of winning independence' (Associated Press, Beijing, 8 March 1992).

The pride with which M.E. Bugra defended his Turkic national heritage in his letters to Chu Chia Hua in 1952 is still felt among many inhabitants of Eastern Turkestan 40 years later. In one of his letters Chu had made the claim that 'China is really a big melting-pot where cultures blend with each other to form one harmonious whole ... The modern Chinese race is, therefore, the offspring of many racial elements, which accounts for the brilliance of the Chinese civilization and the continued vigor of the Chinese people as a nation.' Chu likened the situation to an old Chinese saying: 'The massiveness of Mt Tai is the result of the accretion of small particles of soil; the immensity of rivers and oceans is due to the confluence of small streams.' Bugra's response came in a following letter:

> Your Excellency Dr Chu Chia Hua;
> You stated that Chinese culture is in a position to enlighten the entire world and would absorb unto itself as the sea absorbs the small streams, or the Tian Shan mountains do not discard its grains of sand ... No matter how large and splendid the Chinese culture is, the cultures of the Turkic and other nations pose a very large history and are of great importance, and cannot be compared to a small brook or grains of sand, as you have quoted. (Bugra, 1954, pp. 27–8)

CONCLUSION

The necessity of dealing with the 'minority problem' is not new to the current Chinese administration. For millennia this problem has occupied an important place in Chinese policy-making and has been linked to a myriad of factors ranging from Confucian principles of hierarchical order to territorial and economic gain. Since the 1950s the official policy concerning the rights of minority nationalities has steadfastly maintained that the minority regions are unquestionably part of China. The Constitution adopted on 4 December 1982 again clearly states that 'all the national autonomous areas are inalienable parts of the People's Republic of China'.

Current land and resource claims of the Chinese state are legitimized in terms of 'national unity' and 'natural boundaries'. Although these ideologies have been rather successfully incorporated into the identities of several national groups in China (including the Han majority), increasing international contacts with a rapidly changing world are fuelling disillusion among several of China's already sceptical national groups. The issues surrounding Tibet are fairly well known. The Dalai Lama, who (much to the chagrin of the Chinese government) was granted the Nobel Peace prize in 1989, has brought the concerns of his people to the attention of the world community. While groups other than the Tibetans, such as the Mongols and Uighurs, are without such well-known international representatives, they do, however, have something the Tibetans lack: a direct cultural link with kindred peoples across the border.

As in the case of the USSR not long ago, policies issued by the Chinese central government – including those meant to deal with the 'minorities problem' – are backed by the resources of a vast country and an infrastructure which includes a railway network spanning the entire East–West axis of the country. Unlike the former Soviet Union, however, one national group, the Han, comprises approximately 92 per cent of China's citizenry. Also unlike the former USSR, China has few minorities serving in its military. Considering these overwhelming statistics, minority discontent among individual groups of Tibetans, Uighurs or Mongols hardly seem to be among the major 'factors which could lead to instability' in China.

On the other hand, a handbill was distributed in 1984 to foreign students at Beijing University which proclaimed the formation of an 'Asian Republics' Confederation'. It stated: 'The final objective of our struggle is the elimination of the existing solution of the question of nationalities and forming an independent Confederation of Asian Republics, independent of the People's Republic of China' (Asia Watch, 1991, p. 6). Two years later, in 1986, exiled ethnic leaders from the territory of the People's Republic of China began publication of a new journal called *Common Voice (Journal of the Allied Committee of the Peoples of Eastern Turkestan, Mongolia, Manchuria, and Tibet presently under China)* (Asia Watch, 1991, p. 6). In addition to these recent developments within the non-Han population of China, it should also be noted that the leader of the 1989 Tian'anmen demonstration was not a Han but a Uighur from Xinjiang. Uerkesh Daolet (his Uighur name being approximated in Chinese as 'Wu'er Kaixi' and thus recorded by the Western press), was 'listed number 1 on the state's 21 most wanted list' after the Tian'anmen crackdown. As Gladney (1991, p. 295) points out: 'The Uighur student leader was joined by a host of other minority nationality students and teachers, far disproportionate to their population. At least three students from the Nationalities Institute were killed during the incident, including a Muslim Kazak and a Tujia minority.'

Thus, despite the unequivocal position of power held by the Chinese state, discontented non-Han peoples in China none the less played a key role in orchestrating what was 'perhaps the greatest pro-democracy demonstration to have ever taken place in China' (Gladney, 1991, p. 295). Today, the conjunction of dissident minority groups within and beyond China is growing. This situation, combined with the rapid changes occurring across China's state boundaries, have the potential to fuel national movements in China and create a situation in which political fracturing within the 'Chinese motherland' is just possible.

REFERENCES

Asia Watch 1991: 'Crackdown in Inner Mongolia'. *Asia Watch Report* (July).

Bugra, Mohammed Emin 1954: *Eastern Turkistan's Struggle for Freedom, and Chinese Policy*. Istanbul: Osmanbey Mat. T.L.S.

Chang, Kwang-Chih, 1986: *The Archeology of Ancient China*, 4th edn. New Haven, Conn.: Yale University Press.

Chiang Kai-shek 1947: *China's Destiny*. New York: Roy Publishers (1st edn, 1943).

Eberhard, Wolfram 1982: *China's Minorities: yesterday and today*. Belmont, Calif.: Wadsworth Publishing Company.

Fei Xiaotong 1981: 'Ethnic identification in China'. In *Towards a People's Anthropology*, Beijing: New World Press.

Franklin, Bruce (ed.) 1973: *The Essential Stalin: major theoretical writings, 1905–52*. London: Croom Helm.

Gladney, Dru 1991: *Muslim Chinese: ethnic nationalism in the People's Republic*, Harvard East Asian Monographs no. 149. Cambridge, Mass.: Harvard University Press.

Holley, David 1991 : 'Beijing credits its stability to '89 use of force'. *Los Angeles Times*, 10 April.

McMillen, Donald 1979: *Chinese Communist Power and Policy in Xinjiang, 1949–1977*. Boulder, Col.: Westview Press.

Moser, Leo 1985: *The Chinese Mosaic: the peoples and provinces of China*. London: Westview Press.

People's Republic of China 1983: *Constitution* (adopted on 4 December 1982 by the Fifth National People's Congress of the PRC at its Fifth Session). Beijing: Foreign Languages Press.

Peters, William 1991: 'Central Asia and the minority question'. *Journal of the Royal Society for Asian Affairs*, XXII, 152–7.

Schwarz, Henry G. 1971: *Chinese Policies Towards Minorities: an essay and documents*, Occasional Paper No. 2, Western Washington State College Program in East Asian Studies.

Wang, William 1982: 'The Chinese Language'. In *Human Communication: language and its psychobiological basis*, (Scientific American), Oxford: W.H. Freeman.

Xue, Muqiao (ed.) 1982: *Almanac of China's Economy, 1981*. Hong Kong: Modern Cultural Company.

Yu Ying-shih 1967: *Trade and Expansion in Han China: a study in the structure of Sino-barbarian economic relations*. Berkeley, Cal.: University of California Press.

Yuan, Qing-li 1990: 'Population changes in the Xinjiang Uighur Autonomous Region (1949–1984)'. *Central Asian Survey*, 9(1), pp. 49–73.

Zhongyang Minzu Xueyuan Shaoshu Minzu Yuwen Xi (Central Nationalities Institute: Department of National Minority Languages) 1984: *Uyghur Tili Darsligi* (Lessons in the Uighur Language). Beijing: Central Nationalities Institute.

PART II

Long-submerged Identities

10

Edgar Kant and Balto-Skandia: *Heimatkunde* and Regional Identity

Anne Buttimer

Geography has never flowered alone; it has always been tied in with the general advances of science, trade, and technology; it literally has no history separate from the intellectual and cultural currents of the time.

F. Lukermann

Issues of national prestige and identity stirred among the sponsors and audiences of geography during its early years as an academic field. For the first two generations of disciplinary effort in Europe at least, imaginations continued to fly towards exploration of the few remaining *terrae incognitae*, but they also turned to the local environments of the homeland, where landscapes were read as indices of cultural belonging, as 'medals struck in the image of civilization', tangible evidence of historical relationships between human cultures and milieux. *Heimatkunde* (literally, knowledge of home areas), which became a required item in the apprenticeship of young recruits, was a style of practice intimately linked with questions of local, national and regional identity. It held a special appeal in the lands of Norden.

The term *Heimatkunde* had, literally and metaphorically, diverse connotations, ranging from school exercises and field excursions designed to promote a sense of patriotism, on the one hand, to training in map-making techniques and the delimitation of local, regional and national boundaries, on the other. At one end of the spectrum it might connote a Weimar-style *Bildung*, including studies of history, biology, culture, folk traditions and sense of place; at the other end it could refer to· hard-nosed empirical inventories of land use, resources and landscape morphology. The goals and conceptual design of home area studies also varied widely among schools: imperial nations apparently keen to bolster their particular self-images, world views and expansionary geopolitics, emerging or colonized nations keen to affirm the integrity of their own *Lebensräume*.

Philosophical underpinnings for home area studies can be traced to Pestalozzi who believed that schoolchildren should gain some comprehension, at a very early age, both of planetary systems and of the immediate environment of their own homes. Home area studies formed part of primary and secondary school curricula throughout the lands of Norden, but their status and rationale varied. *Hembygdforskning* in Denmark and Sweden, for example, was conducted at *Folkhögskolor* (folk highschools), founded by the Danish clergyman Gruntwig to provide education particularly for rural adults in the latter part of the nineteenth century. In Sweden also the folk highschools addressed the interests of rural areas, but ecclesiastical connections were not so explicit. Helge Nelson, once Rector in a Swedish folk highschool, and later Chairman of Lund University's geography department regarded *hembygdsforskning* as direct heir to the Linnaean topographic surveys, and key promoter of patriotism: knowledge of home areas was seen as a precondition for love of the fatherland:

> For one likes to know what it is one ought to care for. And as one has got to know it, then it has usually grown in value, it has received a richer content and greater importance, for oneself. Thus the increased knowledge of the home-area will strengthen the feeling for it and make it warmer and richer. But the increased knowledge will also widen the eyes and let the home-area emerge as the small part in the big whole, in fatherland. Then the love of home area can grow to include all our land and people.
>
> (Nelson, 1913)

To the east, in Finland and in the Baltic countries, *Heimatkunde* was an integral part of university curricula, and its civic importance was explicitly affirmed. Scientific knowledge and understanding of home areas could foster a sense of local and national identity, just at times when the homeland's cultural and territorial integrity was being questioned geopolitically.

Apprenticeship in geography in the early twentieth century must have inevitably involved tensions between the 'home-making' interests of place and regional identity on the emotional and pedagogical side, versus the 'bread-winning' interests of economic and administrative rationality on the other. In the language of Alfred Schütz, one might suggest that horizons for 'home' and 'breadwinning/husbandry' were situated within very different zones of reach and relevance: their conceivable maps of restorable and attainable reach scarcely isomorphic (Schütz and Luckman, 1973, pp. 36–40). Competing claims on space, resources and territory thus gave wings to the field of geopolitics. Geography, and specifically *Heimatkunde*, contributed towards the elucidation of such questions. There were substantial differences, of course, among models inspired from German, French or British sources; questions of *scale*, contextual relevance, cognitive style and ideological implications varied too. Scandinavian geographers tended to follow German precedents, but there were fascinating exceptions, especially on the periphery. In fact, one might claim that the most intimate links between geography and regional identity were forged in the peripheral lands of Eurasia from the Balkans to Catalonia, the Celtic fringe to Karelia.

Balto-Skandia offers an ideal arena in which to observe connections between the practice of *Heimatkunde* and issues of national and regional identity. Despite the diversity of physiographic base and the dramatic record of its geopolitical history, this region may appear to be, relatively speaking, more culturally homogeneous than others in the twentieth century. The term itself, Balto-Skandia, has something of a geo-poetic ring, appealing perhaps more to teachers of world geography or definers of cultural regions than to realists in political geography. For the historian of ideas it might well be considered as a flight of geographical imagination, evoking memories of the Greek sailor Pytheas who travelled to the North during Alexandrian times (Pliny the Elder, 4: 94–5; 37: 35–6). It also evokes images of the Roman (and later Christian) province of Dacia, of the Al-Idrisi (twelfth-century) map, or late mediaeval networks of Hansa ports. Closer to memory might be the royal family connections between Sweden and Poland, the holy wars of Gustav II Adolph during Sweden's golden era in the seventeenth century, and Oxenstierna's geopolitical strategies which included, *inter alia*, the strategic location chosen for building or taking over universities on the fringes of the empire: Turku, Riga, Tartu, Greifswald and Lund. The appeal of a unified Balto-Skandia has obviously had different meanings for folk at central and peripheral locations. And here, as in other regions, historical attempts to create a unified geo-political region stretching from the North Sea to the Baltic have generally been initiated by imperial interests from within or from without. Most complex, historically, has been the identity of people residing on the eastern shores of the Baltic Sea.

It is to the ideas and career history of one notable pioneer on this frontier, Edgar Kant (1902–78), that this paper is addressed. Born into a merchant family in the Baltic port of Tallinn, Estonia, Kant's early career coincided with his native land's short-lived autonomy as a nation state (1920–40) (see figure 10.1). Staunchly patriotic yet liberal in politics, the atmosphere of his childhood was one which encouraged study of international affairs, science, a rational attitude toward economic and social matters and a strong love of nature, folk traditions, and culture (Buttimer, 1987). Having volunteered for military service at the Russian front at the age of sixteen he returned to Tallinn and witnessed the Tartu (1920) treaty which established the independence of his native land. He studied botany, chemistry and climatology at the University of Tartu and was an eager reader of philosophy, history and social science. Quite early he became a disciple of the famous Finnish scholar Johannes G. Granö, whose *Reine Geographie* seemed indeed appealing. Conceptually, this showed that research on the human dimensions of geographic enquiry could be conducted with a rigour and precision comparable to that of physical geography; methodologically, it opened doors for detailed empirical observations at various scales; and, practically, it offered an integrated perspective on population and milieu and an ideal framework for *Heimatkunde*.

It may indeed have been *Heimatkunde* which initially inspired Kant's choice of geography as a university subject. Herein lay a practice with great potential for promoting the educational goals of his country. The horizons of his research interests were eventually to transcend those of local neighbourhoods, territorial niches, and even Estonia itself. At the peak of his early career, in the mid-

Figure 10.1 Baltic States, 1917–45

(*Source*: reproduced with permission from the Pennsylvania State University Press, 1978)

1930s, few issues occupied his attention as compellingly as that of Balto-Skandia, the vision of a cultural-political unit surrounding the Baltic Sea. In the context of his career as a whole, of course, curiosities about the relationships between geography and national or regional identity and those associated with *Heimatkunde* should be regarded as the preoccupations of youth; they would lose their salience after his involuntary emigration to Sweden in 1944. Anthropogeographical interests, and issues of culture and local identity would thereafter yield place to more functional and theoretically based research on urban-economic systems and the grand promises of regional science. His creativity would find ample challenge and support in the Lund School during the 1950s and 1960s.

It was Kropotkin, one of Kant's favourite authors, who claimed:

> the purpose of geography ... is to teach us ... that we are all brethren, whatever our nationality. It must show that each nationality brings its own precious building stone for the general development of the commonwealth .... (Kropotkin, 1885, p. 942)

Heimatkunde offered an ideal course for someone committed to promoting the educational interests of the fatherland. Students would volunteer for summer projects all over the country: mapping physiographic, historical, bio-ecological and ethnic features of Estonia's diverse local areas (Granö, 1922; Rumma, 1925). Kant was placed in charge of field-work in towns and cities, and his first thesis on Tartu (Kant, 1926), with its graphic illustrations of everyday life, rhythms of light and dark, warm and cold, circulation in and out of the city, was a masterpiece. From 1927 to 1936 Kant was editor-in-chief for the entire collection of *Heimat* studies, *Eesti*; from 1938 to 1940 he was chief editor of the *Eesti Atlas* (Atlas of Estonia).

In the broader context of this interwar period, the goals and academic status of *Heimatkunde* were not uncontroversial. Geography's concern about its own identity as an academic discipline was not always compatible with the multidisciplinary concerns of national and cultural identity. Relationships between anthropogeography, *Reine Geographie* and *Heimatkunde* thus posed a number of conceptual and methodological problems. While remaining an admirer and life-long friend of Granö, Kant dared to disagree with his mentor on a number of points (Kant, 1932), such as geography's relationship to biology and social science, regional identity and appropriate relationships between pure and applied science.

A key tension within early twentieth-century regional geography was that between chorological and ecological orientations, in other words between the contrasting claims of formism and organicism. While *Reine Geographie* was methodologically more akin to the former, Kant was sympathetic to the latter. He was, of course, a brilliant cartographer and always an advocate of empirical field research; he also appreciated the analytical and communicative value of maps. But he wished to complement the predominantly formistic (Hettnerian) approaches of the day with at least three further research directions: temporal, social and ecological. Ecological insight seemed imperative for him in studies of population and resources, agriculture and rural life; he even used organic analogies in describing the *Lebensräume* of towns and cities. His summary presentation of the Tartu study was built around organicist metaphors: (a) the heart (centre); (b) the lungs (green areas, hospitals, sanatoria); (c) the limbs (suburbs); (d) the torso (basic ground plan); and (e)–(h) the extremities (Kant, 1926, pp. 261–2).

Granö was cautious about human ecology. The integrity of geography as an academic discipline demanded an avoidance of such environmental determinism as had been associated with anthropogeography. While open to exploring humanist aspects of landscape aesthetics and environmental experience, he felt that geography should define a specific core for its own research domain and also be clear about the boundaries separating it from other fields (Granö, 1929). Human ecology, in Granö's view, was simply part of sociology, admissible perhaps in *Heimat* studies, but not an integral part of geography. Kant, on the other hand, impressed with the writings of Kropotkin, Le Play, Fleure, Geddes and Mukerjee, argued that ecological and social dimensions should be an integral part of geographic enquiry (Kant, 1932, 1948a). His linguistic versatility (he spoke at least eight European languages and could read fifteen of them), coupled with his extensive travels, encouraged him to position Estonian work within the wider framework of international scientific endeavour. Reviewing debates about ecological and social facets of geographic enquiry as articulated by Schlüter, Hassinger and Klute in the German literature, and debates between advocates of *sociografie* and *sociaal geografie* in Holland (Kant, 1948a), he developed his own style of practice, one which was to ascribe a much more important place to history. It was a style far closer in spirit to the French school than to any of his nearer neighbours. He was especially fond of Maximilien Sorre's bio-geography, and was a life-long friend of Demangeon.

Questions of disciplinary boundaries were never really a major concern for Kant. But he was concerned about the integrity of knowledge and was keenly aware of the vacuum which had been created by the conceptual war between anthropogeography and spatial or chorological science. Schooling himself in diverse fields of natural and social science he strove towards more comprehensive and logically defensible modes of disciplinary practice. On matters of ecology, for instance, he drew a critical distinction between what he called *auto-ecology*, or study of the direct effects of climate or any element of the physical environment on human feelings, health and spiritual life, and *synecology* (ecology of communities) which would seek macro and general patterns of mutual adaptations between groups and their milieux (Kant, 1932, 1934a). In this he took Vidal de la Blache's map of world population as a paradigmatic illustration.

> It must be remembered, however, that the significance of environment and its influence on man and population changes with the force of collective achievements over nature. This is not so well accomplished by the individual, as by a social entity. Even where man seems to act as an individual he is really utilizing spiritual and material social aid.
>
> (Kant, 1934, p. 3)

Anthropo-ecology was the term he used to describe his own conceptual approach to the study of population and *Lebensräum*. It would enable him to observe and document the complex drama of people and localities, society and space, bearing in mind both the bio-physical contexts and historical experiences. Fully aware of the hazards of environmental determinism on the one hand, and those of *ceteris paribus* science on the other, he was always a firm advocate of empirical field research as well as attunement to history.

Anthropo-ecology thus offered a valuable framework for the multidisciplinary efforts of *Heimatkunde*, opened windows on to the French tradition of geography and social history, and eventually afforded possibilities for the extension of curiosity beyond the boundaries of Estonia to include the whole of Balto-Skandia.

HORIZONS OF HOME: ESTONIA AND BALTO-SKANDIA

It was from an anthropo-ecological perspective that Kant approached questions of national and regional identity. In typically organicist fashion, he sought to position his own land, and all of its components, within a broader regional and global setting. Hence his fascination with Balto-Skandia, a cultural-political unit including Scandinavian and Baltic lands, a region characterized by a certain internal homogeneity of geophysical, biotic and cultural-historical features.

Ideas about a Northern European 'natural region' were expressed already in the nineteenth century. In 1871 the Finnish botanist Norrlin published a thesis on the 'Limits of Finland and Scandinavia from the Viewpoint of Natural History' based on exhaustive field observations in the Onega-Karelia region (Norrlin, 1871). It was a Finnish geologist, Ramsay, who defined Fenno-Skandia as the 'well-defined region which is attached to the rest of Europe only through the isthmian land connections between the Gulf of Finland, Ladoga, Onega and the White Sea' (Ramsay, 1898). While the Swedish geographer Högbom and foreign geologists such as Suess preferred to confine the term Fenno-Skandia to the Archaean shield of crystalline rock (Suess, 1901; Högbom, 1913), Nordic geographers generally preferred to use the term in a geographical sense (Sederholm, 1928; Sömme, ed., 1960, p. 14). What seems to have appealed especially to Kant, however, were the ideas of Rudolf Kjellén and Sten de Geer, both of whom attempted a scientific exploration of the physical and cultural geography of the area, and had seriously entertained the notion of a region called Balto-Skandia.

The main criteria on which de Geer's delimitation of the Northern European region was based (De Geer, 1928) included (a) the Fenno-Skandian bedrock itself; (b) the outer limits of Quaternary glaciation; (c) post-glacial landscape phenomena (erosion and deposition); (d) distribution of the Nordic race; (e) distribution of Scandinavian and Baltic languages; (f) distribution of 'Protestantic Christianity'; and (g) territorial extent of 2000-year old Nordic states. Superimposing the boundaries of nine distinct distributions, de Geer had proposed a synthetic line which, in his view, would circumscribe the 'natural region' of Northern Europe, or Balto-Skandia. Kant eagerly adopted this thesis and proceeded to elaborate on these criteria for regional identity, emphasizing their implications for Estonian identity.

(a) The Fenno-Skandian bedrock itself, a crystalline shield constituted the core of the Scandinavian peninsulas. Strata deposited around this shield have been subjected to massive folding, erosion and deposition. Throughout

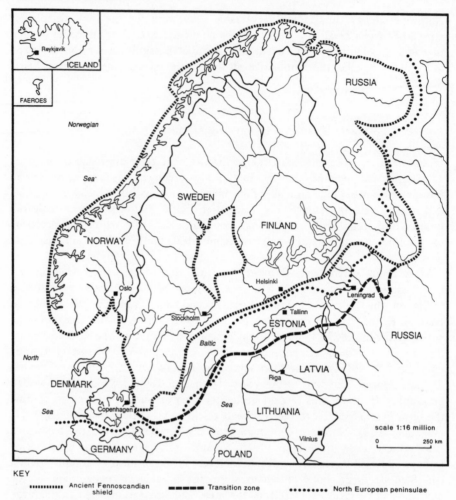

Figure 10.2   The Fenno-Skandian core

(*Source*: after de Geer, 1928, with permission from *Geografiska Annaler*)

the Tertiary era this heartland of Scandinavia remained above sea level but the Baltic and Danish coastlines were frequently inundated. Fossils recorded in the Linnaean surveys of the eighteenth century afford evidence of the extent of marine deposits in Southern and Central Scandinavia (fig. 10.2)

(b) At the approach of Quaternary times the climate became colder and the entire area was covered by a continental ice sheet within a dome-like formation (*Inlandsis*) which extended eastwards across the Baltic into Russia and southwards to cover Denmark, North Germany and Poland. Subsequent glaciation, separated by interglacial phases, left cirques, valleys covered with glacial till and eskers. Lines of moraines mark the retreat of glaciation. The beds of Lakes Ladoga and Onega were formed by glacial erosion; Lakes Ilmen, Peipos and Beloye were all partially formed by morainic dams. Retreat of the ice sheet led to substantial uplift of the land mass, exposing it to erosion especially on the north-western (Scandinavian) side and deposition on the south-eastern (Baltic) surfaces. Northern Sweden is now 12,000 feet higher than it was at the time of glacial retreat and uplift continues today at a rate of 20 inches per century in the Stockholm area, and even 36 inches in the Gulf of Bothnia where presumably the Würm glaciation reached its maximum thickness.

(c) On physiographic criteria, the natural region of Fenno-Skandia could be defined in terms of the outer limits of Quaternary glaciation. Post-glacial events and processes had obviously carved out roughly similar horizons of opportunity for human occupance. Within this region, too, there was a relatively similar climate: with the exceptions of Denmark, Skåne and western Norway, the entire region belongs to the (Köppen) Boreal Forest climatic zone. The distribution of potential resources for agriculture, forestry, energy and industrial development, of course, showed wide differences between Western and Eastern zones. Scandinavia had vast mineral reserves but lacked substantial deposits of coal; the Baltic area benefited from post-glacial soil deposition, but it lacked the 'white coal' of hydro-power afforded by the Fall line in Scandinavia. There were shale-oil reserves in north-eastern Estonia which, as Kant noted, could provide a valuable resource for industrial development (and which indeed has been exploited since then), but basically the Eastern Baltic was a region best devoted to forestry, fishing and agriculture.

(d) Over this glacially differentiated physiographic base successive migration streams had passed, so the actual landscape was regarded as the historical outcome of sequent occupance by people of highly contrasting backgrounds: Stone Age, Roman, Slavic, Teuton and Scandinavian. The same physical environment had held different meanings for these successive groups, depending on their skills, goals, and ways of life. Throughout, however, the fundamental divisions of Estonian national space were those defined by the 'Baltic boundary', or highest marine traces left after the final retreat of Quaternary glaciation (figure 10.3).

Above this line were the fertile landscapes of Central and Southern Estonia, a region once covered with Arctic vegetation and in later times more highly favoured for agriculture and eventually industrial development. Surrounding it on western, north-western and eastern sides, were stretches which had lain under the continental ice-sheet and were later inundated by Baltic waters. These zones showed records of older habitation but younger vegetation; here were the domains of fishing and forestry, domains which eventually suffered population decline and rural to urban migration from the end of the nineteenth century. Ways of life, habitat style and landscape morphology differed strikingly between these two regions:

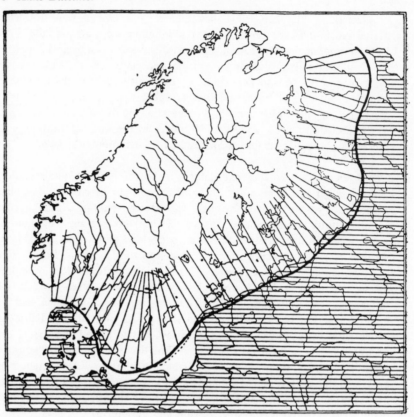

Figure 10.3　The 'Baltic boundary': sketch of boundary between the erosion-territory and the accumulation-territory of the last ice sheet

(*Source*: after Woods, 1932)

> Undoubtedly, the two regions indicated, differing in genetical, mor-
> phological and ecological background, are Estonia's most essential
> and chief natural geographical divisional foundation, in the limits of
> which a more detailed geographical or environmental subdivision
> can be made. (Kant, 1934a, p. 6)

Historically speaking, the critical distinction between 'superaquatic' and
'subaquatic' regions of Estonia had only become salient at the end of
the Iron Age (c. 11–12c AD); evidence of previous occupations (such as
during Roman and Stone Age) showed a more even distribution. Throughout all successive occupations, and especially with the development of
technologically more sophisticated modes of livelihood, the distinction
between these two regions could be documented.

On Huntingtonian criteria of civilization and climate, of course,
Estonia and the eastern shores of the Baltic could scarcely fare well; and
yet, in historical perspective, a remarkable level of civilization had been

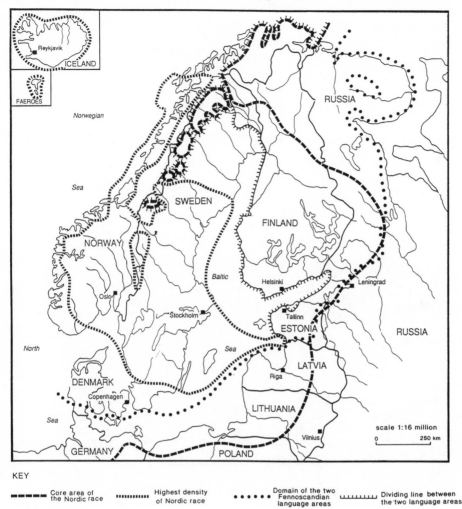

Figure 10.4 Core area of the Nordic race

(*Source*: after de Geer, 1928)

achieved. Like de Geer, Kant sought cultural explanations. Far more important than the conditions of environment was the quality of the Nordic race, a subject regarded as quite legitimate in those days (Lundborg and Linders, 1926). In the region as a whole de Geer had distinguished between the Scandinavian and East Baltic racial types, and Kant emphasized Estonia's greater affinity with Finland rather than with its southern neighbours (figure 10.4).

(e) In terms of language also, de Geer identified three groups: (i) Scandinavian (Danish, Swedish, Norwegian); (ii) Finnish and Estonian; and (iii) Latvian–Lithuanian. For a number of economic and geopolitical

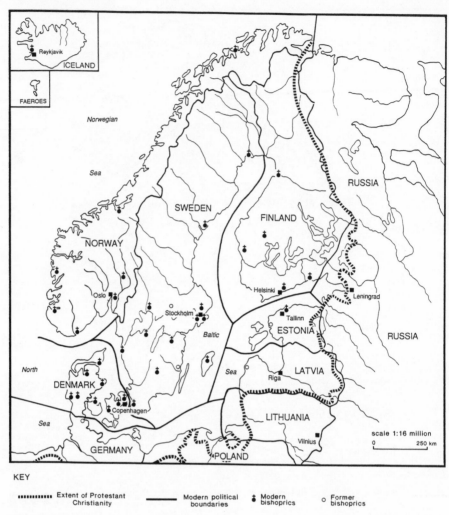

Figure 10.5 Extent of Protestant Christianity

(*Source*: after de Geer, 1928)

reasons Kant showed the extent of interaction which had actually taken place among people of these three language groups and argued on these grounds for their complementariness.

(f) On criteria of 'civilization', Kant had no hesitation in adopting de Geer's criterion of religious affiliation as a potential common denominator. The spread of 'Protestant Christianity' afforded an unequivocal delineator (figure 10.5). He took pains to document the enormous importance of Swedish influence on Estonia's intellectual and economic life ever since the time of Vasa.

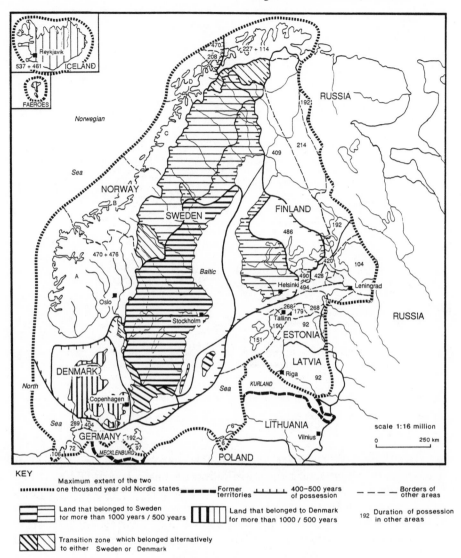

Figure 10.6 Maximal extent of 1000-year-old Nordic states

(*Source*: after de Geer, 1928)

(g) Geopolitically, over 2000 years there really had been only three major Nordic states which at one time or another had held sway over the Balto-Skandian area; of these, he claimed, the era of Swedish occupation had been the most favourable from an Eastern Baltic point of view (figures 10.6 and 10.7).

Superimposing the boundaries defined by these separate criteria, physical as well as cultural, de Geer produced his final synthetic map, and identified

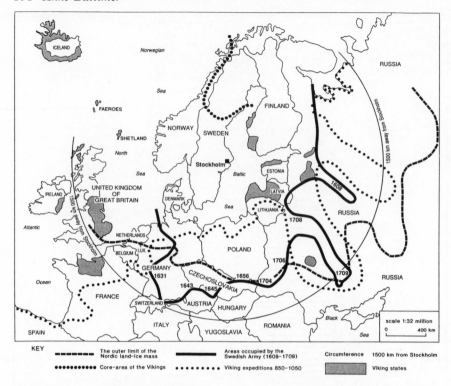

Figure 10.7   Nordic core and periphery in historical perspective
(*Source*: after de Geer, 1928)

the 'mean' geographic surface which, in his opinion, could be regarded as 'Balto-Skandia'. Kant agreed enthusiastically with this judgment (Kant, 1934b, 1935b). On historic as well as on social and ecological grounds, he argued, the interests of all three Baltic states might best be served from participation in such a region. Each state had its own potential contribution to make to Balto-Skandian civilization; his own research therefore was the clarification and promotion of Estonia's economy and society within this region.

REGIONAL PLANS FOR ESTONIA

Anthropo-ecology and *Heimatkunde* thus held more than intellectual appeal for Kant: they afforded a means of rendering the fruits of scholarly research available for societal needs. And Estonian society, in the wake of independence, offered several challenges, all the way from writing a State Constitution to the housekeeping of its domestic economy. Freedom had fostered the proliferation of political parties and folk movements, and Kant joined some of his fellow veterans from the 1918–20 war in an effort to restore some rationality to the political landscape. A growing conservatism among other veterans prompted

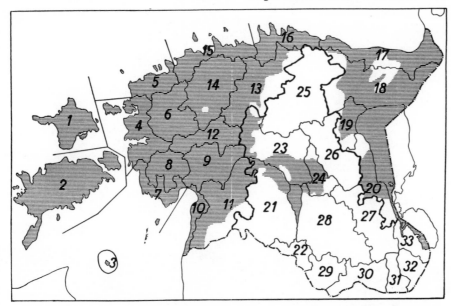

Figure 10.8 Natural regions of Estonia
(*Source*: after Granö, 1922; Kant, 1934a)

Kant and others to initiate the *Eesti Rahvuslaste Klubi* (Estonian Patriotic Club) in 1930. He also helped launch a newspaper *Vaba Sona* (Free Word) in 1933 and a successor *Uus Sona* (Our Word) in 1934. In assuming such editorial responsibilities, particularly with ERK, Kant hoped to create a forum for more rational debate about the political challenges facing Estonian society at the time.

To introduce more rationality to the functional organization of Estonia's landscapes and economic life would be Kant's own special contribution. In his doctoral dissertation on 'Problems of Environment and Population in Estonia' (1934a) he had argued that each advance in the technological mastery of space and resources had involved new interpretations of milieu (Kant, 1934a, 1934b, 1935a). Through an exhaustive inventory of population, agriculture and levels of living, he came up with a proposal for reform of administrative regions.

The fundamental divisions of Estonian *Lebensraum* were, of course, those of 'subaquatic' and 'superaquatic' zones. Superimposing this on Granö's original regionalization of the country (Granö, 1922), Kant endeavoured to harmonize economic and ecological perspectives (figure 10.8). Granö's system was based on chorographic analyses of material elements, combinations of which were used to discern 'landscape unities' (Granö, 1922, 1929). The three-element system seemed most appropriate for Estonia, in Kant's view, and it also allowed for more detailed statistical analysis based on communal statistics. To this formal pattern Kant also wished to add a more functional perspective (figure 10.9).

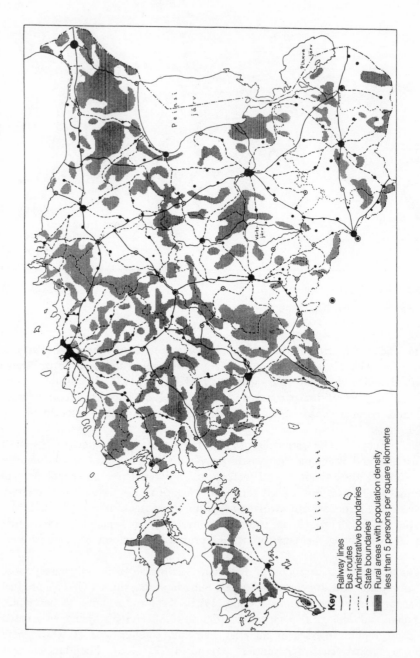

Key

- Railway lines
- Bus routes
- Administrative boundaries
- State boundaries
- Rural areas with population density less than 5 persons per square kilometre

Figure 10.9 The Central Place System of Estonia

His own field surveys had not only included in-depth mappings of population, agriculture and variations in quality of life throughout the land, but also analyses of migration flows, industrial and commercial activities, the origins of all settlements, their size, structure and functions. His most striking innovation consisted in the definition of urban hinterlands and fields of influence (figure 10.10).

Who discovered the notion of 'Central Place Hierarchies'? Kant identified one Carl Brunckman, a Swedish economic historian, who published the idea in 1756. His own discovery actually preceded Christaller's but, in his typically self-effacing manner, he celebrated his Bavarian colleague's *chef d'oeuvre* and introduced him to the Swedes later.

Far more significant than questions of *site*, morphology or history, were those of *situation*, and he noted several instances of the cardinal importance of transportation and accessibility. For a country so dependent on primary production, lacking resources for industrial development, Estonia should invest in trade and commerce: hence the importance of belonging to a Balto-Skandian unit with the access westwards which it afforded. No doubt this was a conviction deeply engrained from childhood.

Quite soon after the defence of his dissertation, Kant became Associate Professor, later Ordinary Professor of Economic Geography at the University of Tartu (1936). One of his first practical applications of his exhaustive field research on population and *Lebensräume* was a plan for the rationalization of administrative units in the land. In a formal presentation to the Ministry of the Interior on 10 September 1935, 'Territorial reform of communes: considerations of economic geography and central place organization', Kant argued that the highest possible degree of rationality in spatial organization would strengthen the nation. Already in this document one could sense the burgeoning of more functionalist attitudes toward spatial organization. He laid out certain imperatives for rationality in the reform of communal boundaries, for example:

- that administrative structures should reflect the optimal hierarchical ordering of central places;

- that taxation should promote rather than hinder rational development of that hierarchy;

- that the development of transportation facilities should be harmonious with the central place system;

- that time should be allowed for research and the acquisition of necessary empirical data; and

- that change should begin with the larger units and only after the overall size and number of regions had been decided should one begin to unify the smaller ones. (*Uus Sona*, 11 September 1935)

The inductive, grass-roots approach of earlier *Heimatkunde* had by now yielded place to a more normative and hypothetico-deductive approach to the func-

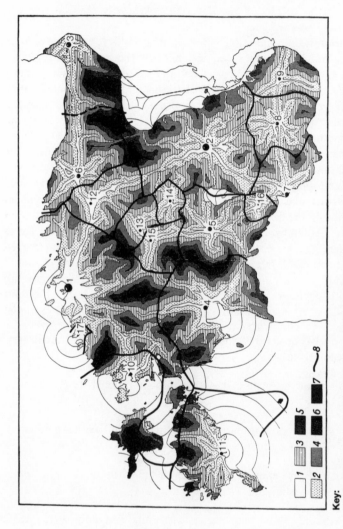

Key:

Urban settlements: 1 Greater Tallinn Tallinn-Nõmme; 2 Tartu; 3 Narva; 4 Pärnu; 5 Viljandi; 6 Rakvere; 7 Valga; 8 Võru; 9 Petseri; 10 Haapsalu; 11 Kuresaare; 12 Tapa; 13 Paide; 14 Põltsamaa; 15 Türi; 16 Tõrva; 17 Paldiski. The heavy curved lines (8) indicate boundaries between urban hinterlands, the shaded zones (1–7) symbolize time–distance intervals from 1–6 hours and 6 hours.

Figure 10.10 Urban hinterlands in Estonia: polycentric isochrones of time–distance to Estonian cities

(*Source*: Kant, 1934a)

tional organization of space. Throughout the 1930s one could sense that economics rivalled both sociology and biology as a key source of insight into matters of population and *Lebensraum*. And he also assumed a more active role in national economic affairs. He chaired the Society for Economic Sciences (1936–9), directed the Tartu Regional Industrial Inventory (1937), led the government-sponsored Labour Market Survey (1940) and served on the Advisory Boards of the Estonian Institute for Economic Forecasting and the Estonian Institute for Natural Resources (1938–40). In 1938 he was appointed Pro-Rector of the University of Tartu and president of the Humanities Section of the Estonian Academy of Sciences. Between 1938 and 1940 he was a member of the Collegium of the Estonian Cultural Foundation.

Many of Kant's ideas on functionalist approaches to the management of economic and urban life found fertile soil at Lund where, surrounded by an eager cadre of young graduate students, he could play a creative role in promoting applied and quantitative geography. On questions of *Heimat* and national identity, however, he remained rather silent. Housekeeping replaced home making, organicism shrunk to mechanism as the characteristic style of his Lund period. But the old penchant for formal classification and mapping could also find expression, as indeed could his genius for conceptual and methodological speculation. Most significant perhaps, was his monumental attempt to produce a *Lexicon* of geographical terms in 13 different languages: a task which alas remained incomplete when he died in 1978.

ANTHROPO-ECOLOGY, NATIONAL IDENTITY AND BALTO-SKANDIA

More than a decade has passed since the death of Edgar Kant. It is tempting to speculate on how he might construe the events in Estonia and Balto-Skandia today. The Baltic region certainly claimed prominence in the Scandinavian press during the 1980s. While formal political boundaries of states have remained stable since the 1940s, there have been many dramatic conflicts and unresolved tensions over functional boundaries within the area. Nordic states have reached a high level of standardization and compromise on matters of economic and social policy, but they still differ considerably in their attitudes toward continental Europe, the Baltic countries, and those of the former USSR. Denmark joined the European Economic Community in 1973 and Sweden is still considering membership; what this may mean for their stances on EFTA and NATO is still not clear. During the 1980s 20 per cent of Finland's foreign trade was still with the USSR. *Perestroika*, however, sent fresh breezes blowing throughout that vast territory, opening up prospects for livelier contacts between western and eastern shores of the Baltic Sea. Finnish television programmes were viewed in Northern Estonian homes, and Eastern Baltic destinations were flaunted as favoured attractions in Swedish tourist ferry-line brochures. Green parties grew in the late 1980s and concern mounted about nuclear matters; differences of opinion certainly not coincident with those of national territories.

The Baltic Sea itself became an arena for other kinds of drama during the 1980s. The 'whiskey-on-the-rocks' episode at Karlskrona in 1981 opened a

decade of submarine hide-and-seek all along Sweden's coastline and archi-
pelagos, and protracted legal debates about the criteria on which transgression
of territorial waters should be defined. As in other marine settings, issues of
military technology and those of fishing territories often become conflated.
The (1958) Geneva Convention, it was assumed, had 'solved' the question
once and for all with its *median* line between the coastlines of autonomous
riparian states; in the case of the Baltic, this is not easy to harmonize with
the 12-mile limit of territorial waters on the continental shelf. There in the
middle lies Gotland, officially part of Swedish national territory. In 1988 a
compromise was reached between Sweden and the USSR on how the 'white
zone' might be shared: an agreement highly favourable to Sweden, but one
which would have been inconceivable if, for example, Gotland and Estonia had
been independent countries (Utrikespolitiska Institutet, 1986, 1988). Ecological
problems, too, transcend those of national territories, and worries about pollution which have reached virtually panic proportions. On 14 February 1988
Svenska Dagbladet featured a cartoon on the Baltic as 'Sewer for Seven Nations',
illustrating the annual dumping of noxious materials recorded by the Baltic
Marine Environment Protection Commission (Helsinki, 1980).

Such ecological conditions may not interfere with submarines, but they were
obviously lethal for fish. Alarming signs of malformation and disease among
pike and other species were recorded and a viral epidemic in 1988 threatened
virtually to exterminate the seal populations of the Baltic. In May 1988 a
Greenpeace bus set out from Copenhagen on a Baltic expedition, equipped with
advanced laboratory equipment for analysing air and water pollution. It headed
northwards through Sweden to Finland and then southwards across the eastern
Baltic countries to participate in a marine science exhibition at Leningrad,
ending the tour in Berlin.

And Estonia? Issues of national identity for all republics within the former
Soviet Union became major news items during the late 1980s. Thoughts turn
to Kant and speculations about his potential comments are unavoidable. One
might suspect, indeed, that he would be thinking *European* rather than *Balto-
Skandian* on subjects of geography and national identity. On the strength of
recent developments, the relatively freer circulation of people and ideas which
air travel and the relaxation of frontier protocol on personal identity, currency
and language have allowed, he might have seen the EC as fulfilment of a dream
for a more rational organization of the European economy as a whole. Where
exactly the easterly boundaries of Europe should be defined would perhaps still
pose an enigma for him. On ecological questions Kant would deliberate, but
on Green Party politics he might well have misgivings. On questions of political
geography one could safely guess that Kant would strongly advocate a *functional*
rather than a geopolitical approach to the spaces of Europe: far more important
than the container-spaces of morphologically defined political territories were
the dynamics of functional systems and the *nodal* organization of accessibility
surfaces. Kant's Europe might have been one such as Kropotkin, Réclus or even
Weber (or Delors) dreamed about: a Europe of identifiable national cultures,
co-responsibly creating and re-creating sociality and economic vitality in the
wider continental and global context of the late twentieth century.

GEOGRAPHY AND NATIONAL IDENTITY IN THE LATE TWENTIETH CENTURY

Connections between geography and human identity were indeed intimate during the opening years of the twentieth century. This is attested in career histories, in texts and in many of the discipline's 'applied' endeavours, for better or worse. In the radically different and constantly changing contexts of the late twentieth century, some reassessment of traditional notions are surely in order. *Identity*, after all, is only one of the many constellations of human interest which are associated with the environment: humans also seek *order*, *niche* and *horizons* for life (Buttimer, 1983, 1993). The priorities assigned to these human interests in geographic research has changed dramatically over the century, and the postwar generation has tended to place primary emphasis on *order*. Its preferred cognitive styles, too, such as those of formism and mechanism, have been best suited for the elucidation of order in space, time and social activities. The latter years of the twentieth century have witnessed at least two major objections to these inherited practices: first, they failed to accommodate the growing environmental crises (tensions between order and niche) and, secondly, they could scarcely handle emerging patterns of regional and national identity, particularly in the so-called 'developing' world.

Kant's Balto-Skandian work occupies a unique position in that perennial challenge of negotiating the distinct interests of identity, order, niche and horizon. Each of these, empirically speaking, projects its own constellation of 'reach', and the nation has, more often than not, provided the umbrella under which this negotiation has been investigated: audiences and sponsors have welcomed 'scientific' evidence about the biophysical, social and historical foundations for national and cultural identity. The fostering and promotion of Estonian national identity indeed offered such an umbrella for Kant's initial endeavours; in fact, his work explored not only issues of *identity* and *niche*, but also elucidated issues of *identity* and *order*. Horizons of scale on the ordering of space and human activities, he demonstrated, had varied through history, each occupying group creating its own appropriate order within the framework of resources (niche) within reach. During the 1930s he argued strongly for an ordering of lifespace which would promote both the integrity of national economic life and the integration of his home nation into a wider regional context. After his move to Sweden, however, issues of 'home' (identity and niche) were probably 'bitter sweet'. Less emotionally demanding were issues of order: the functional organization of space and time offered intellectual puzzles for which he already had welcome suggestions. These were the contributions which apparently were most welcome in Sweden.

Although Kant's early research was conducted at micro-levels of observation, his horizons were never confined to the local scale. Rather, he sought to relate particular cases to the national context and also to relate his own methods and approaches to international scholarly trends. His theoretical imagination and analytical innovations, however, were clearly at the urban and (nodal) regional scale – fields of urban influence and the dynamics of migration flows were his favourite foci. And these would claim further appeal after his move to Sweden. Research horizons could be set at the national scale, models for comparative-national analyses proposed, and eventually the global scale explored in his monumental *Lexicon*.

The career journeys of twentieth-century geographers all reveal the extent to which emotional, aesthetic and moral considerations have entered into apparent choices of substantive focus and epistemological orientation. Nor can such choices be understood without reference to the changing circumstances of audience and sponsorship for their efforts. The growing interest in national and regional identity around the world today demands a reassessment not only of cognitive styles, but also of those world views projected by early twentieth-century European geographers which placed such priority on nationhood as a fundamental building block for world humanity.

The anthroposphere today might more appropriately be conceptualized as an arena of diverse cultural worlds, each attuned to distinct melodies of tradition, values, aesthetic and moral preferences: pockets of order sustained by livelihood which are themselves dependent on some combination of local and global resources. Superimposed upon, and at times quite insensitive to, the traditional values of these cultural worlds, are systems of economic and technological order which hold increasing discretion over the global interactions of humanity and earth. Such systems today support gross inequities in human access to resources and also give rise to serious and possibly irreparable damage to the biosphere itself. How feasible, then, might be those prospects envisioned by Kropotkin and Kant, of a world in which 'each nationality brings its own precious building stone for the general development of the commonwealth'? Already towards the close of his own career, Kant sensed the challenge to discover ways whereby international and cross-cultural dialogue could be facilitated on issues of human co-responsibility about the global interactions of humanity and its terrestrial home.

REFERENCES

Alexandersson, G. 1972: *De nordiska länderna: natur, befolkning, naringsliv*. Stockholm.
Bergsten, K.-E. 1988: 'Geography: my inheritance'. In T. Hägerstrand and A. Buttimer (eds), *Geographers of Norden*, Lund. Lund University Press, pp. 61–70.
Buttimer, A. 1983: *The Practice of Geography*. London: Longman.
—— 1987: 'Edgar Kant, 1902–1978'. *Geographers: Biobibliographical Studies*, 11, 71–82.
—— 1993: *Geography and the Human Spirit*. Baltimore, Md: Johns Hopkins University Press.
De Geer, S. 1928: 'Das geologische Fennoskandia und das geographische Baltoskandia'. *Geografiska Annaler*, I, 119–39.
Granö, J. G. 1922: 'Eesti maastikulised üksused' (*Die landschaftlichen Einheiten Estlands*). Tartu: *Loodus*.
—— 1929: *Reine Geographie. Eine methodologische Studie, beleuchtet mit Beispielen aus Finnland und Estland*. Helsinki: Acta Geographica II, no. 2.
Greenpeace magasin, 1988, no. 2, Editorial.
Hägerstrand, T. 1978: 'Edgar Kant 21.2.1902–16.10.78'. *Svensk Geografisk Årsbok*, 54, 96–101.

—— 1982: 'Proclamations about geography from the pioneering years in Sweden'. *Geografiska Annaler*, 64B, pp. 119–25.

—— 1983: 'In search for the sources of concepts' in Buttimer (1983), pp. 238–56.

Högbom, A. G. 1913: *Fennoskandia*. Heidelberg: *Handbuch der Regionalen Geologie*, IV Band.

Kant, E. 1926: *Tartu. Linn kui ümbrus ja organism* (Tartu: study of an urban environment and organism). Tartu: K.-U. 'Postimehe' trukk.

—— 1932: 'Geograafia, sotsiograafia ja antropoökoloogia' (Geography, sociography, and anthropology). *Sitzungsbereichte der Naturforswcher-Gesellschaft bei der Universität Tartu*, 39.

—— 1934a: 'Problems of environment and population in Estonia'. Tartu: *Publicationes Seminarii Universitatis Tartuensis Oeconomico-Geographici*, no. 7.

—— 1934b: 'Estlands Zugehörigkeit zu Baltoskandia'. Tartu: *Publicationes Universitatis Tartuensis Oeconomico-Geographici*, no. 9.

—— 1935a: *Bevolkerung und Lebensraum Estlands. Ein Anthropoökologischer Beitrag zur Kunde Baltoskandias*. Tartu: Akadeemiline Kooperativ.

—— 1935b: 'Estland und Baltoskandia. Bidrag till Östersjöländernas Geografi och Sociografi'. *Svio-Estonia*, 80–103.

—— 1948a: 'Den sociologiska regionen, den sociala tiden och det sociala rummet. Några kritiska tankar med anledning av en ny regional sociologi'. *Svensk Geografisk Årsbok*, 24, 109–32.

—— 1948b: 'Omstridd Mark. I. Om Bebyggelsens ålder i Lill-Estland i samband med den lokala folktraditionen från den svenska tiden'. *Svio-Estonia*, 1944–8, 5–70.

—— 1952: 'A polyglot glossary of geographic terms'. *Comptes Rendus 16 Congrès International Géographique*, Lisbonne, 1949, pp. 419–24.

Kropotkin, P. 1885: 'What geography ought to be'. *Nineteenth Century*, 18, 940–56.

Lundborg, H. and Linders, F. J. 1926: *The Racial Characters of the Swedish Nation*. Uppsala: Swedish State Institute for Race Biology, esp. p. 126.

Nelson, H. 1913: 'Hembygdsundervisning i folkhögskolan'. In *Svensk folkhögskolans årsbok*, trans. T. Hägerstrand, *Geografiska Annaler* (1982), 24B, 119–25.

Norrlin, J.P. 1871: *Om Onega-Karelens vegetation och Finlands jemte Skandinaviens naturhistoriska gräns i öster*. Helsinki.

Pliny the Elder, *Natural History*, trans. H. Rackham. Cambridge, Mass.: Harvard University Press/Loeb Classical Library.

Ramsay, W. 1898: 'Über die geologische Entwicklung der Halbinsel Kola in der Quartärzeit'. *Fennia*, 16, no. 1.

Rumma, J. 1935: 'Die Heimatforschung in Eesti'. Tartu: *Publicationes Universitatis Dorpatensis Geographici*, no. 4, pp. 1–17.

Schütz, A. and Luckmann, T. 1973: *The Structures of the Life World*, trans. Richard M. Zaner and H. Tristram Engelhardt Jr. Evanston, Ill.: Northwestern University Press.

Sederbohm, J.J. 1928: 'Withelm Ramsay'. *Terra*, 40, 2–3.

Sömme, A. (ed.) 1960: *A Geography of Norden*. Stockholm: Svenska Bokförlaget.

Suess, E. 1901: *Das Antlitz der Erde*. Vienna, III, 1.

Svenska Dagsbladet, 1988: articles published on 14 February, 7 May and 23 June.

Utrikespolitiska Institutet 1986: 'Östersjön – Fredens hav?' *Världspolitikens Dagsfrågor*, rapport no. 10.

—— 1988: *Regeringens proposition 1987–88*: 175.

Vardys, V.S. and Misiunas, R.J. (eds) 1978: *The Baltic States in Peace and War*. University Park and London: Pennsylvania State University Press.

Woods, E.G. 1932: *The Baltic Region: a study in physical and human geography*. London: Methuen.

11

Stateless National Identity and French-Canadian Geographic Discourse

Vincent Berdoulay

Mention is often made of the interrelationships between the development of geographic research, theories and institutions on the one hand and, on the other, state-based nationalism as a particular form of expression of national identity. However, this may not be as systematic as one might think. The case of French language geography in Canada has long been intriguing in this respect: whereas geography and nationalism mutually reinforced each other in the great founding countries of contemporary geographic thought, this was not so apparent in the case of French Canada (Hamelin, 1962–3, 1984).

We may have been too attentive to a geography laid out by a few 'national schools', that is, within a limited number of important Western nation-states. As the sociopolitical phenomenon that the nation-state embodies is, and especially was, far from universal, one should enquire as to what type of geographic thought may unfold where national identity does not coincide with the territory organized by the state.

Even though the contemporary growing affirmation of Quebec as sovereign territory concerns the vast majority of Francophones north of the United States, this population has evolved without a privileged relationship with the state (Brunet, 1957; Monière, 1979). Their case brings to mind the many other populations with a strong national identity but without much political control over their territory. What type of geography may be produced by a minority who cannot rely on its own state? How is its territory thought of and talked about?

This type of question can also shed light on the current Canadian crisis, whereby the move of Quebec towards independence or some other form of sovereignty is the territorial outcome of the failure of a hierarchically conceived federal state to encompass French-Canadian identity. The relatively 'quiet' and peaceful affirmation of Quebec in the contemporary world can be better understood when one takes into account the long process which has led to a minimally conflictual adequacy of national identity and its spatial support. As the following pages will show, this relative adequacy comes in great part from the active work of geographers and from the societal success of their discourse on territory.

It may well be that some other outbursts of national identity in the contemporary world have proved highly conflictual and peace-threatening because of the lack of a long-lasting geographic discourse and practice which could have made compatible national identity, the world view and modernity. This type of geographic contribution, whenever it has been successful, has produced a mature buffer between a people's suddenly freed aspiration for establishing its own state and the realities of the larger inherited territorial context.

In order to approach this question, I shall first propose an appropriate conceptual framework. I shall then show how it helps in understanding the former French-Canadian context. Finally, I shall underline the most specific discursive features related to national identity. This examination refers to the Francophone population in Canada until about the middle of the twentieth century; that is, before Quebec gained sufficient strength to reduce discrepancies between the lived world of this population and its political territorial base.

SCIENTIFIC DISCOURSE AND CULTURAL CONTEXT

We are now better aware of the interrelationships that exist between geographic thought and its societal context. I have tried to provide a balanced view of the nature of these interrelationships by developing a contextual approach to the history of geography, and by showing how it enlightens the emergence of the French school of geography at the turn of this century (Berdoulay, 1981a and b). It is clear that these interrelationships are usually complex and indirect as they are mediated by scientific and scholarly institutions. This is partly related to the very inertia inherent in the forms of scientific (including geographic) discourse. Its specific modalities, which include various genres of expression, languages, rhetorics and so on, provide a definite autonomy in relation to geographic thought (Berdoulay, 1988). This is why the problem with which we are dealing here can be approached in the light of the discursive dimension of geography.

In this discipline in particular there is an incipient realization that one should cast a broader view on the evolution of scientific discourse (which I consider here as the discourse relying on, or claiming to have, scientific forms of reasoning, inference and expression). Accordingly, scientific discourse is viewed as grounded in lay geographic discourses, whether spontaneous or professional, which legitimize the social practices of a society or of a particular population. This phenomenon is at the very root of geographic thought, which grew out of the human activity of measuring, dividing and transforming the earth's surface (Pinchemel, 1981). In other words, scientific discourse entails some dependency on a diverse geographic knowledge, that is, a kind of 'geosophy' (Wright, 1947), which may function as a proto-geography, or

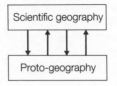

Figure 11.1 Scientific geography and its relation to proto-geography

'*Vor-Geographie*' (Hard, 1982). Obviously, this dependency may be reciprocal. However, one knows little about its ways and means. In schematic form, the problem is summarized in figure 11.1.

The realization of the existence of different, but interrelated, discourses may be placed within the more encompassing theory of culture. Following some research in anthropology and sociology, one may distinguish (at least at the level of cultural production) between a primary, or folk, culture and a secondary, or high, culture (*culture première* and *culture seconde*, in the terms used by Dumont, 1968). The latter refers to the production of intellectuals, artists, novelists, and so on, and it includes scholarly culture, whereas the former refers to productions which depend on everyday practices, to the lived experience of the people, to the routines constitutive of the *genre de vie* (Cette culture ..., 1981). Parallel with the other level of culture, it includes popular culture. Of course, these levels, or types, of cultural activities are strongly dependent on each other because they involve complex iterative processes; but differentiating them helps to set the (scientific and lay) geographic discourses in a broader framework, as simplified in figure 11.2.

Now, the role of national identity can be grasped more fully. It is rooted in popular culture and it also implies a proto-geography. It is via these concepts that its impact on scientific geographic discourse may be assessed. Practically, the problem consists in finding out to what extent there is communication of knowledge between the two types of discourses or, better, in appreciating the discrepancy which exists between them.

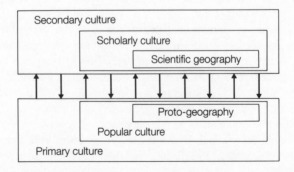

Figure 11.2 The structure of primary and secondary culture

Broadly speaking, the realities dealt with by these discourses are the territories necessary for national identity to unfold. Consequently, the concept of territory that should be used here refers to the set of places which a population organizes and counts on for upholding its own identity. It involves the means necessary for acting on people and things; of these, efficient but not vital means depend on the state.

THE FRENCH-CANADIAN CASE

Before the British takeover, Canada was simply one of the provinces making up the kingdom of France and participated in the latter's cultural life. Scholarly culture and popular culture were in direct communication and dialogue (Roy, 1930; Filteau, 1978). In spite of its small population, New France impressed observers by the high level and quality of its intellectual life (Benson, 1966).

This situation radically changed with the British control of Canada which was established by the Treaty of Paris in 1763. Popular culture became almost completely cut off from scholarly culture: the former continued to draw on the same bases while innovating in its adaptation to the new political and economic environment. But scholarly culture was eradicated because most of the local elite had to leave Canada. It was replaced by a new, British elite, anchored in the military and in trade, whose scholarly culture was part of the English-speaking world and, consequently, was cut off from local popular culture.

Slowly, a French-Canadian elite emerged, producing once more a scholarly culture in tune with the folk experience. But because of Canada's relative confinement within the anglophone world (especially in commercial, economic and political affairs), this small scholarly cultural activity had to meet the considerable challenge of drawing on French-Canadian popular culture while being confronted with the locally dominant models of British and Anglo-American scholarly cultures. Trying to extend the former French cultural heritage and what it could grasp from French-speaking Europe, this little elite achieved early, original and often idiosyncratic syntheses.

Broadly speaking, the increasing number of ties with secondary culture in France and Belgium were still insufficient to draw Canadian popular experience significantly closer to international French cultural production. Attempts to bridge this cultural gap were greatly disfavoured by ideological divergences, notably by the defensiveness and conservative strategies of the local French-Canadian elite (Galarneau, 1970; Simard, 1987). Thus, the strong national identity at the popular level was not fully echoed in high, scholarly culture. Nevertheless, with the increasing national affirmation of French-Canadians, cultural production became more balanced, as is summarized in figure 11.3.

Another challenge for French-Canadian scholarly culture in its attempt to encompass popular experience was totally related to its territorial bases. Broadly speaking, French-Canadians as a group did not have political control over the land they occupied. Even when taking into account their increasing

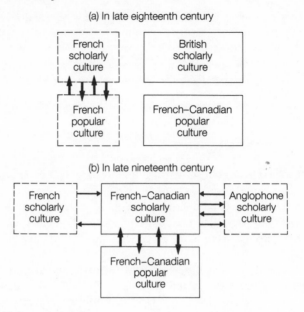

Figure 11.3   Major cultural flows

control of some state institutions, correspondence between these and French-Canadian territory could never be achieved. National identity had to cope with various and often discrete political frameworks, leaving scholarly culture the challenge of trying to harmonize quite different popular experiences and academic contexts.

First, the territorial framework was remodelled by the British in order to establish their domination. A complex set of political, legal, administrative, religious and economic devices served this purpose. Many of these constraints were progressively relaxed when the French-speaking population of Canada finally became a demographic minority (in the middle of the nineteenth century). Only the Saint Lawrence valley remained mainly French-speaking; it was to become the hearth of the emerging political territory of Quebec. But in spite of its long-term efforts, this territory always lacked the full attributes of the State, many of which were reserved for the anglophone-controlled federal level.

Institutions which were not directly dependent on the State had long been devised (such as the parish) and reinforced autonomy *vis-à-vis* state political structures and boundaries (Rioux and Martin, 1964; Savard, 1980; Claval, 1979, 1980; Bélanger, 1984).

At the same time, Quebec did not encompass all the French-speaking population. Its diaspora was important, spreading not only into the Prairie, into northern and eastern Ontario, into Acadia, but also in greater numbers into the United States (mostly New England) (Louder and Waddell, 1983). A culture of mobility resulted from this territorial experience (ibid.; Morissonneau, 1978 and 1981). Many people were involved: net migration from Quebec alone amounted to over half a million, and there are now almost as many descendants of French-Canadians in the United States as there are francophones in Quebec (Lavoie, 1972; Abramson, 1973). Because scholarly culture was firmly rooted in the concentrations of the Saint Lawrence valley and tied to a fully sedentary *genre de vie*, it had difficulty in taking into account the experience of mobility which was constitutive of popular culture. Agricultural colonization schemes associated with forest exploitation (Séguin, 1980) were the major response to this challenge, whereby French Canada was losing population to the United States. Even if the settlers in these colonization areas remained less numerous than those migrating to the United States, their experience, which was frantically hailed by the elite, played a major role in French-Canadian national identity and also revealed the difficulty for the elite of influencing a mobile population (Dussault, 1983).

Consequently, this lack of correspondence between the experienced territory of the French-Canadians and the long-standing Canadian and *Québécois* state institutions was echoed by a tension between popular and scholarly cultures. This accounts for the difficult dialogue between the proto-geography conveyed by popular culture and French-Canadian scientific geography which was confronted with the ideas put forth by the great national schools, themselves quasi-foreign to the mainsprings of French-Canadian experience. A particular type of geographic discourse was to emerge in French Canada.

PARTICULAR FEATURES OF FRENCH-CANADIAN GEOGRAPHIC DISCOURSE

The overarching concern of this discourse was with the survival of French-Canadian culture. Given the weak, not to say antagonistic, action of the State, an overall strategic, sometimes subversive, geography came to the fore among this population. Its originality clearly stemmed from its unfolding at the margins of State authority.

In this respect, it diverged completely from English-Canadian geographic discourse with its dominant political and commercial concerns. Most of all, this discourse implied the national (pan-Canadian) unity of the country (Sénécal, 1989). It evaded the question of the presence of a French-speaking culture. When this was mentioned, rhetoric reduced it to a mere historical event (a phase in the colonization process), whose contemporary impact is limited and highly localized. But most significant was the idea of demonstrating the unity of the country by focusing on its relationships with the outside world (mostly the British Empire) or its environmentally based regional structure.

A most interesting thesis was developed by Harold A. Innis, whose contribution to geography has been remarkable, however misunderstood it may have been (Dunbar, 1985; Berdoulay, 1987 and 1990; Parker, 1988). Put briefly, he argued that the potential for unity is set in the environment. He insisted on the role played by the Canadian shield, which determined a large east–west hydrographic system (based on the Saint Lawrence, the Great Lakes, and the Saskatchewan, Fraser and MacKenzie rivers). As soon as the fur trade was established, man had taken advantage of these communication possibilities to build a country thus organized around the Canadian shield (Innis, 1930). Always attentive to communication technologies, Innis was critically aware of the threat to Canadian unity posed by diverse regionalisms and their close relations with the neighbouring United States.

One cannot find the same interest in French-Canadian geographic discourse for such environmental sources of (real or desired) Canadian unity. One may even note a rising scepticism at the turn of the century. For instance, the most notable French-Canadian geographer at that time, Emile Miller, shed a very different light on the geography of Canada (Miller, 1913). On one hand, he subscribed to the confederative goals (which, for French-Canadians, included equal status with their English-speaking fellow citizens), albeit in a sceptical and unhopeful manner. On the other hand, he made it clear that geography was playing against Canadian unity. The three great regions of British Columbia, the Western plains and the more fragmented eastern part of Canada were quite autonomous in their relationship to each other and, furthermore, were almost irresistibly attracted by the United States. Only one sub-region escaped the strength of this attraction to the southern neighbour: the Laurentian region (*la Laurentie*), as the manifest land of the French-Canadians. Larger than Quebec, its extent remained ill-defined, understandably so because of the above-mentioned French-Canadian pattern of settlement. Thus, even if the Canadian shield were given considerable importance, it stopped short of organizing the whole of Canada.

More specific geographic themes related to this primary concern for survival: they were all aspects of a discourse on spatial strategies usually at odds with the central authority of the state. First, this difficult relationship with the state was characterized by an insistence on the most decentralized forms of organization. It focused on local administrative or religious units, such as the county, the diocese and the parish. This was not surprising because these local levels made up the territories over which French-Canadians felt most mastery.

The second aspect was the growing importance given to the parish. It actually ended up by coinciding spatially with most functions of local government, such as education and municipal services. It preserved cultural and religious values, it promoted the development of a community spirit, and it was the introduction to parliamentary life. In other words, the parish was a 'State in miniature' (Miller, 1913, p. 85). It was considered as the best locus for the unfolding of national identity.

The third aspect concerned the strategy of colonization. Here again the parish played a central role as a means of organizing new territories for the French-Canadians. Given its highly autonomous life, this model of cellular expansion could alleviate anti-French state policies, and it could also minimize the effects of provincial and even international boundaries. Contrary to what one may think, the parish was not considered a completely closed world; it gave to its inhabitants a sense of belonging to a vast cultural diaspora (not locked in a territorial State) and of contributing to universal goals, thanks to the focus on the Rome-based Catholic Church. Colonization involved much organizational back-up: state involvement was then accepted as far as it complied with the ends and means of the ever-present Church organization. Unanimously, it was felt that survival depended on the control of the land. Much debate concerned the areas to be colonized and their (dispersed or continuous) spatial pattern of extension; this was especially the case with the opportunity to colonize the West in view of the increasing anglophone State hostility (Rouillard, 1909; Miller, 1917).

Interestingly enough, environmentalism was mostly called for in association with this concern for survival. More precisely, its use served the purpose of legitimizing the French-Canadian claim to existence and respect. By underlining their anteriority in Canada over other Europeans, they strengthened their cause not only by stressing some historical right but also by pointing to their better adaptation to the environment (Miller, 1913).

What follows from all this is a fundamental ambiguity *vis-à-vis* the province of Quebec. On the one hand, the local level is hailed but, on the other, Quebec is felt to be the hearth of French-Canadian culture and the only territorial unit endowed with some State attributes which may be used by French-Canadians. The growing disappointment with the Confederation as a means of development for French-Canadians made Quebec a more significant territory in their spatial strategy, at the expense of a transboundary territorial realm (Sénécal, 1989 and 1992).

Thus, French-Canadian geographic discourse was cast in very particular forms. As part of a scholarly culture which rested on relatively weak demographic, political, economic and institutional bases, it necessarily favoured genres adapted to this limited support and resources. The foremost is probably the essay: as the best-fitted medium for both conveying scientific information and proposing strategic orientations, it played a major role in Quebec (Fournier, 1985). Arthur Buies's geographic work best illustrates this discursive approach to the realities of his country. It accounts for the otherwise apparent contradictions of the juxtaposition of detailed, accurate and positive environmental analyses with quasi-lyrical developments and practical enticements to promote colonization (for example, Buies, 1880, 1889). This is the logic of the genre.

Albeit less action-orientated, a related genre is the travel account (or journal). It provided diffusion of information and reflexive discussions on remote and foreign areas. By achieving this, it made readers better aware of their own land and of their place in the world (Brosseau, 1987). Strategic considerations were at times clearly present (for example, in the numerous works by Faucher de Saint-Maurice).

In order to spread geographic knowledge and correlated strategies, school textbooks were of great importance in French Canada; in fact, they constituted the main published geographic material in the country for a very long time, and they escaped State control. Their publication history is long, consistent and enlightening about a geographic knowledge which is at the interface of scientific research and popular culture (Berdoulay and Brosseau, 1990, 1992). Most significant is the advanced textbook which was also read by the cultured public among the elite, thus contributing to moulding the geographic view of the world (for example, the many expanded editions of Holmes, 1832–3; Garneau, 1912; Miller, 1924).

Regional monographs constitute another genre; they concerned areas to be developed or to be devoted to colonization schemes. As evidenced by the works of S. Drapeau, J.-L. Langelier and A. Pelland, they were really a genre in the sense that they followed a similarly structured outline and were action-orientated. The topics which they examined tended to be the same, and dwelt on scientific as well as popular information. It may even be argued that these monographs laid out the geographic grid through which the idea of Quebec and its regional make-up have been established and taken for granted ever since (Sénécal and Berdoulay, 1992).

Finally, a quantitatively minor but ideologically important genre concerns toponymy. Its significance comes from the high stake in control over the land. Even when purely symbolic, this type of territorial appropriation or reappropriation was central to establishing a balance between national identity and environmental referents. It was spontaneously started by the elite (as evidenced in the respective *Bulletins* of the Société de Géographie de Québec and of the Société du Parler Français). Support from the state of Quebec began with the creation of the Commission des Noms géographiques in 1912 at the instigation of geographer E. Rouillard. Since then, this geographic genre has been regularly cultivated, including within a broader perspective the university level (Dorion and Hamelin, 1966).

In addition, rhetoric took a particular slant in the French-Canadian geographic discourse. The figure of metonymy seemed the best fitted to glorify some elements of strategy while eschewing difficult aspects, especially those which escaped the French-Canadian influence. In a sense, it was a way of preserving ambiguity. For instance, the synecdoche whereby the part is taken for the whole was consistently used to shift ambiguously from the description of the francophones as a sub-group of the Canadian population to their representation as carriers of the country's characteristics and interests. This figure had the advantage of maintaining the ambiguity of territorial boundaries. For instance, the term *pays* allowed one to shift freely from the local scale to the provincial and then national (that is to say, federal) scales, while varying the degree of affective bonds that it may be wished to attribute to any of these levels.

More particularly, the use of the ellipsis and of hyperbole was widespread. The ellipsis obviously consisted in not mentioning a point which neither fitted the message sent to the reader nor could be openly criticized because of the insecure position held by the author *vis-à-vis* the State or religious authorities. Often in association with this figure was the frequent use of hyperbole. The frequency was a function of the importance of the spatial strategy pushed by the author. For instance, hyperbolic developments qualified the territories to be colonized, the rural scene or natural resources accessible to French-Canadians. Arthur Buies is thus known for his quasi-lyrical descriptions of the Ottawa valley and of the Lac Saint-Jean region (Buies, 1889 and 1890). Authors of essays, travel accounts and textbooks often indulged in this figure of rhetoric. The city of Quebec and Niagara Falls were most often presented in their picturesque or majestic aspects. But it was also the case with 'ordinary' landscapes (forests, fields and rivers), which were carriers of the cultural values held by colonizers. The same epithets tended to be repeated over and over again: they connoted riches, fertility, abundance and especially beauty. Aesthetics was called forth to support a feeling of attachment to, or appropriation of, a territory. For a time, hyperbole would contribute to the construction of a whole geographic myth, that of the North as the Promised Land (Morissonneau, 1978).

The use of metaphor was relatively limited: they mostly concerned the rootedness of French-Canadians in their rural territory. In the same organic vein, the parish was referred to as a 'social cell' or an 'organism'. But this organicism remained very limited and belonged more to the style than to the deep structure of the discourse. This was in strong contradistinction to the importance of the organic metaphor in major national schools of geography, especially in Germany and France (Berdoulay, 1982).

Why was French-Canadian national identity and geographic discourse not dependent on the organic metaphor? Obviously, this must be related to the avatars of the territorial history of this population. Without a clear spatial pattern, the territories of identity could not be easily described in a systems framework; rather, they would form the motive of a constant struggle against various environmental and political odds. Not surprisingly, the geographic presentation of French Canada often tended towards the telling of a narrative (for example, in the above-mentioned books by Buies and Miller, and in the widely spread textbooks of the Marists). The epic dimension of the narrative highlighted the actions taken by the French-Canadian population. Even though these actions were diverse, they all tended towards cultural survival. The French geographer Raoul Blanchard evidently built on this tradition in his five-volume regional geography of French Canada (from 1935 to 1954).

Writing the geography of French Canada was like finding the various evidences of the French-Canadian founding *geste*. Rather than organicism, the fundamental metaphor that should be singled out is that of an epic story about territorial conquest (or at times *reconquista*), and resistance.

CONCLUSION

The example of former French Canada reveals the originality that a geographic discourse may hold when it is backed by a strong national identity and at the same time its minority status entails ambiguous relations with the state and its territory. It is characterized by spatial strategies which favour local, decentralized, sometimes subversive (anti-State) options, balanced by a sense of working in favour of universal goals. It rests on specific genres which can reach a wide public and relies on particular rhetorical figures, such as hyperbole, in order to reinforce the relationship with controlled or desired territory.

The often epic slant of the discourse points to the very important issue in modern epistemological and textual analysis of not opposing both what is descriptive to what is explanatory, and also what is descriptive to what is narrative. These discursive patterns should be viewed as neither hierarchical in status nor fully separable.

In fine, the conclusion is that contemporary geographic discourse is too much marked by the scientific models which have been elaborated in major nation-states. Geographic thought may significantly gain from the study of very different forms of discourse.

REFERENCES

Abramson, H.J. 1973: *Ethnic Diversity in Catholic America*. New York: John Wiley.

Bélanger, M. 1984: 'Le réseau de Léa', *Cahiers de géographie du Québec*, 28(73–4), 289–302.

Benson, A.B. (ed.) 1966: *Peter Kalm's Travels in North America*, vol. 2. New York: Dover.

Berdoulay, V. 1981a: 'The contextual approach.' In D.R. Stoddart (ed.), *Geography, Ideology and Social Concern*, Oxford: Basil Blackwell, pp. 8–16.

——1981b: *La Formation de l'école française de géographie (1870–1914)*. Paris: Bibliothèque Nationale (Comité des Travaux Historiques et Scientifique, Section de Géographie).

——1982: 'La métaphore organiciste. Contribution à l'étude du langage des géographes'. *Annales de géographie*, 91(507), 573–86.

——1987: 'Le possibilisme de Harold Innis'. *Canadian Geographer*, 31, 2–11.

——1988: *Des mots et des lieux. La dynamique du discours géographique*. Paris: Editions du CNRS.

——1990: 'Harold Innis and Canadian geography: discursive impediments to an original school of thought'. In R. Preston and B. Mitchell (eds), *Reflections and Visions*, Waterloo: University of Waterloo, Department of Geography, Publication Series no. 33, 51–60.

——and Brosseau, M. 1990: 'L'ouverture sur le monde dans les manuels de géographie canadiens-français'. *Cultures du Canada français*, 7, 71–8.

——1992: 'Manuels québécois de géographie: production et producteurs (1804–1960)'. *Cahiers de Géographie du Québec*, 36(97), 19–32.

Brosseau, M. 1987: *L'Autre comme apport à notre identité: l'image de l'Arabe dans les récits de voyage au XIXᵉ siècle*, Ottawa: University of Ottawa, Department of Geography, Research Notes no. 55.

Brunet, M. 1957: 'Trois dominantes de la pensée canadienne-française: l'agriculturisme, l'anti-étatisme et le messianisme. Essai d'histoire intellectuelle'. *Ecrits du Canada français*, 3, 33–117.

Buies, A. 1880: *Le Saguenay et la vallée du Lac St-Jean*. Quebec: A. Coté.

——1889: *L'Outaouals supérieur*. Quebec: Darveau.

——1890: *La Région du Lac St-Jean, grenier de la Province de Québec. Guide des colons*. Quebec.

'Cette culture que l'on appelle savante' (1981): *Questions de culture*, l, special issue.

Claval, P. 1974: 'Architecture sociale, culture et géographie au Québec: un essai d'interprétatlon historique'. *Annales de géographie*, 83(458), 394–419.

——1980: 'Le Québec et les idéologies territoriales'. *Cahiers de géographie de Québec*, 24(61), 31–46.

Dorion, H. and Hamelin, L.-E. 1966: 'De la toponymie traditionnelle à une choronymie totale'. *Cahiers de géographie de Québec*, 20, 196–211.

Dumont, F. 1968: *Le Lieu de l'homme. La Culture comme distance et mémoire*. Montreal: Hurtubise.

Dunbar, G. 1985: 'Harold Innis and Canadian geography'. *Canadian Geographer*, 29, 159–64.

Dussault, G. 1983: *Le Curé Labelle. Messianisme, utopie et colonisation au Québec (1850–1900)*. Montreal: Hurtubise HMH.

Filteau, G. 1978: *La Naissance d'une nation. Tableau de la Nouvelle France en 1755*. Montreal: Editions L'Aurore.

Fournier, M. 1985: 'Essai en sociologie: littérature sociale et luttes politiques au Québec'. In P. Wyczynski, F. Gallays and S. Simard (eds), *L'Essai et la prose d'idée au Québec*, Montreal: Fidès, pp. 143–79.

Galarneau, C. 1970: *La France devant l'opinion canadienne (1760–1815)*. Quebec: Presses de l'Université Laval, and Paris: A. Colin.

Garneau, A. 1912: *Précis de géographie*, 2nd edn, 1917. Quebec.

Hamelin, L.-E. 1962–3: 'Petite histoire de la géographie dans le Québec et à l'Université Laval'. *Cahiers de géographie de Québec*, 7(13), 137–52.

—— 1984: 'Destin d'une géographie humaine mal aimée'. In *Continuité et rupture. Les sciences sociales au Québec, 1935–1985*, Montreal: Presses de l'Université de Montréal, vol. 1, pp. 87–109.

Hard, G. 1982: *Lehrerausbildung in einer diffusen Disziplin*. Karlsruhe: *Karlsruher Manuskripte zur mathematischen und theoretischen Wirtschafts- und Sozialgeographie*, vol. 55.

Holmes, J. 1832–3: *Nouvel abrégé de géographie moderne ....* Quebec: Neilson & Cowan.

Innis, H.A. 1930: *The Fur Trade in Canada*. New Haven, Conn.: Yale University Press.

Lavoie, Y. 1972: *L'Émigration des Canadiens aux Etats-Unis avant 1930. Mesure du phénomène*. Montreal: Presses de l'Université de Montréal.

Louder, D. and Waddell, E. (eds) 1983: *Du continent perdu à l'archipel retrouvé*. Quebec: Presses de l'Université Laval.

Miller, E. 1913: *Terres et peuples du Canada*. Montreal: Beauchemin.

—— 1917: 'Où faut-il coloniser?' *Bulletin de la Société de Géographie de Québec*, 11(5), 271–6.

—— 1924: *Géographie générale*. Beauceville: Editeur L'Eclaireur.

Monière, D. 1979: *Le Dévelopment des idéologies au Québec des origines à nos jours*. Montreal: Ed. Québec/Amérique.

Morissonneau, C. 1978: *La Terre promise: le mythe du Nord québécois*. Montreal: Hurtubise HMH.

—— 1981: 'Mobilité et identité québécoise'. *Cahiers de géographie de Québec*, 23(58), pp. 29–38 (reprinted in Louder and Waddell (eds), 1983).

Parker, I. 1988: 'Harold Innis as a Canadian geographer'. *Canadian Geographer*, 32, 63–9.

Pinchemel, G. and P. 1981: 'Réflexions sur l'histoire de la géographie: histoires de la géographie, histoires des géographes'. *Bulletin de la Section de Géographie, Comité des Travaux Historiques et Scientifiques*, 84, 221–31.

Rioux, M. and Martin, Y. (eds) 1964: *French Canadian Society*. Toronto: McClelland & Stewart.

Rouillard, E. 1909: 'L'Ouest canadien'. *Bulletin de la Société de Géographie de Québec*, 3(6), 11–19.

Roy, A. 1930: *Les Lettres, les sciences et les arts au Canada sous le régime français*. Paris: Jouve.

Savard, P. 1980: *Aspects du nationalisme canadien-français au 19ᵉ siècle*. Montreal: Fidès.

Séguin, N. 1980: *Agriculture et colonisation au Québec: aspects historiques*. Montreal: Boréal Express.

Sénécal, G. 1989: 'Les géographes et les idéologies territoriales au Canada: deux projets nationaux contradictoires'. *Cahiers de géographie du Québec*, 33(90), 307–21.

—— 1992: 'Les idéologies territoriales au Canada français: entre le continentalisme et l'idée du Québec'. *Revus d'Etudes canadiennes*, 27(2), 49–62.

—— and Berdoulay, V. 1992: 'Stratégies d'argumentation et aménagement du territoire: le rôle des monographies régionales au Québec (1850–1915)'. *Etudes canadiennes*, 32, 1–18.

Simard, S. 1987: *Mythe et reflet de la France*. Ottawa: Presses de l'Université d'Ottawa.

Wright, J.K. 1947: '*Terrae Incognitae*: the place of imagination in geography'. *Annals of the Association of American Geographers*, 37, 1–15.

12

Nationalism and Geography in Catalonia

Maria Dolors García-Ramon and Joan Nogué-Font

In recent years there has been a revival of ideologies of a nationalist character that give special relevance to the study of nationalism in geography. The international scene is full of signs that suggest that the last decade of this century will be strongly dominated by conflicts of a nationalist character, especially in Europe.

In the making of European nationalisms different factors that we cannot now analyse have played a role in their consolidation. Our aim in this chapter is to show how geography as a discipline, understood in its widest sense, has been a significant agent in the making of some of these European nationalisms. We shall focus upon Catalan nationalism and its close connection with the Catalan school of geography.

The chapter is divided into four sections. The first deals briefly with the study of nationalism from the viewpoint of contemporary political geography. The second presents the study area, Catalonia, with an emphasis on its most relevant cultural, historical and geographical features. The third section analyses the links between geography and nationalism in Catalonia through four relevant indicators: the activity of geographers in hiking clubs – cultural institutions that were very popular and had a marked nationalist orientation; the strong presence of Catalan geography in the projects of the regional planning of Catalonia; the role of the Catalan Geographical Society, founded in 1935; and, finally, the importance given to the teaching of geography in Catalan schools under the influence of movements of pedagogical renewal in which Catalan geographers were prominent. The final section presents some conclusions.

CONTEMPORARY POLITICAL GEOGRAPHY AND THE STUDY OF NATIONALISM

After several decades of decay, political geography has experienced a strong revival in many countries. Traditional political geography centred almost exclusively around the study of the state. This tradition originated during the last century in the work of F. Ratzel who is usually considered the father of political geography. In his *Politische Geographie* (1897) Ratzel clearly identifies political geography with the study of the structure and territorial dimension of the state. Since then, the state has been the main subject of study of political geographers. In fact, many interesting works of political geography that centre directly or indirectly on the study of the state are still being published (Johnston, 1982; Dommen and Hein, 1986; Foucher, 1988). It cannot be denied that the state is one of the politically organized spaces that most require attention from scholars and that have been most influential in the last two centuries. However, it is not the only territorial expression of political phenomena.

The awareness of this has led political geography in recent years to enlarge its scope. A new political geography arises starting from a wider conception of the notion of political space, understood now as collective actions localized in a concrete place, as systems of relationships based on old affinities and reproduced in short-term interactions (Kirby, 1989). In short, the aim is to conceive of a political map of the world centred not only on nation-states as the only political entities, but as a wide array of political spaces, from political, tribal or ethnic alliances to the various neighbourhoods of a metropolitan area.

Such a wider conception of the notion of political space merely represents the translation into the field of political geography of an interesting series of conceptual and methodological innovations experienced by geography during the last decade. We refer, among others, to the reappraisal of the role of culture (Cosgrove, 1983; Thrift, 1983a), to the new emphasis on the meaning of place in the explanation of social phenomena (Agnew, 1987; Entrikin, 1990; Pred, 1984; Soja, 1989; Thrift, 1983b), and to the efforts for reconstructing a regional geography capable of linking the general with the specific (Cooke, 1988; Johnston et al., 1990; Massey, 1985).

This thematic and at the same time theoretical and methodological reconstruction is the setting of the current interest of geography in nationalist issues which earlier political geographers scarcely recognized, with some exceptions (Knight, 1971). But in recent years a number of outstanding contributions have appeared, such as Agnew (1984, 1987, 1989); Anderson (1986a, 1986b); Blaut (1986); Boal and Douglas (1982); Bureau (1984); George (1985); Johnston et al. (1988); Knight (1982, 1984); Lacoste (1986); MacLaughlin (1986); Orridge and Williams (1982); Williams (1982, 1985); Williams and Kofman (1989); and Zelinsky (1984, 1988).

A growing interest in the nationalist issue is also observed in Spanish geography, although this is so recent that few research works refer to the territorial dimension of the nationalisms within Spanish territory; this is quite surprising since Spain is a real benchmark for nationalist conflicts. Among the works by Spanish geographers one can mention Aymà (1981); Lluch and Nello (1983, 1984); Nadal (1990); Nogué-Font (1991) and Vilà i Valentí (1989).

THE TERRITORIAL DIMENSION OF CATALONIA

Catalonia is a nation, without being a state, in the north-eastern corner of Spain (see figure 12.1). Its distinctive identity, based on its own language, culture, history and territory, was recognized to a certain extent in the Spanish Constitution of 1978, and it was considered to be one of the 'historical nationalities' recognized in different degrees during the Second Republic (1931–9). The Catalan Statute of Autonomy was approved in 1979 by the vast majority of the population.

The integration of Catalonia into what was to become Spain started in the fifteenth century with the marriage in 1489 of King Ferdinand of the Kingdom of Aragon (of which Catalonia was a part) with Queen Isabella of the Kingdom of Castile that led to an irreversible unification of both Kingdoms. In spite of this dynastic unification, Catalonia (as well as the other parts of the Kingdom of Aragon) kept intact its specific political, administrative, financial and legal institutions until the second decade of the eighteenth century. An attempt to erode this specificity under the reign of Philip IV led in 1640 to the open rebellion of Catalonia against the King, complicated by the intervention of France. Despite military defeat, Catalonia kept its distinct institutions, although several Catalan territories (Roussillon and the northern part of Cerdagne) were ceded to France in the Peace of the Pyrenees in 1659.

During the War of Spanish Succession (1705–15), Catalonia (together with other parts of the old Kingdom of Aragon) fought against the new Bourbon dynasty. The armies of Philip V conquered Catalonia only after prolonged and fierce resistance in 1714, and all the traditional institutions of Catalonia were wiped out by the Decree of Nueva Planta in 1716. Spanish was also imposed as the only official language.

In spite of the loss of its own political institutions Catalonia experienced intense population and economic growth during the eighteenth and nineteenth centuries. An enterprising commercial and industrial bourgeoisie emerged – in marked contrast to a mostly agrarian Spain – which favoured the rise of the Catalan movement at the end of the nineteenth century. But Catalonia did not enjoy real political autonomy until the short democratic experience of the Second Republic, after which the Franco dictatorship abolished the Statute and tried by every means to dilute the Catalan national identity.

Under the Constitution of 1978, the *Generalitat* (that is, the Catalan Government) enjoys a degree of political autonomy that allows it to manage many aspects of the daily life of the six million inhabitants of Catalonia. This population is very unequally distributed across an area of 32,000 square kilometres comprising the provinces of Barcelona, Tarragona, Lleida and Girona (see figure 12.1). Seventy per cent of the Catalan population is concentrated in the metropolitan area of Barcelona, which covers no more than 10 per cent of the total area of Catalonia: clear evidence of the urban character of contemporary Catalan society. Together with Spanish, Catalan is the official language of Catalonia and it is spoken in all the region by a majority of the

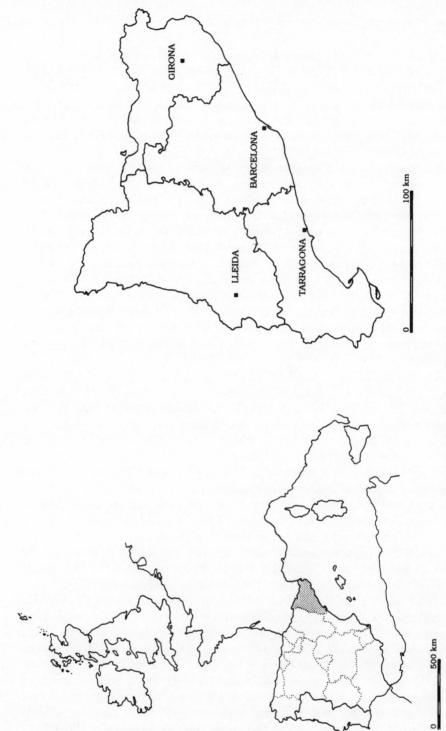

Figure 12.1   Catalonia and its provincial divisions

Figure 12.2 Boundaries of the Catalan language

population, although most of the immigrants from other regions have Spanish as a first language. Catalan is also spoken in Valencia and the Balearic Islands (see figure 12.2), regions with strong cultural and historical ties with Catalonia.

GEOGRAPHY AND NATIONALISM IN CATALONIA

Hiking, geographical knowledge and nationalism

In Catalonia hiking was never merely a sporting activity as in some other countries. From its beginnings it was inspired by a will to give to a scientific dimension as well as a political and a nationalist one (Lluch, 1961). Hiking contributed to the awakening of patriotic and nationalistic sentiments which moved beyond intellectual circles to embrace the aspirations of the Catalan people.

This peculiar notion of hiking was born in Catalonia at the end of the nineteenth century within the framework of a cultural movement of nationalist inspiration (the *Renaixenca* or Renewal) supported by the influential Catalan bourgeoisie of the time (Marti-Henneberg, 1986). Enlightened young members of Barcelona's bourgeoisie founded in 1876 the first hiking club in Catalonia with the name of *Associació Catalanista d'Excursions Cientifigues* (Catalan Association of Scientific Hiking). Two years later, in 1878, the *Associació d'Excursions Catalana* (Catalan Hiking Association) was created and both clubs merged in 1891 to give birth to one of the most influential societies of civic and cultural character, the *Centre Excursionista de Catalunya* (Catalnn Hiking Club), still alive today and loyal to its original ideals.

The *Associació Catalanista d'Excursions Cientifigues* brought together a wide array of people from different disciplines. Geographers, geologists, botanists, historians, archaeologists, ethnologists and even writers and poets had their place in an association whose statute established as its first objective: 'with the aim of investigating everything that deserves attention in our beloved country from a scientific, literary and artistic viewpoint, a Society is founded under the name of *Associació Catalanista d'Excursions Cientifigues*, comprising under this term the diverse branches of human knowledge'.

In 1904 Joan Maragall, one of the most outstanding Catalan poets and a member of the *Centre Excursionista de Catalunya*, summed up the spirit of the Catalan hiking movement in sentences that are worth quoting:

> our hiking movement is neither a sport nor a pleasure or a work: it is love and not merely an abstract love for nature as a whole but for our nature; it is difficult to love the whole world if one does not start by loving the land in which one has been raised; within the love for the motherland is contained the live love for the entire world and who on behalf of the latter abhors the former he or she has none of them. Our love for nature lives through the love for Catalan nature; Catalonia is for us a summary of the world. Hence we can proudly say that the soul of our hiking movement is the love for Catalonia. (Maragall, 1960, p. 860)

All those poets, scholars and scientists had something in common that somehow led geography to an outstanding position in the Catalan hiking movement: the need to know the land intimately, to make an inventory of it and to map it. This task was a top priority, and it had a clearly nationalist interpretation, in agreement with one of the basic principles of the *Associació*: 'to love the land one has first to know it'.

Thus Catalan geographers were active from the beginning in the hiking movement and played a conspicous role within it. In the first half of the twentieth century geographers such as Norbert Font i Saguer, Flors i Calcat, Goncal de Reparaz, Miquel Santaló Pau Vila and later Josep Iglesies, among others, were significant members of the *Centre Excursionista de Catalunya* or belonged to hiking clubs in other Catalan towns, many of them branches of the *Centre Excursionista de Catalunya* in Barcelona. They frequently published articles and notes in the Bulletins, gave lectures at the hiking clubs and were often members of the executive boards as in the case of Pau Vila, who was the chairman of the *Centre Excursionista de Catalunya* from 1932 to 1935.

After the Spanish Civil War and several years of inactivity at the beginning of the Franco dictatorship, Catalan geographers again held important positions in the Catalan hiking movement. During this period, the work of the geographer Salvador Llobet deserves special mention. He founded a publishing house, *Alpina*, that in recent decades has been a reference point for the Catalan hiking movement. Specializing in the publication of maps and hiking guidebooks, during those difficult years *Alpina* satisfied a steady demand for such publications. It is relevant to point out that the Boy Scout movement in Catalonia had very nationalist leanings during the dictatorship and was instrumental in the diffusion of this ideology in a regime that had suppressed political parties and the free press.

It must be pointed out that the symbiosis between geography, hiking and nationalism in Catalonia was favoured during the first 60 years of this century by the fact that the dominant geographical paradigm was the Vidalian one. It is well known that regional French geography was 'imported' into Catalonia at the beginning of the century by Pau Vila, the father of Catalan geography (García-Ramon and Nogué-Font, 1991), and it soon became the model to be followed by other Catalan geographers. The emphasis of French regional geography on the holistic approach to the relationship between society and environment in a given area – the exceptionalist character of which was strongly marked – together with the emphasis on fieldwork and a direct knowledge of the land made this method the most suitable geographical response to the aspirations of the hiking movement and to those of the nationalist project as well.

The debate over territorial division: the comarca *versus the* province

The hiking movement was not the only connection between Catalan geography and nationalism. The link was also established through the active participation of geographers in the projects of territorial division of Catalonia. It has frequently and rightly been asserted that if there exists a Catalan school of geography, one of its most distinctive features is a very strong interest in regional planning (Cassasas, 1979). It is important to point out that Catalan nationalism has always urged the abolition of provincial divisions existing today and imposed by the central government in 1833. In the various periods

OLD DIVISION (1936): 38 *COMARQUES*   NEW DIVISION (1987): 41 *COMARQUES*

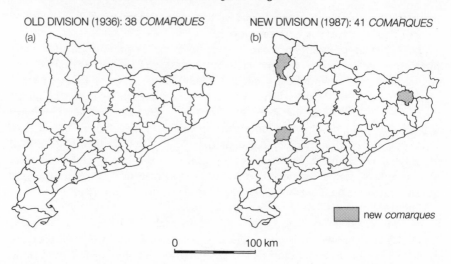

new *comarques*

0 _____ 100 km

Figure 12.3   The territorial division of Catalonia into *comarques*: (a) old division (1936): 38 *comarques*; (b) new division (1987): 41 *comarques*

during which Catalonia has enjoyed some degree of political autonomy, geographers have been sought by the regional governments to elaborate and implement alternative projects of regional planning. This was the case during the Second Republic (1931–9) and it has again been the case since 1978, the year of the proclamation of the democratic Constitution.

In 1932 the *Generalitat* of Catalonia, the institutional form of Catalan self-government which had been re-established with the advent of the Second Republic in 1931, wished to set up a new territorial organization for public administration. Therefore, the Catalan Government asked the geographer Pau Vila (García-Ramon and Nogué-Font, 1991) to prepare a project of territorial division. In establishing the new territories, the main considerations were physical relief, the effectiveness of the communications network, and the territorial attraction of each area's market. Catalonia had to be divided into the smallest possible number of areas, with some degree of equality of physical size and population. Popular sentiment had to be respected as much as possible, provided that it did not produce excessive territorial partitioning. The proposal presented by Pau Vila divided Catalonia into 38 *comarques* (similar to French *pays*) but, for various reasons, it was not approved until 1936.

Due to the work of Vila, the public became sharply aware that the Catalan territory should be politically and administratively restructured with the highest regard for those territorial units with a natural and/or historical character that were *comarques*. Such was the feeling of Catalan nationalism too, as is clearly shown in the *Bases de Manresa* of 1892 (a manifesto with a strong independentist bias) and, to cite another example, in the founding Assembly in 1929 of the *Partit Separatista Revolucionari Català* (Catalan Revolutionary and Separatist Party) in which the abolition of provinces was urged together with the establishment of the *comarques* as the basic administrative unit of the Catalan nation.

The debate is still alive and geographers are conspicous participants in it (Drain, 1990; Lluch and Nello, 1984). The present regional government has recently (1987) implemented a territorial division that has virtually the same number of *comarques* as that designed by Pau Vila 50 years ago (41 *comarques* instead of 38; see figure 12.3). Nevertheless, the Constitution establishes that the four provinces must be kept as administrative units in spite of the opposition of the nationalist circles and of a number of geographers too.

The Catalan Geographical Society

It is well known that the Catalan Geographical Society has played a role in building up the nationalist identity of Catalonia, especially in organizing cultural resistance during the worst period of repression of the Catalan identity, from the end of the Civil War in 1939 to approximately 1970.

The Catalan Geographical Society is one of the 22 societies affiliated to the *Institut d'Estudis Catalans* (an institution that functions as the Catalan Academy of Sciences), which was created in 1907 to promote scientific research and, in particular, Catalan culture. The Catalan Geographical Society was founded in 1935, counting among its founders such well-known geographers as Pau Vila, Josep Iglésies and Lluis Solé-Sabaris. From its beginnings to the present day it has brought together not only academic geographers but also other scientists as well as hikers and geography 'amateurs'. Most of its regular activities (lectures, field trips and so on) have had a nationalistic or patriotic connotation, especially up to the early 1970s.

The Society was very active in 1935 and 1936, when Pau Vila was the president. Among other activities, the Society tried to create a degree in geography within the Catalan University which during the Second Republic (1931–9) enjoyed a high level of autonomy. The Society also tried to reinforce the traditional links between Catalan geography and the Vidalian school of geography through the organization of a field trip to Catalonia in the autumn of 1936 for geographers of the *Association de géographes français* and the *Institut de géographie de Paris*, presided over by E. de Martonne. But the outbreak of the Spanish Civil War (1936–9) put a stop to all kinds of activities and the *Societat* was seriously affected by the war. Afterwards, most of the leading Catalan geographers had to go into exile and the Society – like many other Catalan institutions – suffered the repression to which Catalan culture, in general, was subjected during the Franco dictatorship.

Nevertheless, the institution survived, thanks to its spirit of resistance, its loyalty to its original purposes and the role played by some individuals, such as its first secretary, Josep Iglésies. From 1947 to 1963 – a period of Catalan geography that is referred to as the 'geography of the catacombs' (Iglésies, 1989) – Josep Iglésies started again to arrange the meetings and organize the lectures of the Society; these were usually held at his own home in Barcelona (Cuadros and Durá, 1988). The intense – although underground – relations between the Society and the French Institute at Barcelona (its director was

Pierre Deffontaines) also played a significant role in the survival of Catalan geography and its nationalistic bias during those difficult years. He organized discussions and lectures – some given by invited foreign geographers – that were attended by young Catalan geographers such as Salvador Llobet, Joan Vilà i Valentí and so on. Pierre Deffontaines was even awarded by the *Institut d'Estudis Catalans* the prize given to foreign scholars who have played an active role in favour of Catalan culture (Solé-Sabaris, 1989).

Later, the institutionalization of academic geography in the University in the late 1960s and the slow opening up of the Franco regime made the Catalan Geographical Society change and adapt to the new intellectual and political context. The year 1972 could be considered a turning point when Lluis Solé-Sabaris – also a founder of the Society – became its president; he was a professor of Physical Geography at the University of Barcelona and he made a real effort to strengthen the links between the Catalan Geographical Society and academic geography at the University (Riudor, 1988). From then on a large number of the members of the executive board of the Society also held teaching positions in the Geography Departments, either at the University of Barcelona or at the Autonomous University of Barcelona. (García-Ramon et al., 1993).

At the present moment the Catalan Geographical Society is quite active and has more than 400 members. It continues to organize activities: the monthly lectures, the field trips, the seminars on specific subjects offered by foreigners or local geographers and the publication of the Bulletin (that since 1984 has been called *Treballs de la Societat Catalana de Geografia*). In March 1991, during the presidency of Lluis Cassasas i Simó, the Society organized the First Catalan Congress of Geography that attracted more than 300 participants. One of its objectives was 'to discuss all the aspects related to geography and to its insertion into Catalan society', and one of its features was its peripatetic character (sessions being held in eight Catalan medium-sized cities, besides Barcelona). This is probably related to one of the early objectives of the Catalan Geographical Society: to reach all the regions and *comarques* of Catalonia and make the people feel that geography can play – and even more, must play – an important role in the building up of the Catalan national identity and the Catalan culture.

National identity and the teaching of geography in schools

It is a well-known fact that primary schooling was made compulsory during the second half of the nineteenth century in most European states. It is also well known that geography was one of the disciplines included by the administration in almost all the curricula of primary education in public as well as private schools.

There are several explanations for this choice (Capel, 1981), though one reason seems particularly relevant in the context of this essay: geography was considered by the administration of the nation-states as a key factor in the diffusion among citizens of a sense of national identity. As Nogué-Font (1991) has explained in more detail, the nationalist ideology of the European nation-state saw in the teaching of geography a highly suitable vehicle to spread its political discourse. Knowledge of the geographical space of the country was considered necessary in order to grasp the idea of being a nation, the idea of the motherland. The future citizen had to learn to link an abstract idea (the nation) with a concrete and tangible reality, that is, the physical and spatial setting of the nation. This was also the reason for the great emphasis given to cartography in schools. The teaching of geography was closely linked to the illustration and construction of the feeling of national identity.

Our argument is that this was not only the case with nationalism in nation-states but also in nations which were not states. They obviously had less influence on the educational system but, whenever circumstances allowed (a certain degree of political and administrative autonomy, for instance), the nationalist goverments of these nations which were not states also used geography with the same nationalist purpose.

This is the case in Catalonia during the Second Republic as well as today. What is remarkable and to a certain extent specific is that, in both historical periods, the teaching of the geography of Catalonia in schools has not merely been a government decision but has been closely linked to the movements of pedagogical renewal in which Catalan geographers have been very conspicuous. This combination of nationalism, geography and pedagogical renewal has been a distinctive element of the Catalan geographical tradition (Casassas, 1979).

It is particularly important to note that the majority of the founders of the 'Catalan school of Geography' were teachers and educators: J. Palau-Vera, Miquel Santaló and Pere Blasi, among others, were teachers besides being geographers. The teaching of geography, including field trips, fieldwork and map-making presented to those geographer-educators an exceptional chance to implement the principles of pedagogical renewal and it allowed them at the same time to introduce their pupils to knowledge of the Catalan land and territory. Thus, the study of geography had great importance in the teachers' training school founded in 1906 by Joan Bardina and also in two exemplary educational institutions: the *Escola Normal* (a teacher's training school) of the *Generalitat* during the Second Republic where Pau Vila was a teacher, and the *Institut Escola* (a high school) of the same period where the geographer Lluis Solé-Sabaris taught.

During the last years of the Franco regime and up to the present, the most influential movement of pedagogical renewal in Catalonia – known as *Rosa Sensat* – has followed exactly the same path: geographers deeply concerned with education – among them Enric Lluch and Pilar Benejam – have played a decisive role in the design and development of this movement.

CONCLUSIONS

From the beginning, the Catalan geographical tradition has been deeply involved in the defence of the national identity of Catalonia. The close historical links between nationalism and geography in Catalonia can be traced through different indicators: the role of geographers in the development of the hiking movement, a very popular activity with a marked nationalist bias, the conspicuous presence of Catalan geography in the several regional planning projects which had clear nationalist implications; the task of the Catalan Geographical Society in organizing cultural resistance during the worst period of repression of Catalan identity; and finally, the importance of the teaching of the geography of Catalonia in schools closely related to movements of pedagogical renewal in which Catalan geographers played a remarkable role.

From the beginning of this century to the end of the 1960s – more than 50 years – French regional geography has been the reference paradigm for Catalan geography. This paradigm was perfectly fitted to the nationalist goals of a large number of geographers, as we have seen through the analysis of the hiking movement and in the case of the claim for a new territorial division based on historical national entities – the *comarques*; in a number of cases these *comarques* have been the subject of regional monographs to be presented as doctoral dissertations.

In the same way, geographical research on Catalonia as a whole has often helped to strengthen the nationalist project. Among these, the outstanding example is the *Geografia de Catalunya* published by Aedos and edited by Lluis Solé-Sabaris. The three volumes, written in Catalan, were published between 1958 and 1968 and became a very important reference point for the nationalist cultural resistance. The best Catalan geographers of the time contributed to this initiative, and it was also highly successful from a commercial point of view. The work not only placed the geographical analysis of the land in the service of Catalan national identity but it also constitutes the last piece of work of the regional school in Catalonia and is its masterpiece (a sort of Vidal's *Tableau* for Catalonia).

Since the 1970s, Catalan geography has followed new theoretical and methodological paths more inspired by Anglo-Saxon geography than by the French regional tradition. At the same time the changes in the political setting, especially the granting of a certain degree of autonomy, have led to a less unanimous appraisal of the national issue than in the years during which very basic national rights were repressed by the dictatorship. However, there are links with nationalism but nowadays it is more appropriate to speak of 'nationalist geographers' than of 'nationalist geography'. The links between geography and nationalism are established through an individual political decision rather than through a coincidence of the aims of methodology (the regional approach) with those of an ideology (nationalism), as was the case during the first 60 years of this century.

BIBLIOGRAPHY

Ainaud, Joan et al. 1989: *Catalunya: De l'any 1000 a l'any 2000.* Barcelona: Fundació Enciclopèdia Catalana.
Agnew, John A. 1984: 'Place and political behaviour: the geography of Scottish Nationalism'. *Political Geography Quarterly,* 3, 191–206.

—— 1987: *Place and Politics: the geographical mediation of state and society.* Boston, Mass.: Allen & Unwin.

—— 1989: 'Nationalism: autonomous force or practical politics? Place and nationalism in Scotland'. In Williams and Kofman, 1989, pp. 167–93.

Anderson, James 1986a: 'On theories of nationalism and the size of states'. *Antipode*, 18(2), 218–32.

—— 1986b: 'Nationalism and geography'. In Anderson, 1986c, pp. 115–42.

—— (ed) 1986c: *The Rise of the Modern State.* Brighton, Sussex: Harvester Press.

Aymà I Aubeyzon, Josep Maria 1981: 'Els desequilibris territorials, les ideologies de la nació i la consciència territorial'. Barcelona: Departament de Geografia Humana, Universitat de Barcelona, unpublished.

Blaut, J. 1986: 'A theory of nationalism'. *Antipode*, 18, 5–10.

Boal, F.W. and Douglas, J.N. (eds) 1982: *Integration and Division: geographical perspectives on the Northern Ireland problem.* London: Academic Press.

Bureau, Luc 1984: *Entre l'Éden et l'Utopie. Les fondements imaginaires de l'espace Québécois.* Montreal: Éditions Québec-Amérique.

Capel, Horacio 1981: *Filosofia y ciencia en la Geografia contemporánea. Una introducción a la geografia.* Barcelona: Barcanova.

Casassas i Slmo, Lluís 1979: 'Pau Vila en l'evolució de la geografia catalana'. In *Divisió Territorial de Catalunya, La. Selecció d'escrits de geografia de Pau Vila*, Barcelona: Curial, vol. 1, pp. 5–24.

—— 1986: 'About the Societat Catalana de Geografia'. *Treballs de la Societat Catalana de Geografia*, special issue, August–September, 17–20.

Cooke, Philip 1988: *Locality, Structure and Agency: a theoretical analysis*, paper presented at the 84th Annual Meeting of the Association of American Geographers, 6–10 April, Phoenix, Arizona.

Cosgrove, Denis 1983: *Social Formation and Symbolic Landscape.* London: Croom Helm.

Cuadros, Ignasi and Durá, Antoni 1988: 'Josep Iglésies i Fort'. *Geographers. Biobibliographical Studies*, 12, 107–11.

Divisió Territorial de Catalunya, La. Selecció d'escrits de geografia de Pau Vila 1979. Barcelona: Curial, vol. 1.

Dommen, E. and Hein, P. (eds) 1986: *States, Microstates and Islands.* London: Croom Helm.

Drain, Michel 1990: 'L'identité territoriale de la Catalogne dans l'Espagne contemporaine'. *Revue géographique des Pyrénées et du Sud-Ouest*, 61(1), 113–37.

Entrikin, J. Nicholas 1990: *The Betweenness of Place: towards a geography of modernity.* Baltimore, Md: The Johns Hopkins University Press.

Foucher, Michel 1988: *Fronts et frontières. Un tour du monde géopolitique.* Paris: Fayard.

Garcia-Ramon, Maria Dolors and Nogué-Font, Joan 1991: 'Pau Vila i Dinarès, 1881–1980'. *Geographers. Biobibliographical Studies*, 13, 133–40.

—— and Albet, Abel 1993: *La práctica de la Geografía en España (1940–1990). Innovación metodológica y trayectorias individuales en la geografía académica española.* Vilassar de Mar (Barcelona): Oikos-Tau.

George, Pierre 1985: *Geopolítica de las minorías.* Vilassa de Mar (Barcelona): Oikos-Tau.

Gregory, Derek and Urry, John (eds) 1985: *Social Relations and Spatial Structures.* London: Macmillan.

Iglèsies, Josep 1989: 'Els quaranta anys de la Societat Catalana de Geografia'. *Treballs de la Societat Catalana de Geografia*, 19, 11–29.

Johnston, R.J. 1982: *Geography and the State: an essay in political geography.* London: Macmillan.

210 *Maria Dolors García-Ramon and Joan Nogué-Font*

Johnston, R.J., Hauer, J. and Hoekveld, G.A. (eds) 1990: *Regional Geography: current developments and future prospects*. London: Routledge.

Johnston, R.J., Knight, David and Kofman, Eleonore (eds) 1988: *Nationalism, Self-determination and Political Geography*. London: Croom Helm.

Kirby, A.M. 1989: 'Tiempo, espacio y acción colectiva: espacio político y geografía política'. *Documents d'Anàlisi Geogràfica*, 15, 67–88.

Knight, David 1971: 'Impress of authority and ideology on landscape: a review of some unanswered questions'. *Tijdschrift voor Econ. en Soc. Geographie*, Nov–Dec., 383–7.

—— 1982: 'Identity and territory: geographical perspectives on nationalism and regionalism'. *Annals of the Association of American Geographers*, 72, 514–31.

—— 1984: 'Geographical perspectives on self-determination'. In Taylor and House, 1984, pp. 168–90.

Lacoste, Yves 1986: *Géopolitique des régions françaises*. Paris: Fayard.

Lluch, Enric 1961: 'La ciència geogràfica'. In Soldevila, 1961.

—— and Nello, Oriol (eds) 1983: *La gènesi de la Divisió Territorial de Catalunya (1931–1936). Edició de documents de l'Arxiu de la Ponència*. Barcelona: Diputació de Barcelona.

—— 1984: *El debat de la Divisió Territorial de Catalunya. Edició d'estudis, propostes i documents (1939–1982)*. Barcelona: Diputació de Barcelona.

MacLaughlin, Jim 1986: 'The political geography of "nation-building" and nationalism in social sciences: structural vs dialectical accounts'. *Political Geography Quarterly*, 5 (4), 299–329.

Maragall, Joan 1960: *Obres Completes de Joan Maragall*. Barcelona: Selecta.

Marti-Henneberg, Jordi 1986: 'La pasión por la montaña'. *Geocrítica*, 66. Monograph.

Massey, Doreen 1985: 'New directions in space'. In Gregory and Urry, 1985, pp. 9–19.

Nadal, Francesc 1990: 'Los Nacionalismos y la Geografia'. *Geocrítica*, 86. Monograph.

Nogué-Font, Joan 1991: *Els Nacionalismes i el Territori*. Barcelona: El Llamp.

Orridge, A.W. and Williams, C.H. 1982: 'Autonomist nationalism: a theoretical framework for spatial variations and its genesis and development'. *Political Geography Quarterly*, 1, 19–39.

Pred, Allan 1984: 'Place as historically contingent process: structuration and the time-geography of becoming places'. *Annals of the Association of American Geographers*, 74, 279–97.

Ratzel, Friedrich 1897: *Politische Geographie*. Munich, Oldenburg. French version by Pierre Rusch and Charles Hussy, *Géographie Politique*, Paris: ERESA.

Riudor, Lluís 1988: 'Lluís Solé-Sabarís (1905–1985)'. *Geographers. Biobibliographical Studies*, 12, 167–74.

Soja, Edward W. 1989: *Postmodern Geographies: the reassertion of space in critical social theory*. London: Verso.

Soldevila, Ferran (ed.) 1961: *Un segle de vida catalana, 1814–1930*. Barcelona: Alcides.

Sole-Sabaris, Lluís 1989: 'Sobre la naixença i el desenvolupament de la moderna geografia catalana'. *Treballs de la Societat Catalana de Geografia*, 19, 31–45.

Taylor, J. and House, J. (eds) 1984: *Political Geography: recent advances and future directions*. London: Croom Helm.

Thrift, Nigel 1983a: 'Literature, the production of culture and the politics of place'. *Antipode*, 15(1), 12–23.

—— 1983b: 'On the determination of social action in space and time'. *Society and Space*, 1, 23–57.

Vilà i Valentí, Joan 1989: 'La població i el territori avuin'. In Ainaud et al., 1989, pp. 89–99.

Williams, Colin H. (ed.) 1982: *National Separatism*. Cardiff: University of Wales Press.

—— 1985: 'Conceived in bondage – called into liberty: reflections on nationalism'. *Progress in Human Geography*, 9, 331–55.

—— and Kofman, Eleonore (eds) 1989: *Community Conflict, Partition and Nationalism*. London and New York, Routledge.

Zelinsky, Wilbur 1984: 'O say, can you see? Nationalistic emblems in the landscape'. *Winterthur Portfolio*, 19(4), 277–86.

—— 1988: *Nation into State: the shifting symbolic foundations of american nationalism*. Chapel Hill, N.C.: University of North Carolina Press.

13

Two Geopolitical Concepts of Poland

Józef Babicz

Józef Babicz

BEFORE THE POLEMICS STARTED

The modern geopolitical concepts of Poland in the nineteenth and twentieth centuries are deeply rooted in the geography and history of the country shaping primarily the national identity. Two events that played a significant role in that long process were Poland's acceptance of Christianity in 966, and its acceptance by Lithuania in 1387 followed by the union of the two countries which lasted four centuries. The merged state, whose territory extended from the Baltic to the Black Sea and was the largest in Central Europe, had its own uniform culture and was conscious of its national identity despite ethnic differences and relationships with the dynasties of other states, such as with Italy and Austria in the sixteenth century, France in the seventeenth century and Saxony in the eighteenth century. When in the late eighteenth century, as a result of military pressure from three neighbouring countries, Poland lost its statehood and was erased from the map of Europe for more than 100 years, it was history and geography that the Poles referred to in their struggle for independence. Their firm will in this respect was expressed in repeated insurrections and also in a rich literature, including geographical writings. Among the numerous viewpoints contained in the latter, particular prominence is given to the concepts expressed at the turn of the nineteenth and in the twentieth centuries. While the question of Polish national identity was indisputable, the discussions dealt with the shape of the future territory of the state within which that identity had to be finally realized. Wacław Nałkowski and Eugeniusz Romer, the most eminent geographers of that time, were the main proponents of these discussions which lasted more than 50 years. Because of the opinion that the varying geopolitical concepts are relative to their different general viewpoints and research interests, it is worth analysing the distinctive features of the personalities of the two scientists.

Wacław Nałkowski (1851–1911) was the most eminent geographer in the part of Poland annexed by Russia. He was a reformist in teaching geography, the author of numerous handbooks and a popularizer of new achievements in the field of geography. He was the author of a model regional monograph of Poland, *Geograficzny rzut oka na dawna Polske* ... (A brief look at the geography of former Poland) published in 1887. He was also a publicist representing the view of the radical intelligentsia and he co-operated with democratic and socialist circles.

Eugeniusz Romer (1871–1954), geomorphologist, climatologist and cartographer, was the creator of the Polish school of cartography. He implemented his ideas relative to the teaching of geography and to geographical studies in the Galician education system and his role in this respect was similar to that of Nałkowski in Warsaw. Besides many polemic dissertations dealing with the geographical problems of Poland, collected in a volume *Ziemia i Państwo* ... (The Land and the State ...) published in 1938, he also elaborated the *Atlas geograficzny i statystyczny Polski* (Geographical and Statistical Atlas of Poland) in 1916. This was of great importance in establishing the boundaries of the country as it recovered its independence: Romer was an expert at the peace negotiations in Paris (1919) and Riga (1921) and then in the plebiscite in Upper Silesia.

Since the principles of their geographical standpoints have not as yet been subject to thorough analysis, the following subjects deserve attention now: (a) the political context of the discussed problem; (b) the origin of the key idea of 'geographical individuality' present in the history of geography; (c) the essential differences between the two geographers in understanding Poland's 'geographical individuality': common views and differing opinions; (d) historical verification of the concept of 'geographical individuality' and of the concepts under discussion.

THE POLITICAL CONTEXT OF THE DISCUSSION

The problem of geographical individuality and of the territorial shape of past and future Poland was discussed by Nałkowski and Romer at the end of the nineteenth and in the twentieth centuries; that is, at the time when Poland did not exist on the map of Europe. Our analysis is exclusively historical and has two aspects: first, the discussion between the two scientists took place in a period characterized by growing conflicts among the greatest European powers; and, secondly, it occurred during a time of increasing aspirations of the Polish nation to regain its independence. That atmosphere of international conflict and the expectation of forthcoming military struggle for an independent state had incited J. Piłsudski to organize in 1910 the *Związki Strzeleckie* (Riflemen Union; in French: *Franctireurs Union*), as the core of the Polish army to be directed against Russia. It was then that the wish to regain territory lost more than 100 years ago revived strongly. The three neighbours of Poland (with its pre-partition boundaries) were not sufficiently united to rule out the possibility of a struggle for freedom on the territory of one of them and simultaneously to claim lost territories from the two others. The subject of the contest was the historical territory of Poland stretching to the Dnieper River in the east and from the Baltic to the Black Sea. In the meantime, the irreversible process of Russianization and the growing national consciousness of other communities, such as the Ukrainians, took place. However, in spite of these processes, the subject of both discussion and controversy was Poland with its past extended boundaries. According to Romer, the leading principle of the controversy was the hundred-year-long 'conflict between Poland's political status and its aspiration to independence'.[1]

HISTORICAL SOURCES OF THE MAIN IDEA OF GEOGRAPHICAL INDIVIDUALITY

In their fight for the state territory, besides historical arguments Polish geographers used physical ones too, following in this respect many precursors in the European way of thinking, well known in Poland both from original works and in Polish translations. Examples of these are Jean Bodin's *Methodus ad facilem historiarum cognitionem* (1566) and Montesquieu's *De l'esprit de lois* (1748) in which the authors postulated the dependence of man and society on physical conditions. Those ideas of physical determinism were revived with new strength in the nineteenth century in H. Buckle's *History of Civilization* (1857) which evoked great interest among Polish geographers. In the first half of the nineteenth century, the prime place held by statistical geography, which dealt with the subject according to categories of states – an inconvenient method because of constant changes in political frontiers, particularly during the Napoleonic wars, was taken over by C. Ritter's concept of geographical regions having natural borders and individual features. Ritter's concepts, after some evolution, were stabilized as the deterministic theory of the state as an organism by F. Ratzel (1895). On the other hand, Vidal de la Blache in his *Tableau géographique de la France* (1903) argued in favour of the possibilistic approach to dependencies between geographical factors and social phenomena.

Before the First World War, Polish geographers in their geopolitical discussions not only knew about the two European models (German and French) for analysing the relations between nature and society but also had their own tradition in this respect. Half a century earlier, Polish emigration literature relied on physiography to prove the Poles' legal right to their territories and, as Romer wrote, 'in the very name "Central Europe" given then to Poland, the thought was suggested to serve not only for the advancement of science but also for the political needs of Poland, geographically justified'.[2]

The old problem of the right of the Polish nation to possess the defined territory was, in the late nineteenth century, presented in a new form concordant with the condition of geographical and political aims. In solving that problem, geographers firmly postulated that the physical 'individuality' of the country is indispensable for the stable activity of state organization. The new formula of the problem was: is former Poland 'a transitionary country' or 'a platform between the two seas' – the Baltic and the Black Sea?

THE ESSENCE OF THE DIFFERING VIEWS OF NAŁKOWSKI AND ROMER WITH
RESPECT TO THE PHYSICAL INDIVIDUALITY OF POLAND

The two geographers accepted the same double task: to define the geographical
individuality of Poland and to justify Poland's rights to the defined territory.
They arrived, however, at different conclusions. It is significant that although
they used the same geographical notions and elements of the geographical
environments, they expressed differing views regarding the question that each
country should have geographical individuality conditioned by the individuality
of its nature. Nałkowski, an expert in both anthropo-geographic and physico-
geographic problems, expressed in the following synthetic way the aim of his
excellent monograph on Poland: 'to define that vague, by nature, physico-
geographic territory of Poland'.[3] He defined that task more precisely:

> We wish to define Poland geographically – to answer the questions:
> what are, first of all, the natural characteristic features assuring Poland
> its individuality and its position as a geographical land; and then, what
> was the influence of those features upon the historical lot and character
> of its inhabitants?[4]

On the basis of his knowledge of the Polish land, Nałkowski, through
quasi-deductive reasoning, came to the conclusion that the '*qualité maitresse*'
of Poland's area is its transitionary character and he inductively confirmed
that statement by giving a number of its important features. According to
Nałkowski, the territory of Poland is a transitionary area which (a) opens
the gate from Western to Eastern Europe and (b) upon which West European
features slowly penetrate the East European ones.[5]

> What is characteristic of Poland and determines its specific character is
> that Poland is a transitionary country, located actually between the East
> and the West of Europe, from the point of view of both communication
> and classification. It is, therefore, the essential geographical feature
> determining that (1) Poland is as if a gate or passage from Western to
> Eastern Europe ... (2) Poland is a region where the features characteristic
> of Western Europe are gradually passing on to the East to penetrate
> the East European features ....[6]

That fully open area of transitionary character is due to the lack of meridian
mountain ranges. Consequently, the Atlantic climate while passing eastward
changed itself into continental. Eventually, respective changes in flora and
fauna took place, as well as ethnic and cultural changes in the physiognomy
of the country. The Baltic and the Carpathian Mountains were the only
natural barriers protecting the state situated in such an open area. On the basis
of these facts, Nałkowski drew the obvious conclusion: upon that transitionary
territory, which had belonged to Poland for several centuries, the recovery of
the state after its 100-year loss of independence could first of all be possible

Figure 13.1 Map of Poland united with Lithuania, by K. Grodecki and
A. Pograbius

(From the Atlas by Ortelius, 1570, 680 × 459mm.) Romer based his arguments on
this map. The river system at the narrowing of the European continent, shown here
distinctly (although not accurately), was considered by Romer to be the factor
influencing the cohesion of the state. Differences between the western and eastern
parts of the map show the different degrees of knowledge about Poland (which
became associated with Europe in 966 by accepting Christianity) and Lithuania (which
did not unite with Poland until 1387 — again by accepting Christianity).

when founded on ethnic and national criteria. S. Lencewicz, Nałkowski's
disciple, defined his view as 'socialistic' though the French geographical
possibilism of Vidal de la Blache also lay at its base.[7]

Romer, who also knew the territory of historical Poland very well, was
primarily a prominent scholar in the field of physical geography. In the
reminiscences of his lectures given for the organization *Macierz Szkolna* in
1907, he wrote that in the patriotic atmosphere 'he set himself to the Polish
geopolitical service'.[8] It was then that he postulated the following: 'Poland is
the land presenting geographical integrity and this is the reason why, in spite
of the political partition, "it is not yet lost".'

The geographical integrity should not only have definite boundaries but
also noticeable and advantageous geographical position and it should

be physically compact. Geographical Poland fully responds to these requirements.[9]

According to Romer, the physical conditions of Poland primarily justified its recovery. He wrote:

In case Poland didn't have physical justification in physical Europe then even its long existence might have been only the result of certain political and cultural assumptions and situations. Eventually the case of Poland is exclusively dependent on similar historic-political constellations that might return. Alternatively, Poland does exist in Europe as an organic member of the physical construction and then the problem of Poland's independence would be much closer since it is not nature but another human will that might be an obstacle in carrying out the postulated human will.[10]

The isthmus area stretching from the Baltic to the Black Sea at the narrowing of the European continent, having a remarkably consistent hydrographic system, determined that place of Poland in the physical construction of Europe (see figure 13.1). The concept justifying the Poles' right to the large territory stretching 'from sea to sea' was called by Lencewicz 'an imperialistic antithesis'[11] in opposition to Nałkowski's socialist concept.

ESSENTIAL DIFFERENCES BETWEEN THE TWO VIEWPOINTS

Romer was decidedly opposed to Nałkowski's geographical synthesis of Poland as a transitionary country, according to which the lack of natural barriers facilitated invasions and, finally, conquest. He called that concept 'a construction of the idea built upon despair'.[12] He did not find in Nałkowski's synthesis any efficient arguments to support the fight for Poland within its former historical boundaries. He also believed that Nałkowski's concept of the 'transitionary' character of the country was not contradictory to his own concept of the Polish territory as 'a platform between the two seas'.

Nałkowski denied the importance of Romer's two arguments that the fundamental features indicating the individual character of Poland are: (a) its hydrographic system (because analogous phenomena are observed in many parts of Europe), and (b) Poland's being 'the platform between the two seas', the Baltic and the Black Sea, upon the continental narrowing. The latter view does not negate the fact that this territory is gradually becoming transitionary, and it may even confirm it. Moreover, Romer's assumption was disadvantageous because it implicated Poland's close alliance with other nations, and nobody had asked for their opinion in this respect; this specifically referred to the communities living on the southern part of the platform. Nałkowski argues further:

Contrary to the stage character, the transitionary feature is applied to the whole of Poland and not just to a part. In the face of various historical events it is a rather flexible feature that may be adapted both to historical and to proper – ethnographic – Piast Poland. It neither forces anything nor resigns from anything.[13]

The standpoint of transitionary character – contrary to inter-Pelagianism – allows the comprehension of Poland and not only of a defined region of the country; it makes possible the adjustment to various historical circumstances because it is flexible and as such it may be adjusted to the Polish territory in the broadest historical sense, as well as to Poland *sensu stricto* – ethnic Poland – Poland of the Piasts. It does not accept force in relation to anything and it does not resign from anything.[14]

Nałkowski considered the Baltic and the Carpathians to be natural boundaries to the north and the south and he considered that the western boundary on the Oder river and the southern one on the Dnieper river to be the boundaries of maximum flexibility.

HISTORICAL VERIFICATION OF THE TWO CONCEPTS

S. Leszczycki, an expert in the modern history of geography and an organizer in this branch of science in postwar Poland, has a high regard for the polemic between Nałkowski and Romer. He wrote that 'the discussion was of great importance in the time before Poland's recovery because it postulated for politically non-existent Poland its place in Europe'.[15] However, the results of the peace conference in Paris (1919–21) proved that Romer – thanks to his work at the Congress Office – greatly influenced the decision about Poland's boundaries. His *Geograficzno-statystyczny Atlas Polski* (Geographic-statistical Atlas of Poland; 1916) and *Polski Atlas Kongresowy* (Polish Congress Atlas; 1921) supplied data supporting his position. His widely known polemic with Nałkowski was of importance mainly in inspiring patriotic feelings in the country.

However, at the time of the peace conference in Paris, both syntheses discussed above had adherents among outstanding geographers, with S. Lencewicz and L. Sawicki supporting the concept of Nałkowski and S. Pawlowski and M. Janiszewski being adherents of Romer. A. Sujkowski was among those who tried to find some common aspects in the two opposing standpoints.

The recovery of Poland has confirmed the correctness of Nałkowski's position, a fact which was later stressed by Lencewicz when he wrote that 'the transitionary character corresponds to Poland in its present boundaries'.[16] Also, Sawicki acknowledged the correctness of Nałkowski's standpoint on the basis of Poland's boundaries shaped after the First World War. He wrote:

Poland is generally understood to be the land historically Polish, stretching to the Carpathians and the Baltic, mostly inhabited by Poles, being the place of cultural activities of the Polish nation. Regarding, however, the concrete boundaries, the concept of Poland becomes evolutionary. Poland will be such as we want it to be through our proper activities, through accumulating social energy on our borderlands to defend the country and to solve our borderland problems.[17]

After the Second World War, the arguments of neither of the two geographical concepts of the territory of Poland – Nałkowski's possibilistic concept and Romer's physical-deterministic one – was acknowledged, although Nałkowski's statement that ethnic and national criteria should be decisive in determining state boundaries remained of importance. However, during the Second World War all those criteria were dominated by political imperatives. Technical progress has also lessened the importance of natural boundaries. While geographical factors still remain an element worth consideration in politics, the geographical-political theories, even those of the early twentieth century, belong now to the history of geographical thought.

At present, after regaining full independence from Russia, there remains on a part of the former historical territory of Poland the national problem of many Poles living in the post-Soviet republics of Lithuania, Byelorussia and Ukraine. These Poles have for a long time been occupying a large area of former Poland and they do not want to leave their homes. Their present fight for the rights of national minorities or for the autonomy of the areas in which they constitute a majority should be supported by Polish society and by its organizations such as *Towarzystwo Miłośników Wilna* (Society of Lovers of Vilnius) and *Towarzystwo Miłośników Lwowa* (Society of Lovers of Lwów). This vital problem for the Polish nation results both from its geography and its history.

NOTES

1 E. Romer, *Ziemia i państwo. Kilka zagadnień geopolitycznych* (Earth and State: some geopolitical problems). Lvov, 1939, p. 15.
2 Ibid.
3 W. Nałkowski, *Polska: Obraz geograficzny Polski historycznej. Słownik Królestwa Polskiego i Litwy* (Poland: geographical picture of historical Poland. Glossary of the Polish Kingdom and Lithuania), vol. 8, 1887, p. 601.
4 W. Nałkowski, *La Pologne, entité géographique*. Varsovie, 1921.
5 Ibid., p. 9.
6 Ibid., p. 12.
7 Stanisław Lencewicz *Wspomnienia o Wacławie Nałkowskim* (Memoire about Wacław Nałkowski). *Przegląd Geograficzny* (Geographical Review), IV (1936), 7.
8 Romer, *Ziemia i państwo*, p. 6.
9 Ibid., p. 11.
10 Ibid., p. 16.
11 Lencewicz, *Wspomnienia o Wactawie Natkowskim*, 7.
12 Romer, *Ziemia i państwo* p. 18.
13 Nałkowski, *La Pologne*, p. 30.
14 Ibid., p. 56; W. Nałkowski, 'Polska jako kraina przejsciowa' (Poland as a transitionary land), *Ziemia* (Earth), (1910), 723.

15 Stanisław Leszczycki, 'Rozwój polskiej geografii w sześćdziesięcioleciu 1918–1978' (Development of Polish geography in sixty years, 1918–1978), *Przegląd Geograficzny* (Geographical Review), (1979), 425.

16 Lencewicz, Wspomnienia o Wacławie Nałkowskim, p. 8.

17 Ludomir Sawicki, *Zarys ogólny geografii ziem polskich* (An Outline of Geography of Polish Lands), Part I. Cieszyn, 1920, p. 15.

14

The Image and the Vision of the Fatherland: The Case of Poland in Comparative Perspective

Ladis K. D. Kristof

The original idea of Fatherland is deeply traditional: it is the land of the fathers; it is the *sacred* land where their bones lie buried. We cannot repudiate or abandon it without repudiating and abandoning them. In other words, without the Fatherland we are fundamentally fatherless; a nobody; the scum of the earth.

The Fatherland is, of course, also the Motherland. Written large or on a smaller, more intimate scale, it mothers us, emotionally and physically, exactly as our mother did. It feeds us, it consoles us, it is our refuge where we seek protection and where we can recharge our emotional batteries to go and venture once again beyond our home and homeland to face the challenges of the unfamiliar in the frontierland. The strength and energy we draw to act beyond home and homeland originates at home. Without this umbilical cord we risk becoming powerless as well as rootless; at best janissaries whose strength and direction originates in foreign traditions, in their very essence alien to those of our ancestors. In other words, as individuals we are in a sense little microcosms of empires: the empire is only as strong and vigorous as the mother country.

We should bear in mind, however, that the concept of Father*land* and Mother*land* are expressions which reflect the more recent history of mankind, namely, that of societies *settled* on land, bound to and rooted in a particular piece of territory, in contrast to the *nomadic* societies which are bonded together not by the community of territory but by the community of blood. The contrast between these two types of societies is underscored in the classic distinction made in Roman times between those governed by *ius territorialis* – the law of the land – and those governed by *ius sanguinis* – tribal law. The transition from the primacy of the blood relationship to the primacy of the territorial relationship – which of the two is the glue that plays the more important role in bonding people into a community and society – has been a very slow and painful process, not yet fully consummated even in the Western world. Ambiguities and tensions between the two bonding forces remain everywhere, ethnic irredenta being merely one manifestation of them. Characteristically, these ambiguities are reflected in the Russian language which was moulded in that great empire-prone space of (northern) Eurasia, first dominated by tribal-nomadic peoples from a centre in the East (Mongolia) and then by agricultural-settled peoples from a centre in the West (Muscovy), the two ways of life clashing but often forming a symbiotic relationship. The Russians have two words for Fatherlands: *otchizna* – heritage of the father (sometimes used also to indicate individual hereditary land holding) – and *rodina* – homeland. However, *rodina* in the sense of homeland is a newly acquired meaning because the root of the word is *rod* which means family in the broad sense of blood relationship, that is, of clan or tribe. Briefly, *rodina*, which is more frequently used than *otchizna*, harks back to the times when the home of the blood – the glue provided by the blood relationship – took precedence over the glue emanating from the home*land*.

With the creation of the modern Fatherlands – the nation-states – the old idea of the Fatherland has largely survived.[1] Thus, the first modern political geographer, Friedrich Ratzel, speaks of the state as a piece of land and a piece of humanity, bound together by a state idea.[2] Similarly, international law defines the state as a territory, a population and an authority holding the two together. Thus the main difference between the old and the new Fatherland is that the latter has been expressly politicized: its very essence – its soul – is not an immutable, inherited tradition, symbolized by the sacred burial grounds which could be neither moved nor abandoned, but a politicized tradition which by the very fact of it having become political has been imbued with life and the dynamics of change. Now it can selectively shed some 'skins' of traditions while assimilating and legitimizing quite novel ways of doing things.

The political dynamics of the modern idea of the Fatherland underscore its geopolitical nature, for the essence of geopolitics is that it is politics and not geography – not even political geography and perhaps not even a sub-division of political science, although of course, a legitimate subject of enquiry by political scientists. While political geography is either regional geography in which the region is politically defined or the study of the impact of political activity on the landscape, geopolitics is, as the etymology of the term indicates, geographical politics, that is, politics in which concepts and ideas about a geographical area (for instance, the territory of the fatherland) play an important, often decisive role. Political geography tends to be descriptive; it is 'geography as an aid to statecraft'.[3] Geopolitics, on the other hand, tends to be prescriptive, in other words, it wants to be statecraft itself. If we review the history of geopolitics – from Ratzel through Mahan, Kjellén, Mackinder, Haushofer, De Gaulle and Kissinger – we notice that most of those who contributed to the store of geopolitical thinking were themselves deeply politically committed, whether professionally or not.

Turning to the component elements of the modern idea of the Fatherland, it is useful to distinguish between the image and the vision of the Fatherland. The former has its roots in the past. It is a mental image of the Fatherland as we know it, as we remember it. It is a series of concrete images of our homeland as recorded by our own eyes or of those close to our heart. It is the *typical* landscape that *we* associate with our Fatherland. Hence it tends to be either the image of our little corner of the Fatherland we know and love best projected on the entire country, or the image of some area of the Fatherland which our ideology or preferred historians and literary writers have designated as the heart and core of the country. Thus the area we consider the original nucleus, the birthplace of the Fatherland, without which we could not even imagine its existence, is always a decisive geopolitical element within our image of the Fatherland.

The image of the Fatherland is, of course, idealized and polished. We see in it the Fatherland as we want it to be. But since it is rooted in our memories and various tales of what has been, it is mostly an image of an idealized past, not an idealized future. Thus even in industrialized countries the image of the Fatherland tends to be rustic, even pastoral.[4] It is the image of the Fatherland of our ancestors, of their idealized life in an idealized landscape. The enemies of today – smog, pollution, crowding – are replaced by the enemies of yesterday – wolves, bears, and other natural or imaginary beasts; and, characteristically, even these beasts are in various ways idealized to make them fit into the bucolic imagery of harmony that reigned, *must* have reigned in the land of our valiant ancestors.

The Fatherland of our vision, on the other hand, is forward orientated. It is not an idealized past but an idealized future, not what our idealized ancestors built but what we the idealists, worthy sons of our fathers, are going to build. Clearly, the vision of the Fatherland is a modern idea. It assumes that the Golden Age is ahead of us, not in the past. It is a child of the idea of progress, of our growing feeling of strength which can transform the Utopia of our dreams into reality. The supernatural powers of individual giants and heroes of the past are replaced by the powers of reason, mass effort and ideological commitment. This tendency to emphasize organized collectivity, science and the role of ideology in the realization of a dreamed-up future is quite characteristic of the visionary Fatherland and visionary leadership, and contrasts sharply with the much more easy-going Fatherland image of the past. The romantic element plays a crucial role in both, but in the visionary Fatherland it is much more a romanticism of the deed, while in the other it is rather a contemplative romanticism that comes to the fore. And since it is the human, not the natural or supranatural, deed that is to play a crucial role in the realization of Utopia, it is man who dominates the scene. While in the past-orientated image of the Fatherland it is nature – the natural landscape – which dominates the scene, in the future-orientated vision of the Fatherland man and his deeds are in the forefront.

There is, however, no inevitable incompatibility and conflict between this past image and future vision of the Fatherland. Quite the contrary: the latter can, even must, rely for strength on selectively choosing past images and fitting them into the visionary future. Polished and refurbished, the heroes of the legendary past can become examples to follow for the collectivity of the present. If they, our ancestors, could achieve singlehandedly that much, why cannot we collectively, in this modern age, achieve much more? Thus while the deeds of the past are glorified they are seen as merely *relative* achievements, not the ultimate in what can be achieved. In other words, the sky is the limit and it is at the skies that this and the future generations must always aim. Thus the attitude of the visionary leaders towards the ancestors, and their Fatherland, real or imaginary, is always ambiguous. Yes, the Fatherland of the ancestors may have been a dreamlike wonder, but essentially was a children's Garden of Eden. An adult man, not to mention the superman of tomorrow, neither could nor would want to fit into such a paradise, even if it were God-given. He wants something better and more grandiose; he is sure he can improve on the works of God and nature; he does not want simply to inherit a world, he wants to build it himself exactly to his own specifications for he is *the* master of this world.

Although the distinction and contrast between the image and vision of the Fatherland must be emphasized, it is often difficult to make in a particular case. Thus in the case of Poland the two often merge. This is due to the fact that having lost their Fatherland the Poles' vision of a rebuilt Poland tended to be confounded and merged with the image of Poland as it was, or was imagined to have been, in 1772 or even way back in the fourteenth century. Other nations have the same problem. For instance, the Armenians, having also lost their Fatherland, sometimes visualize in their dreams the ideal of a future rebuilt Armenia by going almost two millennia back into history, to the reign of Tigranes the Great when their Fatherland reached the apogee of political power. However, the image of the Fatherland so distant in time is not an image living in the oral tradition and imagination. It is not rooted in memories, however coloured and embellished, that have a historical continuity. Rather, it is a discovery – sometimes even an outright fabrication – of a scholar-patriot which is then taken over and diffused by a group of ideologues-patriots. Consequently, it is an idea that originated with a group of intelligentsia, not one that is rooted in the heritage of immemorial traditions. Usually the promoters of the idea must compete with another group of intelligentsia who are promoting a different idea, and the idea comes to the mass of the people as if from above as propaganda, often disseminated through school textbooks. If the idea becomes the dominant one among the intelligentsia at large, and the society's elite, then, after several generations, it may sink down into the conscious and unconscious minds of the majority of the people and thus become appropriated by it as if it were its own. This is, however, a long process and seldom fully successful because the mass of the people are likely to be bombarded by a variety of contradictory ideas and because it is not easy to transfer ideas across culturally hardened social class lines.

A good example of elements of an ideology and image of the Fatherland that originated, or was 'discovered', by the intelligentsia is that contained in the Russian epic *The Tale of the Host of Igor*. No doubt it has penetrated the stratum of cultivated Russians and to some extent also the broader educated masses, and to that extent has become an integral part of the Russian heritage and consciousness. However, if we take the Finnish *Kalevala* epic, which has survived through the ages as a genuine folk tradition, always alive in the consciousness of broad masses and picked up by the intelligentsia not from documents but from a widely distributed oral tradition, we can understand why *Kalevala* has played a much greater role in Finland than has *The Tale of the Host of Igor* in Russia. What rises from the bottom of the society, the ideas that are rooted in the folk consciousness and from there penetrate the strata of intelligentsia, are much more likely to be fundamental to the national heritage than those which essentially originate with, or are rediscovered by, the elite.

True, there are exceptions and the United States offers one. The most popular and widespread image of America, with a strong ideological content added to it, is the one that originated in New England. The Pilgrim Fathers on the *Mayflower* create a contractual society before setting foot on Plymouth Rock and then begin to civilize, that is Christianize, the wilderness of nature and the native inhabitants. The Puritans are hard-working and frugal people and they quickly transform untamed nature into a fruitful garden. The Almighty rewards them and we see them on Thanksgiving Day dining peacefully with the Indians. Soon neat villages and towns, with white-steepled churches at their centres, appear everywhere. This is the image of America to this day portrayed throughout the country.

A 'typical' American Christmas card shows us a bucolic New England village with snow, sleds and its congregational church, and this image stands for America even for Americans who live in big cities and have never seen anything even remotely like what they see in the picture, who often have never even seen snow on Christmas Day. Such has been the strength of the cultural influence of New England – originally to a large extent spread by schoolteachers and primary school textbooks that fanned out from this area throughout the country – that most Americans are not even aware that the colonies established in Virginia, not to mention those in St Augustin in Florida, are older than those of Massachusetts. Indeed, in the popular image of America New England remains its birthplace and symbol as well as the core from which the country was created. It is an image which has been created by an intellectual elite – historians, writers, artists and church leaders – originating in or influenced by the New England Puritan traditions and it has been successfully imposed on and absorbed by America as a whole (except in the South) because class differentiation was relatively weak and social mobility relatively high in America. Moreover, the fact that practically all immigrants entered the country through its eastern gate, between New York and Boston, and usually lived in that area for a time, permitted their socialization to the New England image of America before ever they moved further west. Thus, again excepting the South, a regional differentiation of the country's image has developed in America far less than in various European countries of much smaller size.

In Poland the nobility was proportionally two or even three times more numerous than in other European countries. When it became gradually impoverished under foreign occupation in the nineteenth and twentieth centuries it sank socially downwards penetrating virtually all social classes, yet struggling for several generations to maintain its old value system in the new social context. Consequently, Polish national culture, including the popular ways of thinking about the country and its past, present and future, largely reflected the model as set by the nobility. 'Poland has been a nation dominated and led by the aristocracy and the gentry, with the intelligentsia [overwhelmingly of noble origin] taking over the cultural leadership.'[5]

The image of the Fatherland conveyed in Polish art and literature reflected the views of the ruling elite even after this elite *de facto* ceased to rule the moment Poland disappeared from the map of Europe for almost a century and a half. It was an image dominated by the nobility's country lifestyle as well as by its nostalgia for the glorious past of the imperial period of Polish history and sanguine hopes for the future revival of a Poland 'from sea to sea'.[6] The typical Polish landscape was portrayed by the noble's white manor house and *the* Pole was represented by the nobleman, usually on horseback and in some martial pose to underscore that it was he who defended Poland's independence and its liberties as well as Christianity in general. It was the image of the so-called 'Sarmatian' Poland with a touch of oriental decor to emphasize Poland's, that is the Polish nobility's, mission as pioneers on the edge of civilization, defenders of the faith, somewhere on the borderland, Ukraine.

Surveying Polish literature, it is really amazing to see how overwhelmingly it has focused on the historical eastern borderlands of Poland, and since in these areas the ethnic Polish element was represented almost exclusively by the nobility, it is its life and concerns that are dealt with by the authors while the mass of the people, usually Ukrainian peasants, receive short shrift and are presented negatively.[7]

In art, too, the emphasis has been on the landscape and scenes that have clearly an eastern borderland setting. Ploughs drawn by four or six oxen certainly do not depict the reality of central Poland where the soil is sandy and light but the practice in the heavy black soil zones of Ukraine. Similarly, the forests in most Polish landscape paintings from before 1945 do not represent the pines of western Poland but rather the oaks, beech trees and spruces more common in the eastern borderlands. Even more clearly portraying eastern Poland are the popular scenes with horses where the type of horse, the dress of the people as well as the landscape unmistakably betray the location.

Consequently, we can say that the inherited traditional image of Poland, at least as portrayed until 1945 and which still influences the present generations of Poles, has tended to create a geopolitical distortion. The conveyed image of Poland did not correspond to the reality, at least not the reality where the mass of the Polish ethnic core resided. True, the river Wisła was always said to be the principal axis and 'Mother' of the Polish Fatherland, and the old capital Kraków situated on it as the very essence and heartland of Poland. However, the vision of a truly rebuilt Poland as nurtured from the eighteenth to the twentieth centuries when the country was absent from the map of

Europe, and also during the interwar period, was that of 'Poland from sea to sea', that is from the Baltic to the Black Sea. Within such a Poland the Wisła certainly did not play historically the role of main axis; rather, its axis would be a waterway and commercial route that ran from the Black to the Baltic Sea either along the Wisła, Bug and Dnestr rivers, or along the Wisła, Bug, Pripet and Dniepr rivers, leaving Kraków, two-thirds of Wisła and most of ethnic Poland on one side.

Not surprisingly, the post-1945 Polish regime, trying to legitimize the loss of half of Poland's pre-war territory and the acquisition of German-populated territories in the west and north, immediately began to emphasize not only the role of the Wisła (which seemed to be enhanced because Gdansk, at its mouth, was now in Polish hands) but also that of the Warta and Oder rivers.

There was even an attempt by Polish geographers and publicists to justify shifting the Polish boundary further west than the Oder–Neisse line. Both the arguments in support of the latter line and for going beyond it were based primarily on pure geographical determinism[8] and reflected the political moods and hopes of the expansionists in the Polish camp.[9] In general, since the ethnic argument could not be invoked by Poland in favour of the annexations in the north and west, geographical factors, and especially the allegedly uniting and determining factor of rivers and river basins, were widely used to make political points. Indeed, attempts were made to ground the very existence and territorial structure of Poland in geographical terms.[10] This was rather ironic because just a few years earlier some German geopolitical propagandists had resorted to very similar methods and arguments to reach diametrically opposite conclusions. In fact, clashes of German and Polish geographical–geopolitical arguments and counterarguments had already taken place in the interwar period.[11]

Emphasis on the role of rivers and river basins in the Polish past and present also led of course to an emphasis on the Baltic. Historians were able to point out that there was at least one important Polish publicist, Jan Dymitr Solikowski, who in the sixteenth century had argued eloquently for making Poland into a sea power:

Who has a state with access to the sea, and does not use it [the sea] or let others take it away from him, is distancing himself from all benefits and brings upon himself all losses, from a freeman becomes a slave, from a rich man becomes a pauper, from being a master of himself becomes dependent on a foreigner, from a lord is transformed into a peasant, which is a great obscenity and stupidity.[12]

However, Solikowski remained a prophet in the wilderness. A few Polish cities did belong to the Hanseatic League, but except for Kraków these were German-dominated cities on the border of Prussia which, somewhat nominally and not with heart and soul, belonged to the Kingdom of Poland. The only time the old historical Poland built a fleet and used it in naval operations was during the reign of Sigismund III Waza, a Swede who lost the Swedish crown to an uncle and who needed a Polish fleet solely to regain his throne in Stockholm. The Polish throne he was all too ready to cede to the Habsburgs in exchange for 400,000 guldens.[13]

In the interwar period much was written about 'the marriage of Poland with the Baltic' – when Polish independence was restored General Józef Haller drove his horse into the Baltic and threw a ring into the water in a symbolic gesture – and the government did allocate relatively large sums of money to build a port in Gdynia and a small naval force. But to the average Pole the Baltic and the whole idea of being involved in the affairs of the sea, either economical or political, remained distant if not wholly foreign. As in Russia so also in Poland the average man was a landlubber. He neither had contact with nor any feeling or thought for the sea or seafaring; but in both countries, and especially in Poland, that changed quite radically after 1945. While in 1939 the loss of access to the sea would have left most Poles, except the intelligentsia, rather indifferent and materially hardly affected, today every Pole considers the seaboard an essential part of Poland not only in the economic but also in the emotional sense. In every Pole's image of the Fatherland the Baltic is now an essential component, no matter what regime may dominate the country.

The change in the Polish consciousness and in its image of Poland is largely due to the role Gdańsk has played in recent Polish history. The Danzig of 1939 was tied to Poland by various often-disputed historical records but no living memories, and most Poles were at the time aware that it was an indisputably German place and that it was longing to shed its imposed free city status and merge with the Reich. Now, however, Gdańsk has become a living monument to the struggle for the emancipation of the Poles and of Poland, and a symbol of a truly Polish Poland. In the image of the Fatherland of the new generation of Poles, Gdańsk is the anchor that ties Poland to the Baltic and the wide blue waters beyond it which Polish vessels are now regularly plying in search of new opportunities. But may not some Poles of the older generation still nurture an image of the Fatherland that would rather see Poland drop anchor at Wilno and Lwow?

In the immediate post-1945 period Poland aired grandiose plans for transforming the Wisła, Warta and Oder rivers into major water-highways to the Baltic. Since then a lot has been written, and even more said, on the subject but virtually nothing has been done. The Wisła is as unregulated as ever – during the summer even ordinary flat barges cannot move along it – and its waters are much more polluted. On the Oder not even all the installations inherited from the Germans have been repaired and pollution has reached dramatic proportions. However, the new regime installed in Poland in 1945 needed the vision of a great Wisła–Warta–Oder route to the Baltic in order to counter the older idea of the Wisła–Bug–Dniepr route connecting the Baltic and the Black Seas. The former represented a western expansion while the latter represented the eastern expansion. Much touted was the fact that the new route was crossing and uniting territories inhabited by Poles, which of course became true once the Germans were expelled and the Poles settled in their place. This was contrasted to the old axis which tried to bind together territories of a multinational Polish empire. In general, much was made of the fact that the new Poland was a genuinely Polish Poland; the Poland of the nobility that ruled more non-Poles than Poles now became literally a *Respublica Polonorum*.[14]

The latter point was made both in the ethnic and in the social-class sense. While the old Poland was in fact the Fatherland of only the Polish nobility – lands dominated by it regardless of who the majority of the inhabitants were – the new Poland was the Fatherland of the mass of the Polish people. Hence it was called the Polish People's Republic and, to ground it in history, it was said to revive the tradition of the Poland of the Piast dynasty, which from the tenth to the fourteenth centuries ruled only territories that were ethnically speaking essentially Polish. The Piast dynasty had also the advantage that, according to popular tradition, its founder was a simple peasant and that during that time the nobility's position and privileges had not yet generally become entrenched.[15] This contrasted with the period of the next, Jagiellonian, dynasty (fourteenth to sixteenth centuries) which expanded Poland eastwards and, thus, transformed it into a multinational state in which the nobility became *the* dominant ruling class with increased privileges, at the expense of the monarch's authority and to the great detriment of both peasants and town-dwellers.[16] The expansion eastwards into territories much larger than those of original Poland was the start of enormous estates over which the aristocrats ruled with no regard for law or the interests of the commonwealth. The decline and ultimate fall of Poland in the seventeenth and eighteenth centuries is, indeed, usually attributed to the selfishness and lawlessness of the Polish nobility, especially the aristocrats in the eastern borderlands. The present regime also points out that while Polish aristocrats pressed the expansion eastwards, Poland was at the same time slowly losing its hold on large portions of its western lands into which German political and cultural influences were encroaching. Hence the argument that present-day Poland is in both the geopolitical and social sense a restoration of the original truly Polish Poland, while the Poland dominated by the nobility down to 1945 was a geopolitical and imperialistic aberration.[17]

What is certainly true is that the Poland of today is geopolitically much more compact and that it has a core. It has shed the eastern provinces where national minorities prevailed, and while it has not yet fully absorbed the northern and western German territories, today there certainly is a Polish heartland where Polish cultural, economic and political life is concentrated. Moreover, Warsaw is today not merely the political and administrative capital of the country but also truly its heart. Although it became Poland's official capital well over three centuries ago it did not genuinely gain that status until 1945. Kraków has lingered in Polish hearts and minds as the more truly Polish centre because of the historical, cultural and intellectual traditions tied to it which Warsaw could never match: after all, the Polish kings, even those who reigned in Warsaw, were brought for burial to Kraków. Moreover, there were provincial capitals such as Lwów and Wilno which meant much more than Warsaw to the local Poles. However, this changed during the Second World War. Warsaw's resistance in September 1939, after the rest of Poland had collapsed, and then the Warsaw uprising of 1944 raised the city to the rank of hero and symbol of Polishness and Polish patriotism. At the same time the quietness of Kraków throughout the German occupation has rather diminished its prestige, and Lwów and Wilno are today outside the Polish borders.

As indicated, among many Poles, especially of the older generation, the loss of Lwów and Wilno has not been compensated by the acquisition of German territories in the west. Sentimental ties to Danzig-Gdańsk in the pre-1939 Poland were weak, and to Breslau-Wrocław practically non-existent among the mass of Poles. Stettin-Szczecin did not even figure in Polish minds. Thus the questions of where is Poland, what are its essential territories, without which Poland is not really what it ought to be, linger on. Undisputed only is that the Wisła is Poland's basic axis. Some Poles have even accepted a zigzag line as a schematic representation of the Wisła, with a short side bar indicating Kraków, which together represent the primordial essence of Poland.[18]

Poland is not the only country in such a quandary. In Romania the choice between Transylvania (and Bucovina) on the one hand and Bessarabia on the other has been a cruel one, especially so during the First World War. For a long time the partisans in favour of joining the war on the side of the Central Powers and the partisans in favour of joining the Allied Powers put forward various conflicting arguments but essentially it was a question of what was dearer to them – what did they consider a more essential part of the Fatherland – Transylvania and Bucovina or Bessarabia? To some extent at least the same problem arose in 1944. Because for various reasons in both 1916 and 1944 Transylvania was seen as much more a part of the heart and core of the Romanian fatherland than was Bessarabia, the latter was sacrificed. Of course, in 1944 there was really no choice, but to the extent that a mental choice was made it because easier to accept because the attachment to Transylvania was greater. Without the Transylvanian Carpathians, the birthplace and home of the 'Mioritsa' – its myths, yearnings and music – the very essence of Romania would be truncated.[19]

An even better comparison with the case of Poland is that of Greece. Throughout the nineteenth century the Greek patriots were dreaming, and maybe still dream, of a rebuilt Greece, not in the modern nation-state sense but as the core of an empire – the Byzantine Empire. Hence they did not look to Athens as their capital but to Constantinople. After all, ancient Athens was merely a city-state, and pagan at that, while Constantinople was the capital of an empire *and* centre of unalloyed Christianity. The Greeks expected all Christians of the Eastern Orthodox faith inhabiting the Ottoman Empire to support enthusiastically the rebuilding of a modern Byzantine Empire – a Greece written large, a multinational state led by Greeks and their Church. Bulgarians, Macedonians, Romanians and Serbs were all supposed to join in and not seek to build their own little states. This was an idea similar to that of the Poles who could not comprehend why the Lithuanians, Belorussians or Ukrainians would spurn the embrace of the Polish Commonwealth. Were not the Poles, and that meant the Polish nobility, the natural leaders of this part of the world? Was not their culture self-evidently superior to that of the little Slav brothers to the east and even to that of the big Slav brother? And so many Poles (and Greeks) remain unsatisfied with the Fatherland as it is. They would like it to be big, with a large mission, and they feel crowded and unfulfilled in the little one that fate has bestowed on them.

NOTES

1 There is, however, one important difference between the traditional Fatherland and the modern nation-state. The latter does not tolerate wishy-washy frontiers. It imposes, or at least tries to impose, clear cut boundaries. Up to these boundaries its legitimate authority is practically absolute and its effective power ought to be absolute. Within its sovereign territory every nation-state attempts to maximize the centripetal forces generated by its state idea. Where the bonds of blood, ethnicity and/or ideology overlap the territory of two nation-states the boundary cuts through living flesh, and the pain it imposes on the borderland people can be mitigated only if the principle of absolute territorial sovereignty – of impenetrability of boundaries and territory – is relaxed first, if not in principle than at least in practice.

2 Friedrich Ratzel, *Politische Geographie oder die Geographie der Staaten, des Verkehrs und des Krieges*, 2nd edn. Munich and Berlin: R. Oldenbourg, 1903, pp. 4–9.

3 I am borrowing here the subtitle of W.H. Parker's *Mackinder: geography as an aid to statecraft*. Oxford: Clarendon Press, 1982.

4 David Lowenthal and Hugh C. Prince, 'English landscape tastes'. *Geographical Review*, 55 (1965), 187ff.

5 Jan Szczepański, *Polish Society*. New York: Random House, 1970, p. 19.

6 The tenacity of the image of Poland extending from the Baltic to the Black Sea is really amazing. Marshal Piłsudski attempted to realize it in 1920 when he briefly conquered Kiev and as a consequence almost lost all of Poland. Yet even after this unhappy experience the idea remained alive in the heads of many ardent Polish patriots. This despite the fact that, historically speaking, Poland touched the Black Sea only extremely briefly and rather ephemerally, and the poles and their governments showed little interest in the Baltic prior to 1918 and began to really use it only after 1945.

7 The best, and most influential, example of such writing is the historical novel *Ogniem i Mieczem* (With Fire and Sword) by the famous Polish writer and Nobel Prize winner, Henryk Sienkiewicz. First published in 1883, always in print since required reading in Polish schools and to this day very popular with both the intelligentsia and people in general, also made into a movie, it has had an enormous impact on how the Poles view their history and the nobility's heroic and patriotic role in it, and at the same time prejudiced them against the Ukrainians, making a reconciliation between Poland and Ukraine so much more difficult. The Ukrainians have of course immediately spotted the bias of the novel and recognized the harm it was doing to Polish–Ukrainian relations. See Volodymyr Antonovykh, 'Pol'sko-Russkiia otnosheniia XVII veka v sovremennoi pol'skoi prizme'. *Kievskaia Starina*, 1885, v.

8 See the following writings of the well-known professor of geography at the University of Poznań, August Zierhoffer; 'Rola odry w terytorialno-państwowym organizmie Polski', in *Sprawozdanie Sesji Rady Ziem Odzyskanych*, Kraków, 1947, pp. 21–30; 'Problem zachodniej granicy Polski w świetle geografii politycznej', *Przeglad Zachodni*, III (3) (1947), 203–12; and his 'Le Cadre géographique du territoire de la Pologne',

in *Les Fleuves et l'évolution des peuples*, Paris: Centre Internationale de Synthèse, Institut Internationale d'Archeocivilization, 1950, pp. 15–24.

9 Thus in 1945 the Instytut Zachodni, which was in charge of publications justifying the annexation of formerly German territories, was declaring the Oder–Neisse line as 'the best Polish boundary', but a year later was backing the annexation also of the left bank of the Oder river. Maria Kiełczewska and Andrzej Grodek, *Odra-Nisa, najlepsza granica Polski*, Poznań-Warsaw: Instytut Zachodni, 1945; M. Kiełczewska, M. Glucka, and Z. Kaczmarczyka, *O Lewy brzeg Odry*, ed. Z. Wojciechowski, Poznań: Instytut Zachodni, 1946.

10 Stanislaw Leszczycki, *Geograficzne podstawy Polski wspólczesnej*, Poznań: Instytut Zachodni, 1946; Maria Kiełczewska, *O Podstawy geograficzne Polski*, Poznań: Instytut Zachodni, 1946.

11 Władysław Semkowicz, *Geograficzne podstawy Polski Chrobrego*, Kraków: Krakowska Spółka Wydawnicza, 1925; Robert Maxmann (ed.) *Die Tendenz der polnischen Geographie. Eine Zusammenstellung von vier polnischen geographischen Arbeiten in deutscher Übersetzung*, Danzig: Westpreusischer Verlag, n.d. [1932?].

12 Jan Dymitr Solikowski, 'Rozmowa kruszwicka' (1573), in Edmund Kotarski (ed.), *Kto ma państwo morskie. Problemy morza w opinji dawnej Polski*, Gdański: Wydawnictwo Morskie, 1970, pp. 169–86; on p. 179.

13 Eugeniusz Koczorowski, *Flota polska w latach 1587–1632*, Warsaw: Ministerstwo Obrony Narodowej, 1973, p. 28.

14 Edmund Osmańczyk, *Rzeczpospolita Polaków*, Warsaw: Państwowy Instytut Wydawniczy, 1977, p. 5.

15 The identity of the first Piast ruler underwent revision as the Polish Workers' (Communist) Party became entrenched. At first, as long as the regime tried to woo the peasant away from the Polish Peasant Party led by Stanisław Mikołajczyk, it was propitious to call the first Piast ruler a peasant. Later, however, when the emphasis was on Poland as a workers' land, the regime remembered that the first Piast was allegedly a wheelwright, which was much better than a peasant. In Marxist eyes a peasant is just a petty bourgeois. A wheelwright, on the other hand, could be assimilated into the working class. Thus the first Polish king was a worker, that is, a member of the proletariat.

16 It is certainly true that under the Jagiellonian dynasty Poland was transformed into a multinational empire, but it is at least debatable whether the Piast dynasty was non-imperialistic or merely less powerful and less successful at empire-building. After all, the second Piast king, Bolesław I, had already led an expedition eastwards which reached Kiev, and the last of the Piasts, Casimir the Great, laid the foundation for the Polish expansion eastward when he annexed the so-called Red Russ (the present day Western Ukraine).

17 The contrast between Piast and Jagiellonian Poland began to be emphasized immediately after 1945. See Maria and Zygmunt Wojciechowski, *Polska Piastów, Polska Jagiellonów*, Poznań: Drukarnia św. Wojciecha, 1946. The best-known treatment of the problem is Paweł Jasienica, *Polska Piastów*, Warsaw: Państwowy Instytut Wydawniczy, 1966; see also his *Rzeczpospolita obojga narodów*, 3 vols, Warsaw: Państwowy Instytut Wydawniczy, 1967, 1972.

18 Edmund Jan Osmańczyk, *Matka Boska Radosna patronka Polaków spod znaku rodła*, Paris: Editious du Dialogue, 1989; and his *Niezłomny proboszcz z Zakrzewa*, Warsaw: Czytelnik, 1989; Bolesłzw Czajkowski, *Wszystko o rodło*, Warsaw: Krajowa Agencja Wydawnicza, 1975.

19 Lucian Blaga, *Zum Wesen der rumänischen Volksseele*, Bucharest: Minerva, 1982; and his *Zări si etape*, ed. Dorli Blaga, Bucharest: Ed. pentru Literatură, 1968.

15

National Identity of Ukraine

Ihor Stebelsky

Recent Gorbachev-inspired developments in the USSR have highlighted the importance of national identity. In the Ukrainian SSR, the 'Movement in support of restructuring' (*Rukh*) arose from a groundswell of opposition to Soviet policies that had suppressed Ukrainian nationhood. Indicative of this popular pressure, the newly elected council of deputies of the Ukrainian SSR proclaimed (16 July 1990) sovereignty and then (24 August 1991) independence of Ukraine.

National identity consists of a number of common attributes that strengthen the bond of a people: their language, culture, history and political aspirations. Geography provides the context from which some of the national markers are drawn and in which they are manifested. For example, the physical features of Ukrainian landscapes, the ecology of agriculture imprinted in traditional rituals and art forms, the historic struggle against the invaders, expressed through Ukrainian folk songs and literature, form part of what Jean Gottmann termed iconography: a system of symbols that reinforce national identity.

Human components of the landscape are symbolic, for they represent the historic continuity of a nation. For example, the blue and yellow national flag, the trident emblem and the national anthem, symbols of Ukrainian independence that were banned by the Soviet regime, were propagated by the *Rukh* activists and are now official in Ukraine. Historical monuments, associated with Ukraine's political aspirations for independence, have become sites of rallies. Ukrainian villages, targeted for reconstruction into rural cities with multistorey apartment buildings to symbolize Soviet progress, have recently been revitalized with the construction of private homes and the restoration of churches. The Soviet shaping of the human landscape, part of an attempt to alter Ukrainian national identity to a common Soviet image, was halted.

Group identity may vary in intensity and hence in its political significance. Groups sharing common attributes may be called a people, a nationality or a nation, depending on the strength of their integration. According to Karl W. Deutsch (1953, pp. 78–9), a people is a larger group of persons linked by a complementary division of labour and facilities of communication by means of a common language and other non-verbal attributes of culture. However, a people attaches little significance to its cultural markers in pursuit of its social, economic and political demands. Only when the cultural distinctiveness of national identity becomes important in a people's social, economic and political demands do the people become a nationality. Once a nationality has acquired the power to compel its members to cohesiveness and attachment to group symbols, it becomes a nation. By placing a state organization at its service, the nation becomes sovereign and thus forms a nation-state.

Groups may share multiple, competing identities. Such identities may be at different levels (clan, tribal alliance, state, empire) or categories (social, cultural, racial, religious, territorial), and may constitute a hierarchy of allegiance. Identities have changed through time, with national identity gaining prominence in the nineteenth century.

The purpose of this paper is to investigate the relationships between geography and national identity in Ukraine. The subject will be treated chronologically in geographical context: the formation of the proto-Ukrainian people, the identity associated with tribes, principalities and empires, the emergence and suppression of Ukraine, the rise of Ukrainian nationalism and the struggle for Ukrainian statehood in the twentieth century.

THE FORMATION OF THE PROTO-UKRAINIAN PEOPLE

The patterns of migration, settlement and cultural evolution of prehistoric societies on the territory of present-day Ukraine were closely related to the physical environment of the post-Pleistocene period. About 8000 years ago the Neolithic tribes in Ukraine adopted agriculture. Originating from the Middle East and the Mediterranean by way of the Danubian basin, agriculture diffused eastward along the forest-steppe from the Dniester to the Dnieper, but failed to penetrate the marshy forests of Polisia to the north.

By late Neolithic times, the Trypillia culture emerged in the Ukrainian forest-steppe zone. Distinguished by its advanced farming, elegant painted pottery and clay female figurines, this culture arose some 6000 years ago and lasted some 2000 years. Later, it was overwhelmed by waves of pastoralist incursions from the left-bank Ukrainian steppe: the Old Pit culture, the Bronze-age Catacomb culture and, finally, the Timber Frame culture (the Cimmerians). Despite this turbulence, farming persisted in the right-bank forest-steppe, where the survivors of the Trypillian culture developed impressive lustrous black pottery and built elaborate earthen walls in defence against the Cimmerians.

Subsequently, the Cimmerians were overpowered by the Scythians from the east. Herodotus referred to the ruling pastoralists, who lived on the Ukrainian steppe, as the 'Royal' Scythians, and the farmers of the forest-steppe as the Scythian ploughmen. Much of the time the farmers, the nomads and the Greek colonists of the Black Sea littoral engaged in complementary trade in which the Scythian ploughmen supplied the ports with grain for export to Greece.

By the second century BC, the 'Royal' Scythians were in turn overrun by the Sarmatians.

Meanwhile, from the forested west began to spread the practice of cremation and the placing of the ashes into urns. The Pszeworska culture of southern Poland which involved this practice was associated with the earliest Roman reference to Slavs – the Wends. Its eastern derivative, the Zarubintsi culture, which appeared along the Prypiat', the Desna and the Middle Dnieper, also retained some cultural attributes of the Scythian ploughmen who survived the Sarmatian incursion. The Zarubintsi culture evolved and spread throughout the forest-steppe and parts of the right-bank steppe to become known as the Cherniakhiv culture (second to fifth centuries AD). It may have included not only the Slavic Antes, who inhabited the Ukrainian forest-steppe, but also the Thracian tribes south-west of the Dniester and the Scytho-Sarmatian (Iranian-speaking) tribes, as well as the Germanic Goths (who migrated from the Baltic to the lower Dnieper, challenging the Antes for political control). After the nomadic Huns had dislodged the Goths (AD 375) from the Black Sea Lowland, and the brief rule of the Huns had waned (Attila's death, 453), the Antes (in Iranian this means the border people) prevailed throughout Ukraine.

FROM TRIBAL FEDERATIONS TO EMPIRE AND PRINCIPALITIES

Political organizations such as tribal federations, city states, empires and principalities have imparted group identity to people. The first historically documented Slavic political entity in the territory of Ukraine was the Antes tribal federation. The Antes, together with their closely related western neighbours, the Sclavines, fought the Roman legions and, in the sixth century, launched raids and even settled beyond the Danube. Following the Avar onslaught from the east, their federation disintegrated, but their remnants regrouped to establish Kiev.

In the seventh century a new ethnic and political name emerged on the territory of Ukraine: Rus'. Although the Normanists relate Rus' to the Varangians (who came later), leading Soviet scholars ascribe it to a tribe which inhabited either the Donets or, more likely, the Ros' river basin south of Kiev. According to Rybakov and Tretiakov, the Rus' tribe played a key role in uniting the Polaine and Severiane to form the Rus' federation well before the arrival of Varangians in Kiev. This Rus' federation imparted its name to its immediate territory which, despite subsequent territorial expansion of the Kievan state, remained known as the land of Rus' until the twelfth century.

After the Varangians captured Kiev (Prince Oleh, 879), empire-building commenced. Kiev became the hub of the trade route between the Baltic and the Black Seas. It also became the main centre for recruiting men and collecting furs, mostly from its tributary regions to the north, as it traded and sometimes fought with Byzantium.

The tribes incorporated by Kiev were not all anthropologically or linguistically related. According to Alekseeva, the most closely related to the Poliane were the Slavic tribes to their immediate west and south: the Derevliane, Volyniane, Ulychi, Tyvertsi and the White Croatians. These tribes once belonged to the Antes federation which formed the basis of the Ukrainian ethnos. The Severiane, however, were transitional to the Belorussian tribes of Radimichi, Drehovichi and Krivichi. Still more remote anthropologically were the Sloveny (in and around Novgorod) who imparted the northern Russian dialect, and the Viatichi (in the upper Oka River) who gave rise to the southern Russian dialect. To the north and east were the northern Finnic tribes, the Ves and the Meria (subjected to assimilation by the Sloveny and the Krivichi) and the eastern Finnic tribes, the Muroma and the Mordva (by the Viatichi).

Empire building was not easy. Tribal unions, based on ethnicity, harboured their own loyalties. Since tribute-collecting generated alienation, rebellions were ruthlessly suppressed. Loyalty was rewarded by co-opting the tribal leaders into the prince's administration and by promoting the rank and file in the military. A common bond was sought through the creation of heroic sagas, but a pantheon of the various tribal gods in Kiev proved inadequate. Prince Volodymyr, for this and other reasons, accepted Christianity from Byzantium as the state religion. Sons of the Kievan princes were appointed to administer major towns, whose territorial borders were designed to dismember the dissident tribal areas.

Except for a common dynastic heritage and an elite that had the same written language and religion, the Rus' principalities lacked cohesion. The common people identified more strongly with major towns, principalities or tribes.

Fragmentation of the Rus' empire could not be avoided. Since leadership in Kiev became contentious among the successors, for the sake of peace, Prince Monomakh introduced hereditary succession within each subordinate principality. After the death of Prince Monomakh, however, the dynastic dispute exploded. Prince Andrei of Suzdal, who claimed seniority, plundered Kiev (1169) and removed the crown jewels to his new capital, Vladimir on the Kliazma.

Meanwhile, in what is now western Ukraine, the powerful principality of Galicia united with Volhynia and acquired Kiev. In so doing, Galicia-Volhynia incorporated most of the ethnically Ukrainian tribes under one administration. Its prince was crowned '*Regnum Russiae*', thus enshrining the name Rus' in the subject population. To oppose the Mongol invasion, King Danylo sought help from the Pope, but backed off because the cost would have been conversion to Roman Catholicism. By retaining the Byzantine rite, Galicia-Volhynia set itself off from Poland and Hungary, both of which had accepted Roman Catholicism.

By the fourteenth century a new term, *Russia Mynor*, arose as a result of a split in Church jurisdiction. The Metropolitan of Kiev was lured (in 1299) to Vladimir on the Kliazma (later to Moscow), while the king of Galicia-Volhynia obtained from Byzantium (in 1303) a replacement Metropolitan for Kiev. Ecclesiastical jurisdiction was thus split between what Byzantium called *Russia Major* (outlying Rus', but later translated into English literally as Great

Russia) in the north and *Russia Mynor* (inner Rus', or literally Little Russia) in the south. This political and ecclesiastical division of Rus' along ethnic lines initiated the emergence of a nation-state on the territory of Ukraine. The process, however, was interrupted by the dismemberment of Galicia-Volhynia between Lithuania and Poland. Following a dynastic union (1385), Poland joined with Lithuania to form the Polish-Lithuanian Commonwealth (1569). Within that polity, the people of Ukraine continued to identify themselves with Rus' and the Byzantine (Orthodox) faith.

FROM RUS' TO UKRAINE

The Ukrainian identity originated with the sixteenth-century Cossack movement on the steppe frontier. It culminated with the Cossack wars of liberation and the emergence of the Cossack state whose territory became known as Ukraine. The word *Ukraina* means, in Slavic, 'borderland'. In the twelfth century it referred to the southern borderland of Rus'. By the sixteenth century it attained general usage for the land along the middle Dnieper south of Kiev. Following the Cossack wars of liberation, Ukraine became synonymous with the newly established Cossack state which symbolized both freedom and democracy. This new Cossack polity paved the way for a switch in national identity from Rus' to Ukraine.

National identity was complicated at this time by several competing identities. The old Rus' princes and gentry, rejected by the Cossacks, had become Polonized. While the national identity of people who spoke Ukrainian was Rus', in the Cossack wars with Poland the rebels emphasized the identity of religion (Orthodox, as opposed to Roman Catholic) and class (Cossack or free man, as opposed to lord or serf). In the westernmost provinces, which were not included in the Cossack statehood, Rus' identity remained. Their four Orthodox bishops found accommodation with the Poles by proclaiming union with the Catholic church (Brest, 1596) while retaining the Byzantine rite, and thus established the distinct Uniate Church. This caused a religious split and heated debate in Ukraine. In the long run, however, the Uniate Church (the Ukrainian Catholic Church) helped preserve the nation from Polish (and later Russian) encroachment.

Although the official name of the Cossack state was the 'Army of Zaporozhia', the unofficial name Ukraine was commonly used. After it switched allegiance from Catholic Poland to Orthodox Muscovy, was partitioned and lost sovereignty, the Cossack identity (free man, as opposed to serf in an autocratic Muscovite state) became paramount in eastern Ukraine. Ethnic differences between the Russians (descriptively called *Katsapy* – bearded, like goats – by the Ukrainians) and the Ukrainians (called *Khokhly* – tufts of hair as worn by the Cossack – by the Russians) were very strong. Indeed, by the beginning of the eighteenth century, Hetman I. Mazepa, who governed the Hetmanate (a protectorate of Russia), considered both the land and the people Ukrainian. By contrast, Russian officialdom renamed the Hetmanate, after its absorption into the Russian Empire, *Malorossiia* (Little Russia), and its inhabitants *Malorossiia* (Little Russians).

CRYSTALLIZATION OF THE UKRAINIAN NATIONAL IDENTITY

The growth of modern Ukrainian national consciousness occurred during the nineteenth and twentieth centuries. The spread of romanticism and nationalism from Western Europe spurred interest in the use of the vernacular language. The struggle against economic exploitation and political repression provided the motive for the growth of socialism, nationalism and Ukrainian national consciousness.

In the nineteenth century the Ukrainian ethnographic territories were divided between the empires of Russia and Austro-Hungary. Conditions for political development differed between the empires. Austria, a constitutional monarchy with federal and provincial parliaments, allowed for representation from different nationalities. While German remained the official language of the empire, Hungarian became official in Hungary, while other languages gained local recognition. Russia, by contrast, was an autocracy with a centralized bureaucracy and the obligatory Russian language.

Yet the elevation of vernacular Ukrainian to a literary language (1798) occurred not in Austria but under Russia, in the former Hetmanate, where well-to-do Cossack families had access to education. Some even promoted the revival of the Cossack state and funded the establishment of the first university in Ukraine (Kharkiv, 1805), which stimulated Ukrainian national consciousness.

The right-bank Ukraine (west of the Dnieper), by contrast, lacked the well-to-do Cossack class. The nobility was Polish or Polonized. Yet some gentry developed a local Ukrainian patriotism, and wrote about Ukrainian Cossacks and folklore. Although their writing stimulated Ukrainians, it also created the perception among the Poles that Ukraine was part of Poland. The university in Kiev (established 1834), while dominated first by Poles and then by Russians, also helped incubate Ukrainian intelligentsia.

In Austria-Hungary, the Ukrainian literate class consisted of the Uniate clergy. They continued to use Old Church Slavonic and, lacking the historic experience of Cossack statehood, maintained their ethnic identity as Rusyn (Ruthenian). A breakthrough came, though, when a young clergyman published his lyrical verses in vernacular Ukrainian (1837).

Political developments in Austria-Hungary also stimulated Ukrainian involvement in politics and education. After the 1848 uprisings, to counter the Hungarian and Polish separatist movements the Austrian government granted the Ukrainians emancipation, participation in government, schooling and publications in Ukrainian. By 1851, however, reaction set in. The Hungarians began to assimilate Ukrainians on their territory, while the Poles in Galicia retained their privileged position. In dismay, some Ukrainian intelligentsia, known as Russophiles, turned to Russia and its Slavophile agents for support.

Meanwhile, in the Russian empire, although the printing of Ukrainian was forbidden, it was allowed in Russian transcription as a literary device to express a peasant 'dialect'. Using this device, the writer N. Gogol enriched Russian literature with Ukrainian imagery and Cossack glory, thus stirring Ukrainian national identity among some, but for others binding Ukraine to Russia. Others, using folklore themes, enriched Ukrainian literature. Moreover, the Ukrainian political historian in Kiev, M. Kostomarov, contrasted the Ukrainian tradition of liberty and individualism with the Moscow tradition of authoritarianism and the subordination of the individual to the collective. He believed that the break-up of the tsarist imperial monolith into a federal system of free and equal states would release the creative forces of all peoples, including the Russians.

The critical point which in fact defined Ukrainian national consciousness was reached with the poetry of Taras Shevchenko. His *Kobzar* (1840) at once captivated and joined in spirit what remained of the Ukrainian upper class, the young and numerically small intelligentsia and the serfs. Shevchenko's 'Little Russia' symbolized subservience, whereas 'Ukraine' was the land of Cossack freedom. In his 'Testament' he called for the liberation of Ukraine. As a result, the authorities first censored his poetry, then banned it from publication. Shevchenko was arrested, exiled, deprived of writing, and thus became a Ukrainian national martyr.

With political repression, the Ukrainian intelligentsia in the Russian empire focused on non-political Ukrainian studies, publications and schools. In Kiev, the Ukrainian intellectuals established (1873) the South-western Branch of the Imperial Russian Geographical Society, which provided legitimacy for research on Ukrainian topics. Because the Society's publications led to a greater appreciation of Ukrainian distinctiveness, its members were soon accused (1875) of fomenting separatism. Some, like the political historian, M. Drahomanov, who advocated Ukrainization as a democratic political process, were exiled, while the South-western Branch of the Imperial Russian Geographical Society was closed. The tsar's decree (signed at Ems, 1876) denied the very existence of the Ukrainian language, banned Ukrainian books and prohibited the use of Ukrainian in schools. Denied the right to publish in Ukrainian, the authors smuggled their manuscripts to Austria, where their publications, along with the poetry of Shevchenko and the political writings of Drahomanov (now in Geneva), stimulated Ukrainian national identity.

Under Austrian rule, conditions for the Ukrainians improved. Competition with the Poles and others honed their political skills. Ivan Franko, a brilliant young poet and scholar, attacked the Russophiles and weakened their influence. He participated in the formation (1890) of the first Ukrainian (socialist) political party and became a charter member of the Shevchenko Scientific Society. Meanwhile, an accommodation between the moderates and the viceroy of Galicia yielded political concessions. New electoral laws (universal and equal franchise, 1907) shifted more political power to the Ukrainians. Various Ukrainian institutions (co-operatives, schools and youth organizations) appeared almost overnight.

Ukrainian scholars on both sides of the Austro-Russian border played important roles in honing Ukrainian identity. They provided the ethnographical, historical and geographical interpretations that countered the views of the Russian Slavophiles with concepts of Ukrainian nationhood. The Ukrainian anthropologist and ethnographer, F. Vovk (Volkov), identified specific cultural markers that characterized the Ukrainian ethnos and mapped their extent. M. Hrushevsky, a prolific and respected scholar, dissented (1904) from the official Russian historiography, treated the history of Kievan Rus' as part of Ukrainian rather than Russian history, and outlined the ethnogenesis of the Ukrainian nation. The Ukrainian geographer, S. Rudnytsky, used Hettner's approach in regional integration to argue for the coherence of Ukrainian ethnographic territories and the logic for their formation into a Ukrainian nation state.

Meanwhile, in the Russian empire intensified repression radicalized the political parties among the Russians and delayed the formation of Ukrainian political parties until the turn of the century. Only the 1905 Revolution forced temporary concessions from the imperial government, including the publication of Ukrainian-language newspapers and journals. Another tsarist concession, the elections (1906) to the first *Duma* (Parliament), enabled the Ukrainian deputies to argue for greater autonomy. In anger the tsar dissolved the *Duma* and, after imposing voting restrictions on non-propertied classes, excluded the Ukrainians from succeeding *Dumas*. The ensuing oppression forced some Ukrainian separatists to flee to Galicia where, upon the outbreak of the First World War, they formed the Union for the Liberation of Ukraine.

The First World War imposed great physical losses on the Ukrainian population. The trauma of occupation made people aware of their vulnerability without the protection of a nation-state. Attempts by Russian troops in Galicia (1915) to impose the Russian language and to force the conversion of the Ukrainian Catholic (Uniate) Church to Russian Orthodoxy reduced support for the remaining Russophiles. As the Austrian forces regained ground, members of the Union for the Liberation of Ukraine politically educated some 400,000 Ukrainian prisoners of war in German and Austrian camps, thus raising their national consciousness.

By 1917 the imperial armies were exhausted. While the Russian empire began to disintegrate in March 1917, the Austrian empire collapsed in October 1918.

THE STRUGGLE FOR UKRAINIAN STATEHOOD AND ITS IMPACT

The collapse of the tsarist regime led to the formation in Kiev of two rival Russian political bodies: an Executive Committee (serving the provisional government in Petrograd) and a Soviet of Worker's and Soldiers' Deputies (serving the radical left). The Ukrainians in Kiev, however, established their own organization, the Central *Rada* (*rada* means 'council' in Ukrainian; the Russian equivalent is *soviet*), which provided for political representation of all the Ukrainian parties. Successive congresses of soldiers', peasants' and workers' deputies in Kiev provided mass political education.

Encouraged by the congresses, the Central *Rada* established itself as the highest political authority in Ukraine. The President of the Central *Rada*,

M. Hrushevsky, feared a Russian backlash, for the Russians not only outnumbered the Ukrainians in the Russian empire by 71 million to 35 million but also dominated the cities and industrialized areas of Ukraine. Therefore, instead of independence, the Central *Rada* proclaimed Ukrainian autonomy in Russia. Although this infuriated the Russians, the setbacks at the front forced the provisional government to concede to the Central *Rada* the administration of five Ukrainian (mostly rural) provinces; Kiev, Poltava, Podilia, Volhynia and Chernihiv. Having been promised broad cultural autonomy, the Russian and Jewish parties in Ukraine also agreed to join the Central *Rada*.

The building of a nation-state was hindered by the lack of Ukrainian institutional structures. Moreover, the young socialists who dominated the Central *Rada* placed low priority on the administration and the military. Some even viewed the bureaucracy and the army as instruments of repression that the revolution had made obsolete. Linkage between the countryside and the administration in Kiev was poor and the latter lacked effective military to impose control.

After the Bolsheviks seized power in Petrograd, the Central *Rada* extended its territorial claim to all nine provinces of the former Russian empire where Ukrainians constituted a majority. Meanwhile, the Bolsheviks in Ukraine moved to Kharkiv where they proclaimed a rival Soviet Ukrainian Republic and prepared to overthrow the Central *Rada*. Lacking sufficient force to hold back the Red Army and the Bolshevik agitators, the Central *Rada* proclaimed the independent Ukrainian National Republic (22 January 1918), made peace with the Austrians and Germans at Brest-Litovsk, and sought their military support against the Bolsheviks.

The Austrian and German forces demanded grain from Ukraine for their war effort. Since the Central *Rada* was neither willing nor able to satisfy their demands, the Germans supported a *coup d'état* by Hetman P. Skoropadsky. By drawing on Russian professionals, the Hetman created an effective administration. He established many Ukrainian institutions: high schools, universities, a national archive, a national library and the Ukrainian Academy of Sciences. The Hetman government also extended its territorial claim beyond the nine provinces of the former Central *Rada* to include all Ukrainian ethnic territories. The Germans, however, discouraged the creation of a Ukrainian military force that might challenge their influence.

As the Central Powers appeared to lose the war, Hetman Skoropadsky, having failed to attract the Ukrainian activists to his side, turned to the Russian monarchists for support and as a concession announced federation with Russia. Enraged by this act, the former leaders of the Central *Rada* drove out the Hetman and formed the Directory.

Although the Skoropadsky regime was viewed by the socialists as reactionary and by the nationalists as collaborationist, it provided institutions that promoted national identity. It exposed members of the largely Russified elite to the idea of Ukrainian statehood, and helped expand the concept beyond the (mostly leftist) Ukrainian intelligentsia to landowning peasantry and gentry.

The Directory resurrected the Ukrainian National Republic with goals of both social transformation and civil order. The radical left, led by Vynnychenko, insisted on the expropriation of state, church and private landholdings for distribution among the peasantry, while the nationalistic moderates, led by Petliura, called for more attention to state institutions and the army and pressed for an agreement with the Entente.

Meanwhile, the Red Army launched a devastating offensive. Before the fall of Kiev, the Directory staged a congress (22 January 1919) where it declared the union of the Ukrainian National Republic with the newly formed West Ukrainian National Republic in Galicia. The hope now was to join forces and together defeat the Bolsheviks.

In Galicia, the struggle for statehood also began with the fall of the Austro-Hungarian Empire. In contrast to the Russian empire, however, Ukrainian institutions in Galicia were already developed. As the Austro-Hungarian Empire disintegrated, the subject peoples, including Ukrainians, began to prepare their own independent nation-states. Since the Poles also claimed Eastern Galicia, what resulted was a clash between two nations for the same territory.

The conflict was carried out mostly by regular armies fighting along established fronts. It was a test of strength between 3.5 million Ukrainians of Eastern Galicia and 18 million Poles. Although the Ukrainian military took the initiative, when reinforcements arrived from central Poland, the Poles (supported by France) gained a decisive numerical, technical and diplomatic advantage.

Supportive of this outcome was the prevailing French geopolitical imperative to offset Germany with a strong Poland in the east. Consequently, despite the US President Woodrow Wilson's democratic concept of national self-determination, which the Ukrainians hoped would support their claims, the Entente acceded to Poland's occupation of Eastern Galicia. The Ukrainian Galician army was driven off to the east, where it linked up with the retreating forces of Petliura.

Meanwhile, in eastern Ukraine the Bolsheviks alienated the peasantry with grain requisitions and the Ukrainian radical intelligentsia with their language policy. Bolshevik arrests and executions of 'class enemies' triggered revolts. Taking advantage of this uprising, the anti-Bolshevik Russian forces, led by General Denikin, attacked from the Don. Simultaneously, Petliura's forces, combined with the Ukrainian Galician army, advanced from the west. Squeezed from both sides, the Bolsheviks retreated to the north.

The political goals of Denikin (the Russian empire) and Petliura (independent Ukraine) proved irreconcilable. Although Petliura approached Denikin for co-operation against the Bolsheviks, the Russian refused. Supplied by the French, Denikin launched an offensive against the Ukrainians. Lacking supplies at home and support from the Entente, the remnants of the Ukrainian armies were defeated. The Ukrainian Galician army, reluctant to engage Denikin's forces, surrendered to the latter. Petliura and the remnants of his forces surrendered to the Poles, with whom they later joined in an effort to liberate Kiev from the Bolsheviks at the expense of Western Ukraine.

The common experience of spirited fighting, suffering and defeat left a lasting imprint on the Ukrainians. The conservatives, the nationalists and the socialists blamed one another for the loss of Ukrainian independence. Differences between the Eastern Ukrainians and the Galicians as to which side to choose as the lesser evil at the end of war embittered both parties for at least one generation.

INTERWAR EXPERIENCES AND THEIR IMPACTS

The partition of Ukraine among the Soviet Union in the east and Poland, Romania and Czechoslovakia in the west provided for different paths of Ukrainian national development.

The Bolsheviks chose to placate Ukrainian patriotism by establishing the Ukrainian SSR with all its surrogate statehood trappings. The terror of war communism was replaced with a benign new economic policy. A policy of cultural development 'national in form but socialist in content' allowed the Ukrainian intelligentsia to educate a generation of youth in Ukrainian (albeit socialist) consciousness.

Once the economy had recovered, the policy was changed to build socialism in the USSR. Forced collectivization involved 'class warfare' and induced famine to break Ukrainian peasant resistance. Millions of farmers, the keepers of ancient Ukrainian traditions, had perished. Mass purges decimated the Ukrainian intelligentsia and terrorized the rest into submission. Ukrainian institutions were revamped to praise all things Russian and equate Ukrainian with backwardness or 'bourgeois nationalism'. Soviet agents even assassinated Ukrainian leaders abroad.

In Poland, Ukrainians were dismayed by the loss of their statehood and angered by Polish refusal to grant them autonomy. Anti-Ukrainian policies generated resentment and increased the activity of the Organization of Ukrainian Nationalists (OUN). Despite the repressions of Pilsudski's authoritarian government, some Ukrainian institutions survived in Galicia until the outbreak of the Second World War. Areas with poorly developed Ukrainian institutions, such as Volhynia, or none, such as Polisia, were isolated from Galicia and subjected to intense Polonization.

Romania acquired three different parts of Ukrainian territory: (a) Bukovyna (from the Austrian part of the Austro-Hungarian Empire); (b) Maramarosh (from Hungary of the Austro-Hungarian Empire); and (c) Bessarabia (from the Russian empire). In Bukovyna, where Ukrainian institutions were developed, their closure evoked strong resentment and increased the activity of the OUN. By contrast, in Maramorosh and Bessarabia Ukrainian national identity was undeveloped, and Romanianization provoked little resistance.

Czechoslovakia acquired most of the Ukrainian lands that were formerly under Hungary. Isolated from Galicia and oppressed by Hungary, the Carpatho-Ukrainians (Ruthenians or Rusyns as they still called themselves) were pro-Russophile.

Under Czechoslovak democracy Ukrainian national identity grew rapidly, challenging Ruthenian identity. That change was brought about by young populists, who were impressed by the Ukrainian movement in Galicia. They spread the Ukrainian idea through Ukrainian reading rooms, schools and press.

On the eve of the Second World War, under increasing external pressure, Prague allowed greater autonomy (1938). Then Hitler, who invaded the Czech lands, sanctioned the Hungarian occupation of Carpatho-Ukraine. In response, Ukrainian volunteers both locally and from Galicia joined the 'Carpathian Sich' (a local militia) to offer armed resistance, but were overwhelmed by the Hungarian army. In its last gesture (15 March 1938), the Parliament of Carpatho-Ukraine declared independence, passed its constitution and elected its president. This struggle transformed the youth of Carpatho-Ukraine into patriotic Ukrainians.

CONSOLIDATION OF UKRAINIAN LANDS IN THE UKRAINIAN SSR AND ITS IMPACT

Following the Second World War, the Soviet Union acquired and consolidated most Ukrainian lands within the Ukrainian SSR. Thus, western Volhynia and eastern Galicia (both from interwar Poland), Bukovyna and parts of Bessarabia (from interwar Romania) and Carpatho-Ukraine (from interwar Czechoslovakia) were incorporated into the Ukrainian SSR. Only small strips of historically ethnic Ukrainian lands remained in Poland, Czechoslovakia and Romania, while a larger segment of Polisia and even larger outliers of Ukrainian Cossack-colonized lands were retained, respectively, in the Belorussian SSR and the Russian SFSR.

Integration of the western Ukrainian lands proved costly to the Ukrainian intelligentsia. With the acquisition of western Ukraine (the Molotov–Ribbentrop agreement), much of the Ukrainian intelligentsia fled westward to avoid Soviet purges. Following the Nazi invasion of the Soviet Union, the Bandera faction of the Ukrainian Nationalists (OUN-B) proclaimed Ukrainian independence in Lviv (1 June 1941). The arrest of this 'nascent government' by the Gestapo revealed the true Nazi intentions towards the Ukrainian population. Forced retention of collective farms and recruitment of slave labour generated resistance which was met with arrests and executions. Then OUN-B formed the Ukrainian Insurgent Army, which harassed the Nazi and later the Soviet occupation forces. The returning Soviet forces rooted out the Ukrainian Insurgent Army and thoroughly purged and dispersed the remaining Ukrainian intelligentsia throughout the USSR.

During Khrushchev's de-Stalinization, the authors of western Ukraine, released from concentration camps, were joined by several outspoken writers in Kiev. In the 1960s a new generation of poets, writers and artists began to express Ukrainian patriotism. Publications in the Ukrainian language increased and scholarly journals in Ukrainian studies reappeared.

Brezhnev's renewed policy of Russification generated dissent among the Ukrainian intelligentsia. Arrests or 'unexplained' deaths only drove the protests underground, resulting in clandestine writings that accused the regime of ethnocide (Anon., 1974).

In the post-Second World War period, the increasing proportion of Ukrainians who spoke mainly Russian was interpreted by some to mean that Ukrainian consciousness had suffered a decline. If one accepts Armstrong's

(1968) argument that the principal obstacle to Russification had been the Ukrainian peasantry, then the prospects for nurturing a separate Ukrainian identity could only be pessimistic. Ukrainian bilingual intelligentsia, however, proved to be a bastion of Ukrainian identity. Even those who lived in a predominantly Russian urban environment (as in Crimea or the Donbass) have retained national identity through association with the territory, a Ukrainian way of speaking Russian and some attributes of Ukrainian culture. Competition for better jobs with the Russians stimulated Ukrainian identity among workers and employees (including party officials) in Ukraine. Thus the policy of Russification, reinforced by planned intermixing of labour, have not only failed to bring about assimilation within the Ukrainian SSR but have increased latent discontent.

Gorbachev's *glasnost* allowed national ferment to surface throughout the USSR. Ukrainian intelligentsia demanded the reinstatement of Ukrainian as the official language of their republic. Other concerns, including those of culture, ecology and economy, were raised by a broad segment of Ukrainian society. The trappings of statehood in the guise of the Ukrainian SSR promoted the pent-up desire for an independent state. The incorporation of Eastern Galicia with its intelligentsia into the Ukrainian SSR, together with their penchant for grass-roots organization and parliamentarianism, provided the backbone for *Rukh*.

Presently, Ukrainian passion for independence is dominant in western Ukraine. From there, *Rukh* had concentrated its political activity in Kiev, making efforts to spread its message to the east and south. Recently, even the Donbas coal miners have demonstrated preference for a sovereign Ukraine. This growth of national assertiveness proved to be decisive in the referendum (on 1 December 1991) regarding Ukrainian independence.

SUMMARY AND CONCLUSION

The formation of the proto-Ukrainian people was linked to the natural environment of the Ukrainian forest-steppe. The Ukrainian psyche was shaped by the deeply entrenched agricultural way of life and enriched by contacts with the steppe pastoralists and the Greek colonists on the Black Sea coast. Although the population endured Scytho-Sarmatian, Germanic and Turkic-Mongol incursions, the Slavic element prevailed, consolidating into the Antes tribal federation which formed the proto-Ukrainian ethnos.

Early group identities were associated with tribes and tribal federations. The Kievan princes, who built the Rus' empire, superimposed a universalist Christian identity in the form of the Byzantine (Orthodox) church, headed by the Metropolitan of Kiev. Yet ethnic and economic differences led the Rus' empire to disintegrate into many principalities. Galicia-Volhynia began to emerge into a nation-state, but this process was interrupted by its partition and then absorption into the Polish–Lithuanian Commonwealth.

The steppe frontier of the Commonwealth gave rise to the Cossack society of free warriors. Their war of liberation against Poland culminated in the establishment of their own state, whose territory was known as Ukraine. After its eastern part became a protectorate of Muscovy, Cossack freedoms and ethnicity became the chief identity markers. As Muscovy became synonymous with Russia, in the Hetmanate emerged a clear national identity of Ukraine. The shift in identity from Rus' or Ruthenia to Ukraine occurred by the mid-nineteenth century west of the Dnieper, at the turn of the century in Eastern Galicia, and only in the interwar period in the Carpatho-Ukraine.

The crystallization of modern Ukrainian national identity in the nineteenth century was inspired by the spread of populist nationalism. Its spread was impeded by the division of Ukrainian lands between two empires, and by the oppressive policies of the Russian empire. Although well-to-do Cossack families in eastern Ukraine established Ukrainian as a literary language and rekindled the desire for independent statehood, their efforts were suppressed. It became the task of the populist intelligentsia in Galicia, where a constitutional monarchy allowed the growth of Ukrainian identity and institutional structures.

The Ukrainian poet, Taras Shevchenko, played a leading role in the awakening of Ukrainian passion. Later, Ivan Franko provided the stimulus to fight for an independent Ukraine. Ukrainian social scientists contributed to the definition of Ukrainian identity. They identified national markers and provided a historical and geographical interpretation that challenged Russian imperial views.

The struggle for statehood inflamed the passion of the Ukrainian intelligentsia and infused the military with national consciousness. During the brief existence of an independent Ukrainian state, institutions were created which, retained by the Soviet government, were instrumental in raising a generation of Ukrainians.

When the Soviet government began rapid socialist construction, these institutions were purged or altered to propagate Marxist-Leninist ideology and Russification. In Poland, Ukrainian institutions were restricted while in Romania they were suppressed. Czechoslovakia, by contrast, allowed free development of Ukrainian institutions. Whereas in Poland and Romania authoritarian suppression fuelled the activity of the OUN, in the totalitarian USSR ruthless purges seemingly crushed Ukrainian nationalism.

The loss of Ukrainian independence had a divisive impact on the Ukrainian intelligentsia. Consolidation of the Ukrainian lands within the Ukrainian SSR after the Second World War, by contrast, brought nearly all the Ukrainian intelligentsia together, where mutual understanding developed.

The Soviet regime did not deny the existence of Ukrainian identity, as had the tsarist regime. In keeping with Marxist-Leninist ideology, however, the Communist Party of the Soviet Union considered nationality to be a historic, transitory phenomenon. It 'predicted' that all nationalities in the Soviet Union would merge, sooner or later, into one Soviet nation. For fellow Slavs, the Belorussians and Ukrainians, this merging was expected to occur sooner, because of linguistic affinity. Indeed, the official historiography taught that all three were, at one time, one Rus' people who spoke one language. Scholars who disagreed with such a view were branded bourgeois nationalists.

With the advent of Gorbachev's *glasnost* the imperative for merging nationalities was no longer imposed. Ukrainian classics began to be published in full editions. Following the failed coup against Gorbachev (19–21 August 1991), the political initiative shifted from Moscow to the Union Republics. In this changed political milieu, Ukraine set an irrevocable path to political independence.

REFERENCES

Alekseeva, T.I. 1973: *Etnogenez vostochnykh slavian po dannym antropologii.* Moscow: Izdatel'stvo Moskovskogo universiteta.

Anon. 1974, transl. 1976: 'Ethnocide of Ukrainians in the USSR'. *Ukrainian Herald* (Baltimore, Md: Smoloskyp Publishers) issue 7–8, 35–161.

Armstrong, J.A. 1963: *Ukrainian Nationalism*, 2nd edn. New York: Columbia University Press.

—— 1968: 'The ethnic scene in the Soviet Union: the view of the dictatorship'. In E. Goldhagen (ed.), *Ethnic Minorities in the Soviet Union*, New York: Praeger, pp. 14–21.

Braichevs'kyi, M. Iu. 1968: *Pokhodzhennia Rusi.* Kiev: 'Naukova dumka'.

Deutsch, K.W. 1953: *Nationalism and Social Communication.* Cambridge, Mass.: Technology Press of M.I.T.

Dmytryshyn, B. 1956: *Moscow and the Ukraine, 1918–1953: a study of Russian Bolshevik nationality policy.* New York: Bookman Associates.

Farmer, K.C. 1980: *Ukrainian Nationalism in the Post-Stalin Era: myth, symbols and ideology in Soviet nationalities policy.* The Hague: Martinus Nijhoff.

Gottman, J. 1952: 'The political partitioning of our world: an attempt at analysis'. *World Politics*, IV (4), 512–19.

Hrushevsky, M. 1904, transl. 1952: 'The traditional scheme of "Russian" history and the problem of a rational organization of the history of eastern slavs'. *Annals of the Ukrainian Academy of Arts and Sciences in the U.S.*, II (2), 355–64.

—— 1911, transl. 1941: *A History of Ukraine.* New Haven, Conn.: Yale University Press.

—— 1913–36, repr. 1954–8: *Istoriia Ukrainy-Rusy*, 10 vols. New York: 'Knyhospilka'.

Korobkova, G.F. 1987: *Khoziaistvennye kompleksy rannikh zemledel' chesko-skotovodcheskikh obshchestv iuga SSR.* Leningrad: 'Nauka'.

Kravets', O.M. 1973: 'Pivdenno -Zakhidnyi Viddil Rosiis 'koho Heohrafichnoho Tovarystva'. *Narodna Tvorchist' ta Etnohrafiia*, no. 2, 60–5.

Krawchenko, B. 1985: *Social Change and National Consciousness in Twentieth-Century Ukraine.* London: Macmillan.

Okhrymovych, Iu. 1922, repr. 1977: 'Rozvytok ukrains'koi national'no-politychnoi dumky'. In H. Vas'kovych (comp.), *Materialy do istorii rozvytku suspil'no-politychnoi dumky v Ukraini XXIX–XX stol.* Brussels: V-vo Spilky Ukrains'koi Molodi, no. 3, pp. 67–170.

Pritsak, O. and Reshetar, J.S. 1963: 'Ukraine and the dialectics of nation-building'. *Slavic Review*, XXII (2), 5–36.

Prymak, T.M. 1987: *Mykhailo Hrushevsky: the politics of national culture.* Toronto: University of Toronto Press.

Rudnytsky (Rudnyckyj), S. 1914: *Ukraina und die Ukrainer.* Vienna: Verlag des Allgemeinen Ukrainischen Nationalrates.

——(Rudnitsky) 1915: *The Ukraine and the Ukrainians*. Jersey City, N.J.: Ukrainian National Council.

——(Rudnytskyi) 1922: *Do osnov Ukrains'koho natsionalizmu*, 2nd edn. Vienna: 'Adria'.

——1924: *Ukrains'ka sprava zi stanovyshcha politychnoi geografii*. Berlin: 'Ukrains'ke slovo'.

——1926: *Osnovy zemleznannia Ukrainy. Druha knyha. Antropogeografiia Ukrainy*. Uzhhorod: Drukarnia O.O. Vasylian.

Rybakov, B.A. 1979: *Gerodotova Skifiia: Istoriko-geograficheskii analiz*. Moscow: 'Nauka'.

——1982: *Kievskaia Rus' i russkie kniazhestva XII–XIII vv.* Moscow: 'Nauka'.

Sedov, V.V. 1978: 'Anty'. In V.V. Kropotkin et al. (eds), *Problemy sovetskoi arkheologii*. Moscow: 'Nauka'. pp. 164–73.

——1978: 'Etnogeografiia Vostochnoi Evropy serediny I tysiacheletiia n.e. po dannym arkheologii i Iordana'. In *Vostochnaia Evropa v drevnosti i srednevekov'e*, Moscow: 'Nauka', pp. 9–15.

——1979: *Proiskhozhdenie i ranniaia istoriia Slavian*. Moscow: 'Nauka'.

Subtelny, O. 1988: *Ukraine: a history*. Toronto: University of Toronto Press.

Szporluk, R. 1986: 'The Ukraine and Russia'. In R. Conquest (ed.), *The Last Empire: nationality and the Soviet future*, Stanford, Cal.: Hoover Institution Press, pp. 151–82.

Tretiakov, P.N. 1953: *Vostochnoslavianskie plemena*. Moscow: AN SSSR.

Vovk (Volkov), F. et al. (eds) 1914: *Ukrainskii narod v ego proshlom i nastoiashchem*. St Petersburg: Izd. M.A. Slavinskago.

Vovk, F. 1926: *Studii z ukrains'koi etnohrafii ta antropolohii*. Prague: Ukrains'kyi hromads'kyi vydavnychyi fond.

Wynar, L. 1985: 'Michael Hrushevsky's scheme of Ukrainian history in the context of the study of Russian colonialism and imperialism'. In M. Pap (ed.), *Russian Empire: some aspects of tsarist and Soviet colonial practices*, Cleveland, Ohio: John Carroll University and Ukrainian Historical Association, pp. 19–39.

16

The Quest for Slovene National Identity

Joseph Velikonja

The plebiscite of 23 December 1990 in support of Slovene independence and the declaration of the sovereignty of Slovenia on 7 October 1991 are the concluding acts of a process more than a century old by which the small 'non-historical' nation of the Slovenes achieved its internal self-assertion; the recognition of its sovereignty by most countries of the world by mid-1993 is the final step for recognition within the family of nations.

The interwoven process of social, political and psychological awakening, the self-discovery and subsequent striving for affirmation has had numerous protagonists, who contributed to leading or hindering the evolution towards the goal of national self-affirmation.

Group identity evolves through communication exchanges between participants within their territorial frame.[1] The space-related identity is affected by the range of the communication network and the intensity of the interacting process which has moulded the Slovene area: in the pre-industrial era the dominance of localized interactions, the almost non-existent ties with people and regions beyond the direct experiences of land-bound peasants created a mosaic of distinct local entities, each with its own dialect, customs, attitudes and identities. The everyday life of the peasant did not require more. Occasional journeys beyond the locality confirmed the foreignness of distant areas and often a belief that not much good can come from them.

The transformation of people's awareness from a locally bound spatial identity of the past to a larger regional and national territorial setting occurred within historical events and an often externally imposed administrative frame.

For centuries the Slovene lands were an integral part of the Habsburg domain, administered from Vienna and managed by a stable apparatus of civilian authorities. The fragmented civilian administration reflected the localized needs of the population. In addition, the Catholic Church with its territorially defined parishes provided extra stability. This frame satisfied the common need to 'be part of' for most people. The civilian authority was often foreign and accepted as a necessary evil. It was foreign by the language it used, by the demands it made and by the rules it tried to impose.

Slovene awareness emerged through written language. The Protestant reformers in the sixteenth century codified the language of the peasants and produced the Bible and other religious writings in Slovene. Although at the time most of the peasant population of the Slovene lands was illiterate, the sacred books in their colloquial language enabled the preachers to proselytize. This common-language bond is the first major external evidence of Slovene identity, still in a primeval stage but potentially significant.[2]

The Napoleonic intervention in 1809 broke the century-old mould of the Austrian administrative territorial organization by creating a new unit, the Illyrian Provinces. But beyond that, it adopted the Slovene language as a legitimate vehicle of official communication and documentation: the colloquial language of the peasants was recognized as a legitimate language of authority. Although the period was very short (1809–14), the Slovene territory and its people were never the same again. The old mould was broken. The recognition of a larger ethnic and territorial entity, Slovenia, became the legitimate goal of ever-larger numbers of people. The demand surfaced in the March 1848 revolution in Austria and gave rise to the first openly formulated programme for a 'United Slovenia'.

Because of the absence of an administrative frame for what had been emerging as 'Slovene areas', the weight of advocacy stayed with the cultural manifestations. The poetry of France Prešeren by its eloquence and quality gave legitimacy to the Slovene language and provided support for aspirations to external recognition.

The territorial frame of Slovene lands (with Slovene place-names) was first traced on a map produced by Peter Kozler in 1853,[3] in the aftermath of the March revolution of 1848. At the time, the Slovene political leaders formulated demands for the unification into one administrative unit of Austrian crown-lands inhabited by Slovene-speaking people. The crown-lands of Carniola, Carinthia, Styria and the Littoral were the recognized historical units to which the population owed their basic loyalties and identities. Of them, only Carniola with the capital of Ljubljana was predominantly Slovene. The contiguous territories of Slovene inhabitants included Southern Carinthia, Southern Styria and adjacent lands of the Littoral with Gorizia and the surroundings of Trieste. The 1866 plebiscite left the Slovene Venetia (Benečija) in Italy; the *Ausgleich* of 1867 established the Austro-Hungarian dual monarchy – and cut off Prekmurje (the Trans-Mura region) from the jurisdiction of Vienna by assigning it to Hungary and Budapest. This meant that at the time when the drive for Slovene unification was under way, significant sectors of territory and their inhabitants, ethnically Slovene, came under the jurisdiction of external authorities. The merging of these territories with the rest of the Slovene lands was only partially achieved in 1918 for the Hungarian part, but not with Slovene Venetia. It has remained in Italy to this day. This administrative division survived to the end of the Austro-Hungarian Empire in 1918, though during the final stages of the First World War the Emperor was willing to accept the unification of Slovene territories in an attempt to save the Empire. The offer came too late: the Empire collapsed in 1918.

For a short period between the end of 1918 and the adoption of the constitution of the new Kingdom of Serbs, Croats and Slovenes in 1921, the Slovene lands were united under a provisional government, expecting to be accepted as a unit in the new Kingdom. The short-lived unification was first affected by the 1920 plebiscite in Carinthia, which left most of Carinthia with Austria; the peace treaty of 1920 gave sizeable Slovene territory to Italy; and finally, the new Yugoslav constitution of 1921 split the Slovene lands within the Kingdom of the Serbs, Croats and Slovenes into two administrative units, further weakening the attempts at integration. The imposed Yugoslav constitution of 1931 finally recognized the Slovene entity by creating Dravska banovina, at a time when the drive for total Yugoslav integration became the guiding principle of 'homogeneous Yugoslavia'. The instruments of a 'nation' and its symbols were hindered by continuous pressures. The constitution of 1931 recognized the 'Yugoslav' language, to replace Slovene, Croat and Serbian; it recognized only 'national' political parties, and restricted the existence and activities of 'provincial' institutions. The symbol of the Slovene flag and particular Slovene celebrations were banned. Nevertheless, the internal consolidation of the Slovene people continued, though the links with Slovenes across international borders weakened, because of the limited interest that the Yugoslav authorities had in 'co-nationals' abroad and because of the restrictive regimes toward ethnic minorities in Italy and Austria.

The Second World War and its aftermath witnessed a combination of raised expectations and crushed hopes. All the conflicting forces of the 1941–5 civil war advocated the postwar establishment of a United Slovenia. The new Yugoslav regime under the leadership of Marshal Tito recognized these demands, and all the constitutions of the new Yugoslavia (1948, 1953, 1963, 1974) retained the Republic of Slovenia as a constituent part of the new Yugoslav Federation. Nevertheless, under the communist system the degree of autonomy was severely restricted. The death of President Tito in 1980 removed from the scene the primary catalyst of Yugoslav unity – and opened the way to disintegration. By the end of the 1980s the demands for autonomy and recognition of the sovereignty of Slovenia were aligned with similar demands by Croatia for a major constitutional revision. Lengthy negotiations within the Federation failed to produce an acceptable compromise. This led to the Slovene plebiscite in December 1990 which, with over 90 per cent approval, agreed to implement the sovereignty of the Republic of Slovenia. By mid-1991 the Federation recognized the *de facto* separation.

Within this historical framework, the Slovenes tried to assert their 'loyalty' and with it their own identity. The century-old discrepancies between the layout of administrative territorial organizations and the ethnic and cultural content have shaped the attitudes of the people and their territorial allegiance.

The noted psychologist Anton Trstenjak[4] analysed the roots of the Slovene character by pointing to (a) the small size of the Slovene lands and their relative proximity to the border: almost everywhere in Slovenia the horizon borders on foreign land; (b) the administrative fragmentation of the crown lands encouraged the growth of regional awareness and loyalties; the persistence of the stereotype is still evident; (c) centuries of common experience with Austria brought the Slovenes close to the attitudes of Austrians, sharing their good and bad characteristics.

The Slovene identity has been associated with a peasant existence up to the First World War. The urbanization of the nineteenth century had only a minor impact on city growth: even the largest city, Ljubljana, only grew from 21,000 inhabitants in 1856 to 41,000 in 1910. The urban population for a long time retained its loyalty to Austria and the Slovene bourgeois population in the cities was in a minority.

The drive for a United Slovenia in the middle of the nineteenth century coincided with the fundamental transformation of the communication system and the economic evolution towards a market economy. The railway connecting Vienna with the port of Trieste was completed in 1857; it provides external evidence of the new system: no more localized and self-contained economic, social and cultural patches but the beginning of a market economy that reached beyond the immediate home territory, and subsequent migration to urban centres, first of all within a one-day journey from 'home', but later to distant centres, even abroad. The once-stable mosaic of local identities began to accept a higher layer of awareness, something between the distant and enormous Austrian Empire and the homely local area: the territory of an emerging ethnic and language region of Slovenes. The period of the 1870s was the time of such ethnic and cultural revival; it also coincided with an enormous expansion of publications and an intensification of the education system with the virtual elimination of illiteracy. The political and literary movement to join this newly awakened Slovene entity to other Slavic 'brothers' in the South, in the framework of the Illyrian movement, did not succeed.

The term 'Slovenia' from that time has referred to two significantly different entities: Slovenia as the territory of Slovene-speaking people, and Slovenia as an administrative entity within Austria and later Yugoslavia. Even the post-Second World War administrative boundaries of Slovenia within Yugoslavia did not coincide with the Slovene lands, or with territory with a Slovene majority or a significant Slovene presence.

This identification and self-awareness[5] is the end result of a long process of conflicting tendencies: retention of local and regional ties on one hand, and interconnectedness beyond the local and regional bound on the other. The interplay of economic links, the formation of a 'national language', the creation of institutions and organizations that went beyond regional boundaries, were in part responsible for the 'discovery' of the larger Slovene national frame; it was contrasted with a striving to preserve the locally and regionally bound identity, organization and specificity. The vast majority of Slovenes in the latter part of the nineteenth century were more inclined to support a regional and fragmented loyalty than risk yielding some of their specificities for the unknown broader elements. This conflict is most evident in the Carinthian plebiscite of 1920, where a large proportion of Slovene-speaking people in Carinthia voted for the 'known' Carinthia and Austria rather than for the 'unknown' Slovenia and Yugoslavia. Furthermore, during the period of mass migration overseas between 1880 and 1920, the predominant flow came from the province of Carniola (Kranjska); numerous organizations in the United States, including the National Federation of Mutual Assistance Societies, accepted the title of KSKJ (*Kranjska Slovenska Katoliška Jednota*), giving provincial identification the highest prominence. It was common for Slovene immigrants from the area to identify themselves as *Kranjci* (Carniolians) or Austrians, and not as Slovenes. The process of national self-awareness was only at the initial stages.

'Slovenia' as a recognized entity did not exist on the map of Europe, nor did it exist in the administrative vocabulary of geography. 'Slovenia' was an aspiration rather than a recognizable entity.

In the mid-nineteenth century – and for the Slovene case afterwards – literary figures played a very prominent role not only as disseminators but also as creators of basic instruments of consolidation and, more recently, of change. Although the roots of transition from local dialects to a 'national language' go back to the beginning of Reformation, when the major written works in the Slovene language, with strong dialect overtones, were published, it was the poetry of France Prešeren that affirmed the qualification of Slovene language as a capable structural frame for artistic expression, superseding the efforts of the Illyrian movement for a merger of Slovene and Croatian languages into a new idiom. The awareness of a unified language occurred at the time when the major innovation, the railway from Vienna to Trieste, linked the dispersed regional entities, and further expansion of the network fostered the beginning of the economic integration of larger-than-regional entities. Increased travel and interaction, combined with a major drive for cultural uplift, began to link people and regions which for centuries had lived separately, even though the external appearance of unification under one political power, the Austrian Empire, had given the false impression of integration. Administrative divisions began to be questioned, civil as well as Church territorial organizations. The persistence of an administrative division of the Catholic dioceses, which reflected the boundaries of the crown provinces, survived to the post-Second World War period and was a major factor in delaying full integration.

The stress on regional awareness was encouraged by the political regime of the Empire, whether intentionally or not; this is the subject of considerable controversy. The administrative division of Austria placed the struggle for ethnic recognition at the administrative centres of the respective provinces, in Klagenfurt, Graz, Trieste and Ljubljana, of which only Ljubljana served as the centre of the mostly Slovene hinterland. In other provinces Slovenes were a minority, aiming to secure some minority rights and protection in a fragmented world of local representations. The system of elections by Curias or classes further aggravated the Slovene struggle for ethnic and later national affirmation. To a significant extent, the Slovenes considered the administrative framework to be the most rigid 'objective' obstacle to gaining equal treatment for the language and culture of the Slovenes living on contiguous territory but formally split among provincial entities.

The position of geography and geographers in relation to this framework of external events can be outlined along two parallel lines: the role of geography and the assessment of spatial constraints in the process of space-bound identities, and the role of geographers in formulating, discovering and emphasizing them.

The spatial fragmentation of the Slovene area, especially the physical barriers of mountains, accounts for the separate life of Carinthia, Styria, Carniola and the Littoral. The internal communication within each region, though not intense, was nevertheless able to respond to the demands of local production and consumption centres; it was able to take advantage of occasional outsiders, such as merchants, missionaries and performers, and not remain totally isolated. But such contacts engaged only a small proportion of the population and the spillover effect to the countryside was minimal. The economic restraint of self-sufficiency supported the 'identity' of the region and its specificity. The language of communication was the local dialect, and this sufficed for everyday needs. On the rare occasions when external communication was necessary, German or Latin was used by those few who were able to use it.

The geography of spatial integration, the economic links as well as social and cultural connections, has made possible the expansion of the region of 'identity' from local to regional, and later from regional to national. It did not succeed in generating a 'Yugoslav' identity or self awareness, in spite of the fact that for external observers the Yugoslav frame was the only recognizable one.

Although 'professional geography' had its beginning in Slovenia after the First World War with the establishment of the Geography Institute and geography programmes at the newly created University of Ljubljana, numerous practitioners played their role as intellectuals in the formation of Slovene awareness and Slovene identity. Like the general population, the people who treated geographical issues were divided: local enthusiasts and defenders, and integralists. They often reflected the broader issues and experiences, their part of political and cultural movements that shaped the Slovene reality. The first modern geography of Slovenia by Anton Melik, published in 1935–6,[6] relegates cultural historical comments to ten pages of the 700-page monograph.

Slovene geographers, influenced by land-bound determinisms and restrained in their free inquiry, concentrated more on devising the most 'accurate' scheme of Slovene regions, based primarily on physical land attributes, and in general refrained from analyses of human cultural behavioural attributes. Even the most recent studies by Gams[7] stress physical layout as the dominant geographical constant for Slovene nationhood. The intense interregional migration that was an integral part of the history of the Balkans, and described by Cvijić[8] as 'metanastasic movements', reached only the eastern fringe of the Slovene territory and had a negligible impact on the Slovene–Croatian border. Cvijić himself recognized the peculiarity of the Slovenes and in general excluded them from his analyses.

A major departure from this sociocultural interpretation is provided by Sperans (Edvard Kardelj) in his study of '*Razvoj slovenskega narodnega vprašanja*', published in 1939.[9] Kardelj frames the Slovene search for affirmation through the Marxist dialectic interpretation of a historical class struggle of 'progressive' forces against the strait-jacket of bourgeois exploitive capitalism. His interpretation became the 'bible' for postwar historians and geographers. The history of the Slovene nation was for him evidence of the unavoidable historical process which would eventually lead to the creation of a socialist, internationally integrated Slovene nation, part and parcel of the new world order of communism. This viewpoint had a major impact on postwar inquirers, more concerned with aligning their view with the doctrinaire blueprint than with a more thorough investigation of the processes themselves. For Kardelj, the importance of class consciousness surpassed any territorial attachment.

The concept of Slovenia as a cultural and social entity has been enlarged in contemporary analyses to embrace the 'Third Slovenia', or the Slovene diaspora of settlers, emigrants and their descendants abroad, especially their clustered communities in Europe, in the Americas and in Australia. Such a united 'Slovene Space' would consist of three components: (a) the Republic of Slovenia; (b) the adjacent Slovene ethnic areas in Austria, Hungary and Italy; and (c) the Slovene diaspora throughout the world. This revised concept of the 'unity of Slovene cultural space' transcends the political and administrative framework and provides evidence of a shift from a territorially defined entity to a functionally linked social complex, where participation is more significant than territorial contiguity.[10] The consideration of this interpretation, however, remained more a political and cultural aspiration than a realistic enforceable entity. The 1991 legislation adopted by the Slovene parliament relating to citizenship, private ownership and participation in the cultural, social and political life of the new entity does not pursue this aim. In the 1970s and 1980s, the dictatorial regime of Tito's Yugoslavia and its aftermath was blamed as an obstacle to the integration of the three components of Slovene cultural space.

In summary, the network and interaction scheme of social and territorial behaviour derived from short-distance interaction produced a mosaic-type pattern of small entities; they existed for centuries and were only occasionally disturbed by external events. In the nineteenth century this system slowly gave way to a larger interactive system, based on the beginning of a market economy, tied to the primary node of Ljubljana, but also linked to the external administrative and economic nodes of Celovec (Klagenfurt), Graz, Trieste and Zagreb. However, these did not serve as catalysts for a Slovene cultural affirmation. In these external centres the Slovene identity was in a minority position, engaged in the struggle for recognition of use of the language, the creation of Slovene or bilingual schools, the rights of community organizations without undue interference from the dominant organization: Italian in Trieste, German in Klagenfurt and Graz, Croat in Zagreb.

Recent events are therefore an external manifestation of ethnic self-awareness. The three levels of space-related identity – local, regional and national – persist to the present day with various degrees of intensity in the minds and attitudes of individual participants: the intensity of association varies from individual to individual, conditioned by life experiences and cultural and social aspirations. The fourth level, the Yugoslav and/or the international identification, did not make major inroads among the Slovenes. It is generally confined to recent immigrants from other Yugoslav republics and is as such present in the Slovene lands.

The long path and its concluding conditions reflect the profound roots of space-related identities and their temporal transformations.

NOTES

1 An extensive review of this topic was first presented by the author at the AAASS conference in New York in 1973, and published as 'Slovene identity in contemporary Europe'. In *Papers in Slovene Studies, 1975*, ed. Rado L. Lencek, New York: Society for Slovene Studies, pp. 1–26.

For a recent comprehensive analysis of space-related identity (place identity), see Peter Weichhart, *Raubezogene Identität. Bausteine zur einer Theorie räumlich-sozialer Kognition und Identifikation*, Erdkundliches Wissens, Heft 102. Stuttgart: Franz Steiner Verlag, 1990.

2 Historical references are derived from Bogo Grafenauer et al., *Zgodovina Slovencev.* Ljubljana: Cankarjeva založba, 1979. For a Marxist interpretion of Slovene history, see Sperans (i.e. Edvard Kardelj), *Razvoj slovenskega narodnostnega vprašanja*, Ljubljana: 1939. Subsequent editions (1957, 1970) include lengthy new introduction by Kardelj, at the time the principal theoretician ideologue of Yugoslav Communism.

3 Peter Kozler, *Zemljovid Slovenske dežele in pokrajin*, 1853. The map was confiscated: the Slovene province did not exist as legitimate administrative unit within the Austrian Empire. The map was later reissued.

4 Anton Trstenjak, '*Vplivi kulturnozgodovinskih razmer na oblikovanje značaja Slovencev*' (The impact of cultural-historical conditions on the formation of Slovene character), *Traditiones* (Ljubljana), 19 (1990), 133–44.

5 Weichhart, *Raubezogene Identität.*

6 Anton Melik, *Slovenija*, vols 1 and 2. Ljubljana: Slovenska Matica, 1935–6.

7 Ivan Gams, 'The Republic of Slovenia – geographical constants of the new Central-European state'. *GeoJournal*, 24(4), (August 1991), 331–40. Also Ivan Gams, 'Slovenija na stičišču srednje, južne in jugovzhodne Evrope' (Slovenia at the junction of Central, Southern and South-eastern Europe), *Traditiones* (Ljubljana), 19 (1990), 6–16.

8 Jovan Cvijić, *La Péninsule Balkanique. Géographie humaine.* Paris: A. Colin, 1918.

9 Kardelj, *Razvoj slovenskega* .

10 The concept emerged in mid-1980. See J. Velikonja. '*Kje domovina, si?*' (Where are you, my homeland?), *Draga * 84*, Trieste (1985), 11–22.

PART III

Newly Emerging Identities

17

Coming to Terms with Australia

J. M. Powell

There are, no doubt, subtle distinctions between 'nationalism' and 'national identity', but it is often more useful to see them as running together. Although it may be argued that the first term always connotes emotion-charged commitments to a peculiar extent while the second may describe a quite dispassionate or intellectual sense of rapport, the interrelationships remain more significant than the differences. Established communities may leave the drawing of fine lines on such matters to their crowding ranks of indigent philosophers, but some younger societies have been drawn towards rather self-conscious quests for identity – and in every case the quest confers at once an improved sense of belonging and opportunities for displays of nationalistic fervour. That has certainly been true of the Australian experience during the past two decades, but the earlier transactions of this immigrant nation with a vast and frequently hostile environment were re-evaluated during the extravagant approach to the bicentennial celebrations in 1988. In the process, the historiography of various styles of geographical thought attracted increasing attention.

Geographical thought in Australia has been built on vernacular and scientific foundations, and in the various forms of practical landscape authorship contributed by pioneer settlers, key bureaucrats and leading government agencies. Furthermore, it has reflected and bolstered imperialistic and nationalistic attitudes towards environmental appraisal and resource management; those attitudes have often resulted in exploitative or otherwise inappropriate modes of development, transforming large areas and severely reducing future options. Seen in that light, geographical thought and geographical action have been comprehensively intertwined since the beginning of European settlement in 1788. A major premise of the following discussion is that an improved understanding of that legacy within Australia will promote more prudent national responses to conservation and development themes, while serving to challenge recently contrived separations of the secondary and tertiary sectors of the education profession. In addition, it may help to establish a case for ventures into the production of more 'accessible' forms of geographical literature, thereby assisting towards a mutually sustaining reciprocity between academic geography and its host society.[1]

GEOGRAPHY IN COLONIAL AUSTRALIA, 1788–1901

From the inception of European settlement, each generation of Australian colonists was confronted with the fact that imperial science and scientific imperialism were but two sides of the one coin. Of course their ties to Europe – specifically to Britain – were very strong. That helps to explain why the need for an emancipatory local science was scarcely given high priority by the majority and why, indeed, the client situation was neither obvious nor uniformly unpalatable. From today's perspective it can be said that the 'metropolitan' approach produced a set of ways of doing science based on theories and conventions of discourse developed in Western Europe, and a social structure dominated by learned societies, small groups of patrons, and key innovators and co-ordinators: in metropolitan perceptions, colonial science was 'low science' – derivative, serviceable, fact-gathering; necessarily following the cues of the great theoreticians at Home. The old terminology is provocative, but a trifle misleading. MacLeod (1982) argues that in Britain, imperial science was seen as 'true' science – imbued with the sense of mission, service, assumptions of promised dominion. Its hegemony in the colonies was facilitated by similar local assumptions cut down to serve the functional needs of 'Britannia in another world': for example, the application of science to more profitable investment in the export-orientated primary industries. 'Imperial' to many dedicated colonial scientists implied a spirit of fruitful, family-style co-operation without any notion of centre–periphery hierarchies, but it is relatively easy to establish the definite association of science with the commercial, humanitarian and ideological motivations for empire-building.

If the development of new knowledge is indistinguishable from the growth of imperial hegemony, at another critically significant level a large stock of basic environmental data was compiled without recourse to formal scientific methods. Even so, in the eighteenth and nineteenth centuries the entire Pacific region presented a fascinating challenge for studies of the relationship between society and nature, and the most celebrated events in the discovery and exploration of Australia were closely assisted by scientific patrons in Britain and Europe. The renowned botanists Sir Joseph Banks and Sir William J. Hooker were the most prominent of these individuals in the early years – understandably, since the simplistic Linnaean system of taxonomy was peculiarly well suited to the rapid scientific investigation of newly discovered territories, and trained botanists were usually assigned to the official exploration parties. The great expeditions produced some of the best of our early literature, but in the scientific realm they were essentially filling in the grosser deficiencies on the maps – and it is still less well known that detailed micro-studies around the initial coastal settlement were conducted by a host of ordinary settlers, convict and free.

And yet, from the first landfall of officially endorsed settlement, independent empirical testing made lasting contributions to environmental appraisal. In official circles remote armchair theorists certainly conjured elaborate schemes to manage settlement expansion, and it is also true that the best and most interesting of these reflected an understanding of some rudimentary principles of regional planning. Until the middle of the nineteenth century, however – when the various colonies were granted more responsibility for their own affairs – the changing geography of settlement and land use envisaged at the metropolitan level was based on a virtually unrelieved ignorance of environmental conditions in the Australian periphery. One result was the underestimation of the long term suitability of livestock farming and the concentration of government interest on small enclaves of intensive settlement. The big ranchers (or 'squatters') were not permitted legal security of tenure, but with a fine disregard for imperial schemata they swept across the south-eastern crescent, and their highly mobile and opportunistic mode of settlement resulted in the first great regional appraisals of land capability. Squatting modes of settlement had been effectively discounted by myopic imperial theorists, and these bold colonial appraisals dramatically improved contemporary understanding of the geography of Australia at every level of officialdom. Furthermore, in each colony the compilation of practical administrative maps and regular statistical returns legitimated the broad regionalizations contributed by the squatters' efforts and incorporated them into subsequent planning policies.

Quite as much could be claimed for the autonomous contributions of small-scale farmers in later years. In the parliamentary rhetoric of the nineteenth century, the complicated machinery of land legislation was expected to create 'little Englands' in Australia peopled with an 'industrious yeomanry'. Scientific and technological research in agriculture, and the need for markets, railways and port development, received insufficient attention from the governments of the day, and so the pioneer farmers had to seek their own solutions to achieve suitable working units. In the absence of imaginative and committed government support, various techniques developed as spontaneous folk contributions to the emerging cultural geography of the frontier regions. Obliged to survive on opportunity's crumbs, families of limited means devised common-sense strategies of micro-scale trial and error, intense local and regional mobility, and intricate webs of inter-family supports and intra-family linkages – all of which remain characteristic of man–land relationships in rural Australia. Where the strategies succeeded there was a veritable honeycombing of the 'big man's frontier' by fascinating nursery colonies of a kind of neo-peasant proprietary. Related advance/retreat sequences established the undeniable existence of 'marginal' regions in every colony – Donald Meinig's (1962) evocative analysis of the South Australians' great 'folk experiment' with the environment is still compelling reading – and underlined the centrality of empirical testing in our uniquely geographical history. In the hearts and minds of the Australian public, settlement expansion and landscape authorship were inseparable and, after the accumulation of such impressive evidence, the lessons of trial-and-error came to be preferred over more 'bookish' interpretations.

For most Australians the formal contexts of geographical learning were clearly subordinate to the informal modes until the last quarter of the nineteenth century. What might be called the informal sector of Australian 'geographical' transactions maintained its vitality despite – or because of – a damaging bifurcation which was not corrected by the introduction of universal schooling in the 1870s. Clear evidence of their community's intimate involvement in profound geographical change enveloped the colonial children, but in the schools they were expected to content themselves with hand-me-downs and bizarre gazetteer 'geographies' which seldom acknowledged the bare existence of Australia. They lived their geography away from school – and little wonder:

> And we learnt the world in scraps from some ancient dingy maps
> Long discarded by the public-schools in town;
> And as nearly every book dated back to Captain Cook
> Our geography was somewhat upside down.
>
> (Lawson, 1924)

In the interim, and in severe contrast, some gifted scientists and administrators addressed a general concern for the elucidation of Australia's primary geographical features in their own private and professional roles. Most of these individuals were senior bureaucrats with heavy responsibilities for land survey and disposal and the production of applied environmental data. A good deal of their work was expressed tangibly in contemporary landscapes and some of that survives – in the location and internal morphology of hundreds of small and large towns; in the road and rail networks; in the public reserves, minute and gigantic. In addition, their publications have contributed massively, directly and indirectly, to recorded geographical thought in Australia, and did something to balance the cultural imperialism which handicapped local education. Thomas Mitchell (1792–1855), George Goyder (1826–98), Ferdinand von Mueller (1825–96) and Henry Chamberlain Russell (1836–1907) provide obvious examples (Powell, 1980a, 1981a, 1981b, 1983a; McKay and Powell, 1985).

Geographical agency takes so many forms. So does environmental appraisal. If it were possible to expand here on that admirable nineteenth-century archetype, the earth-scientist/cleric, it would be shown that William Branwhite Clarke (1798–1878), William Woolls (1814–93) and Julian E. Tenison Woods (1832–89), all of them 'Reverend', occupy honourable niches in the cultural history of geography in Australia. Leading interpreters of environmental patterns and processes, they were also alert to the dramatic impacts of pioneer settlement on the landscape. Woods was, in addition, a noted promoter of elementary school geography, but in that respect a greater contribution was made by William Wilkins (1827–92) towards a more secure sense of place. In remodelling the New South Wales curricula, Wilkins insisted on the production of locally relevant geographical material to balance the importations.

From the 1870s a strong intercolonial, mildly nationalistic movement pursued the ideal of scientific co-operation. Colonial governments showed increasing interest in applied science, and the scientists themselves gradually shed the trappings of vice-regal patronage as they developed more independent forms of professionalization. The establishment of Australia's first major geographical societies was intimately related to these developments, and it mirrored the odd mixture of national and imperial sentiments which permeated so many Australian institutions. Days before Queensland's very popular 'annexation' of south-eastern New Guinea, a Federal Geographical Society of Australasia was mooted in Sydney to collect data on Australasia, Antarctica, New Guinea and Polynesia embracing the 'commercial, political and natural sciences'. Germany had designs on New Guinea and was said to be outstripping Britain and the British colonies in the accumulation of vital intelligence on Australia; and in Sydney itself, former members of the defunct Section F of the Royal Society of New South Wales (geography and ethnology) called for a more independent and thoroughly Australian framework. Branches of the Royal Geographical Society of Australasia were founded in New South Wales, Victoria, Queensland and South Australia in 1883–4.

By the mid-1880s a majority of the population was Australian-born and the inhibiting sense of exile had diminished. Popular writers, balladists, artists and poets became teachers in an accelerated learning process. So many of Australia's immigrants had been (in some cases literally) committed to the place before they had found ways to understand and appreciate it. With the emergence of a distinctively Australian school of painting, and enthusiastic responses to Australian verse and literature, the incubus of European 'standards' was gradually exposed and there emerged a greater sensitivity to the qualities of Australian scenery and to Australian achievements and ways of living. The process was advanced though still far from complete by the early years of the twentieth century, but it was soon to be challenged by the aggressive demands of the wider world. Taken together, however, the literary and artistic evocations strengthened the aesthetic appreciation of place, and assisted towards a promotion of more positive geosophical transactions.

Popular interest in science also increased during the 1880s, but geography's attractive universality captured the ambience of the times more than most – it blended the old and new, the amateur and professional (Powell, 1988a). The keenest promoters of geographical education were simultaneously the founders of other scientific societies and prominent in public affairs. For example, the president of the geographical section at the inaugural Congress of the Australasian Association for the Advancement of Science was the re-doubtable John Forrest (1847–1918), explorer and leading politician. The meeting seemed doubly portentous in so far as Forrest's address underlined at once the continuing centrality of geographical thought in the Australian milieu and its inescapable appeal for the unity of specialized and informal modes of enquiry, pure and applied: 'the term geography covers a very wide area, and embraces or is allied to so many questions of great importance to us, and in which *colonial history and colonial enterprise are connected*' (Forrest, 1888, p. 359; my emphasis).

Agreeable sentiments, echoing some contemporary British views; but somehow geography, despite its experiential base in wider society, had only managed to consolidate that position at the most elementary levels in Australia's schools.

At the higher regions of education there was far less evidence of support in influential Australian circles, whereas academic geography in Britain was attracting powerful champions in and beyond the Royal Geographical Society. Furthermore, although the federation movement as a whole spawned many disappointingly short-lived enterprises, it remains surprising that only the Queensland and South Australian branches of the Royal Geographical Society of Australasia survived in good heart into the present era. The Australian groups were not off-shoots of the Royal Geographical Society as is often supposed and, with the exception of exploration where the imperial institution's lead was happily followed, their chosen fields of activity were far more restricted. Is it entirely fanciful to suggest, in addition, that their very independence in a remote corner of the world scarcely promised to advance the subject in conservative academia?

NATIONALISM AND IMPERIALISM, 1900–1960

If school geography was boosted by the nationalistic involvement of leading bureaucrats, at the research level it was difficult to break the grip of imperialistic assumptions, preferences and pre-emptions. Using the examples of the interwar development of the Imperial Geophysical Experimental Survey and the early activities of the Great Barrier Reef Committee, Butcher (1984) and Hill (1984) have shown that enterprising Australian scientists fought hard to establish independent lines of enquiry. The scientists' mixed success was achieved despite a formidable barrage of patronage in the critical area of research funding and the intricate system of imperial scholarships – notably the Rhodes and the 1851 Science Exhibition – which effectively sealed the imperial bond. And the ideal of imperial cultural and economic 'partnership' was honoured at the highest levels of Australian politics: Prime Minister S. M. Bruce (1883–1967), for example, was very much in favour of a new Council for Scientific and Industrial Research (the forerunner of the internationally renowned CSIRO) acting mainly as an efficient conduit for the receipt and diffusion of advanced British work. Thus in science as in so many other areas, influential Australians had accepted a 'client' status.

The CSIR's dynamic leader, David Rivett (1885–1961), shared the imperial vision, but it clashed with his aspirations for the new national body and he insisted on the right and duty of local scientists to take up their own fundamental work (Schedvin, 1987). A peculiarly utilitarian science emerged, concentrating on livestock vaccines, trace element deficiencies, erosion problems and the like. The CSIR led the way in aggressive biological control campaigns to combat locusts, the rabbit menace, sheep blowfly and pasture weeds, and it shared in the spectacular eradication of prickly pear. Entire regions were defined, transformed, saved, by such practical scientific enterprises. The early lineage of this great scientific institution suggests both a link with the colonial past and the advent of a new era: that is, we can trace its origins to the earlier efforts of pioneering public servants and the underestimated endeavours of government departments engaged in the management of land, forests and water resources during the nineteenth century. Landscape authorship and public service remained inseparable. In a currency minted by national endeavour they were two sides of the same coin.

The relationships between these developments and theory and practice in Australian education still await close investigation. In the case of geography, however, the contributions of a few pioneers became relatively well known in the 1980s.

The most obvious example is provided by analyses of the work of T. Griffith Taylor (1880–1963). Taylor's British origins and early career virtually guaranteed him a place in the inner sanctum of Imperial Science. All the ingredients were there: a graduate assistantship under one of the Empire's most revered geologists, Sydney's T.W. Edgeworth David; an 1851 Science Exhibition to Cambridge University; leadership of the western geological party on Scott's ill-fated *Terra Nova* expedition to the Antarctic; generously defined employment as physiographer to the young Commonwealth Weather Service; a return to Sydney University as foundation head of Australia's first geography department. From the latter base he argued that geography had the potential to offer the Australian student a viable reconstruction of the fragmented curriculum and excellent tools for citizenship training. The well-known description of geography as a 'correlative science', arguably Taylor's most frequently cited claim, is easily traced to British and Australian precedents. It declares a vital relationship with Pestalozzian–Herbartian pedagogical methods of which the author was apparently blissfully unaware (Marshall, 1980), and undoubtedly boosted the study of geography in the schools.

In promoting geography as the spearhead of a 'new' education in the 1920s, Taylor deliberately chose to make a stand on *national* issues and 'nation-planning'. Above all, he challenged the dangerous naivety of nationalists and imperialists who had touted ambitious schemes for railway expansion through the centre of the continent and the intensive exploitation of the tropical north, and had predicted future population capacities of 100–500 million. In explaining the value of cautious resource inventories acknowledging environmental limitations, he offered an alternative prediction of 19–20 million for the turn of the century (see figure 17.1). But Australians were not prepared to swallow Taylor's home truths, and his pessimism and especially his so-called 'environmental determinism' – a term he came to embrace in later years – were judged to be 'negative' and 'unpatriotic'. Indeed, his early text on Australia was banned by Western Australia's education authorities because he dared to describe large areas of that state as 'arid' and 'useless' (Powell, 1979, 1988b).

While it was not the only sharpened focus, the Australian emphasis was certainly strengthened in secondary and tertiary geography by Taylor's unflinching campaigns; in the process, however, the subject acquired a reputation as the haven of non-conforming 'mavericks'. Worster (1979) reports a contemporary parallel in the rise of ecology studies on American campuses, where nationalist–capitalist opposition employed the label 'subversive'. Australian society was numerically small, isolated and in many ways disappointingly parochial, and Taylor probably misread its basic insecurities. But in academia

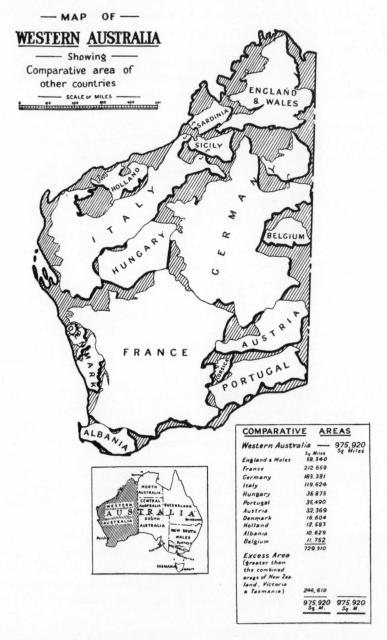

Figure 17.1   Specious spatial reasoning: Western Australia

(*Source*: from T. Griffith Taylor, *The Case of the People of Western Australia* (Perth, 1934), unsuccessfully arguing for secession from the Australian Commonwealth.)

Griffith Taylor *was* Australian geography: he had no senior colleagues, and so the personality factor is crucial to an appreciation of this formative period.

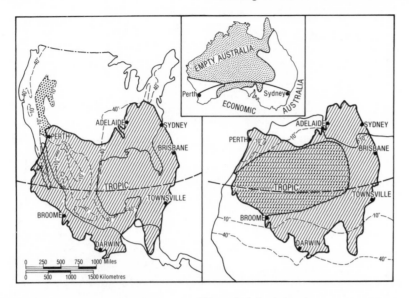

Figure 17.2 Taylor's comments on the 'Gigantic Inheritance'

(*Source*: adapted from T. Griffith Taylor, *Australia* (London, 1940) and his *Environment, Race and Migration* (London, 1937).)

For instance, he never occupied a full chair in Australia, and without question that is a key to his impatient and often truculent demeanour. Yet in addition, although it was a hard-felt injustice, it has often been related to his reluctance to cultivate powerful friends for his infant subject. Unfortunately, the stirrings of nationalistic science reported already offered conspicuously little support: the movement achieved some notoriety in the late 1920s and in the 1930s, but that was too late for Taylor. In 1928 he resigned his Sydney appointment and departed for Chicago, leaving an ambiguous legacy to his Australian successors.

There was no doubt a post-Taylorean 'twilight' for Australian geography, but it can be ascribed too easily to Taylor's disruptive tendencies: he left on the very eve of a lengthy economic depression after all, and in any case his successor, the slightly underrated Scot J. Macdonald Holmes (1896–1966), did produce some valuable practical research in urban and regional planning, conservation and political geography, which was elaborated (and sometimes initiated) by his senior students. Holmes's direct and highly personalized impact on the geography curricula of New South Wales schools, together with his sustained interests in applied geography, have been acknowledged in Australia. Regrettably, on the international scene these achievements have been diminished by Taylor's giant shadow. Throughout the 1930s Holmes's vigorous involvement in conservation and planning brought Australian geographical studies back to traditional local and regional bases. That is not to say that the taint of controversy was removed. Holmes was a very prominent interpreter of popular movements that demanded the creation of new states and his advice

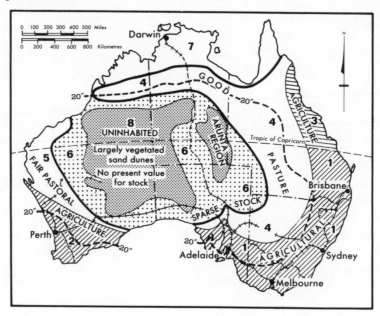

Figure 17.3 Taylor's preferred sequence of settlement

(*Source*: adapted from his controversial summation in the *Sydney Morning Herald*, 28 February 1925. For further discussion on figs 17.1–17.3, see Powell, 1988b.)

was increasingly sought by federal and state governments, particularly on the co-ordination of mapping systems and on the 'regionalization' of administration related to political, environmental and resource management concerns (Powell, 1983b, 1988b).

In the 1930s and 1940s there were some very able workers in regional, historical and economic geography in South Australia, Victoria and New South Wales, and Taylor's interpretations of rural settlement and agricultural expansion (see figure 17.3) were continued in the pure and applied work of the geographer John Andrews (1909–84) and in the writings of a new group of economists and agricultural scientists (for example, Wadham and Wood, 1939). Most importantly, however, in the period under review the demonstration of the application of geographical synthesis to vital national and international issues was quietly and effectively extended by Adelaide's Archibald Grenfell Price (1892–1977).

In the depression years Price was drawn into political affairs and became aligned with anti-socialist factions. Some outcomes were a seat in the federal parliament as a Liberal (that is, conservative) member (1941–3), the establishment of life-long contacts with the imperialist politician Robert Menzies and other leading political figures, and a formal recognition of his involvement in community affairs which began the process leading to his knighthood in 1963. During the same period his political pamphlets on the dangers of inflation and the spread of communism enhanced his standing in the Liberal party, and prefaced the award of a Rockefeller Foundation Fellowship which allowed him to travel extensively in the Northern Hemisphere. *So an Australian geographer returned to the circle.* In 1939 the American Geographical Society published Price's best research monograph, *White Settlers in the Tropics,* which was on the whole an excellent advertisement for geography's synthesizing claims. For much of his long career, Price was regularly consulted on federal–state relationships, Pacific and other international affairs, education and library administration, the management of the Northern Territory, and the writing of geography texts (Powell, 1982).

Holmes's early efforts held the ground at a difficult time and showed the practical utility of geographical approaches to conservation and planning; so he tried, like Taylor and Price, to expose the dangers of mistaken environmental beliefs. Collectively, these three pioneers and a handful of striving colleagues maintained their infant subject's contact with the wider community without, unfortunately, exploiting its older embrace of place-specific art and literature. They built the foundations for an expansion of pure and applied geographical work in the natural sciences, social sciences and humanities; that expansion came with the boom in tertiary education in the 1960s and 1970s.

IMMIGRANTS AND NEGOTIATED CURRICULA, 1960–88

In the early postwar decades, geographical studies were essentially school-based in most states outside New South Wales, but their late introduction at Adelaide and Melbourne Universities did encourage a modest surge in the south-eastern crescent. The regional concept had established a special cachet in wartime planning and in postwar 'reconstruction' programmes; similarly, Taylor returned briefly as a valued consultant to the federal government, and he used the opportunity to make urgent representations on behalf of tertiary geography.

The significance of this interlude is increased by the very fact that it has been comprehensively neglected by most of today's practitioners. Then, as now, geography's image in the full spectrum of national education was largely dependent on what was communicated to thousands of young Australians and what they, in turn, chose to hand on. Until the later 1960s, Australia's broadly based degree structures seemed to answer the extraordinary demands made on the education system by a demographically youthful and rapidly growing immigrant community, but for some subjects they contained the seeds of future strife.

Geography was not the only school subject to be handicapped by the dearth of specialist graduates, but it may have suffered more than most because of its relatively small claim to pastoral care from university departments. History, its synthesizing companion, was particularly well ensconced everywhere. So were English and the major sciences, while sociology and politics were not widely accepted in orthodox school programmes. Something called 'geography' attracted a massive following in the schools, but it was very strange fare. In the 1940s and 1950s it was seldom tended by specialists with a trained basic understanding of the human and physical branches, and regional synthesis, the very anchor of the subject elsewhere, received very little emphasis. In its contents and approaches, the kind of material offered to geography classes was often markedly utilitarian and it occupied an unprotected niche in the curriculum – neither 'arts' nor 'science', but very often pressed into a barren marriage with economics in a dismal 'commercial' neighbourhood. The scale of the problem is well illustrated by the confused Victorian situation: for the most part, school geography was dominated by graduates who had qualified in Melbourne University's unusual 'mongrel major' in geology and a brand of 'economic geography' which occupied a minor position in the giant Commerce Faculty.

When the boom came it drew, inevitably, on immigrants from other English-speaking communities who enjoyed stronger traditions of specialist geographical training: from the United Kingdom, of course, but also from neighbouring New Zealand and to a lesser extent from North America. Given the peculiar bimodal lineage of Australian geography and its somewhat chequered history, there was a clear prospect of an unfortunate dichotomy between native-born geography teachers and those immigrants who chose (or were obliged) to liaise with them: it is difficult to say who suffered the greater cultural shock. The transition was less traumatic where the tertiary input was entrenched and unequivocal (others might say 'relentless and dictatorial'), but although that was not the case in Victoria, some important bridges were built by the subject's foundation professors in that state – at Melbourne (John Andrews, mentioned above) and especially at Monash University (B.L.C. Johnson, 1919–). When the expansion accelerated in the later 1960s and early 1970s, however, a fundamental weakness incident to Australia's 'peripheral' location in the education world emerged with startling clarity. Recruitment for the mushrooming geography departments selected from a narrow band of immigrants flushed with expectations of a bright millennium ushered in by a 'quantitative revolution'. The subsequent sharp decelerations meant that geography's tertiary representation was eccentrically and somewhat precariously placed in terms of its immigrant profile, age structure and intellectual preferences.

That exaggerates, of course. But it is partly true that the recruitment for the other major subject areas was less critically timed and less dependent on our closest cultural relatives in the English-speaking world. Yet it is also vital to stress, once again, the significance of deep-seated cultural change and the daunting profundities of uncertainty in the wider Australian setting. While the optimists celebrated the 'openness' of our national community to outside influences, others warned of its increasing vulnerability. And, in one special sense, endorsement for the pessimistic view was provided by teaching and research developments in geography, wherein the borrowed knife of logical positivism was indiscriminately applied to individual career plans – and to degree programmes and school syllabuses which had long exhibited only the slimmest appreciation of philosophical unity, and were chronically debilitated by native pragmatism (Powell, 1980b).

At the tertiary level that same pragmatism – assisted by (or answering) a measure of tenure-bolstered intransigence and independent philosophical interrogation – led to a promiscuous smorgasbord of course offerings on most campuses. The published expectation announced that students could mix and match with discretion to compile a variety of personalized qualifications. In practice, the institutions invested only the minimum of effort in directing or advising the students on their choices: at its worst, it was a Pontius Pilate approach. While the expansion continued, reliance on the liberal permutations encouraged the (tertiary) 'academics' to pursue their research interests with even greater zeal, for it was taken for granted that there should be a close and very direct link between that research and the specificities of their teaching programme. In geography as in other subjects, the logical short step led to an increasing bifurcation of the secondary and tertiary sectors, and articulate occupants of both areas welcomed their increased independence.

In the 1970s and 1980s it was progressively conceded that the old smorgasbord had encouraged an unhealthy indulgence among staff and students alike: Australian society had reached an age when an adjustment to the national diet was essential. Immigrant academics had been all too ready to ignore the world at their own doorsteps. Undoubtedly, that attitude contributed as much to the estrangement of the two sectors as the temporary infatuation with all things quantitative and any radical move on the part of the teachers' associations. As more of their teaching and research was very deliberately invested in their adopted country, and as their own children moved through the Australian education system, the academics seemed to become more prepared to reconsider their broader professional obligations. And so it was especially unfortunate when the gap suddenly widened, with a move towards professional autonomy in secondary school teaching in the late 1970s. Thankfully, the division is by no means complete, but occasionally each sector is inclined to act in ignorance of the needs, claims and achievements of the other.

In recent years, alerted by the heady return of political-economic perspectives, educators have become more aware of the pervasive influence of unequal exchanges of knowledge production and knowledge transfer within international trading blocs. Seen in this light, higher education in Australia (and, for example, in Canada and New Zealand), may be described as an international transfer agent of British and United States knowledge, directly and indirectly legitimizing and promoting the existing relationships. Notwithstanding the tradition of sanity-saving overseas 'sabbaticals' (which can reinforce professional dependency), the several cultural consequences of the growth of this well-established exchange system now engage urgent attention – underfinanced and minuscule libraries, high student–staff ratios, and a comparative dearth of post-graduate and advanced research schools, indicating the impoverishment of intellectual production in the peripheral societies. (In passing, it may be noted that one of the more useful exercises during a generally hedonistic bicentennial year involved a well-orchestrated legal challenge to the stranglehold established by British publishers over the Australian book trade – a ridiculous anachronism which inflates local prices and nurtures the old imperial habit of 'dumping' excess British production while excluding or discouraging American and other titles.) It has been claimed that cultural elites on the periphery deliberately or unwittingly use the standards of the centre to promote further importations (of ideas, people) from that source, and that certain subjects may be peculiarly at risk in this regard (Powell, 1980b). On this delicate ground, wary academics need to find ways to break down the multiplicity of general principles into proposals for action, and a well-considered involvement with the schools may hold one of the keys.

It is certainly useful and reasonable to examine a simple premise: while status-conscious university personnel became estranged from their local professional contexts by endemic international exchange processes, curriculum reformers and the associated 'teacher autonomy' groups attempted to fill the vacuum by addressing needs, aspirations and experiences in the Australian community. But twentieth-century geography is itself the creature of reform, the child of an earlier 'new education', and to separate ourselves from the radicalism of today is to deny our very roots. Academics and school teachers alike might profit from the realization that geographical thought, whether 'physical', 'human' or 'regional', and whether vernacular, scholarly or intimately related to the landscape authorship of key bureaucrats and public service agencies, has been central to Australia's national experience for 200 years.

For geography's small band of historiographers, the earnest promotion of today's 'new' education has an astonishingly familiar ring. Perhaps the best Australian example is an early statement from the Curriculum Development Centre in Canberra declaring that 'our traditional ways of packaging knowledge into required subjects no longer satisfies society or students' (Curriculum Development Centre, 1980, p. 7). *All* established school subjects, even the sciences, were thenceforth challenged to reorganize in order to address a curriculum based on a Core of Learning and a Core of Knowledge and Experience. The first cited ways of organizing knowledge, dispositions and values, interpersonal and group relationships, learning and thinking techniques, skills or abilities, forms of expression and practical performances. The second was concerned with nine areas: arts and crafts; environmental studies; mathematics; social, cultural and civic studies; health education; science and technology; communication; moral reasoning and values; and work, leisure and lifestyle. Who would seriously deny that the content and approaches of modern geography have been addressing virtually all of these 'core' areas? I suppose what is needed is a better press, not some latter-day Griffith Taylor, to seize this golden opportunity to stake a claim for another kind of 'correlative science'. If that is the way things are likely to go in the schools, then Australia's geographers are further down the track than any of their colleagues in other subject areas. And, even in the late 1980s, the over-blown political emphasis on utilitarian pursuits is to some extent reassuring for many of geography's education professionals. For generations, geography has fostered an interest in conservation, town and regional planning, national and international development, marketing and the like, and its practitioners have done more than most educationists to insist on a *combination* of skills in numeracy, literacy and graphicacy at all levels.

A more obvious illustration emerged in Canberra during the late 1980s, when the federal Labor government served notice that interdisciplinary *Australian* studies should be given clear priority in secondary and tertiary education. School and university geography has probably become more Australian in its choice of content than any other subject area, and it offers a proven potential for combining some of the concerns of the beleaguered denizens of the history, literature and environmental science camps to produce the kinds of comprehensible, practical knowledge now judged appropriate for a young nation with recent immigrant origins. The national heritage of the subject establishes the point and declares a pre-emptive right: geographers already occupy the ground and their substantial improvements were registered long ago. I have argued elsewhere that that is one of the things to be placed boldly on the table, in these days of negotiated curricula.

The sins of tertiary teachers will always be visited on secondary education, no matter how well the schools use their new autonomy. Changes in tertiary programmes still constitute a powerful force to transform the education system as a whole – and it is dangerous to allow it to be portrayed as anything less than a *whole*. The universities are beginning to accept the need for internal reform. The geographers' practical contribution to that end can draw on a national heritage incorporating the experience of a vigorous federation of interests in which the bonding elements are broad educational objectives, rather than tightly defined team-research missions or a guarded public loyalty to some unexamined or unexaminable 'disciplinary' purity. Our collective value as educators is coming under intense scrutiny, and properly so. For comfort, security and plain efficiency, we would be well advised to familiarize ourselves with geography's unimpeachable claims to a distinctive Australian identity.

While no attempt has been made to disguise the nationalistic inclinations in my own approach to the challenges facing our profession, a few strong misgivings must be admitted. Forty years ago, one of Australia's greatest scholars, the geographer Oskar Spate, remarked that our duties to 'core' sectors of the subject are always complemented by the 'accidents of personal inclination', by reactions to foundation works, and by the 'element of opportunism, or adjustment to a geographical environment accidentally given' (Spate, 1953, in Spate, 1966, p. 101). 'Ingenuity' and 'adaptability' might be the preferred terms when we describe modern Australian work in geography, but 'opportunism' is not entirely inappropriate. If geographers like other academics are, in a sense, formed by their responses to their professional contexts, their presentation of their subject must be deeply influenced by those responses. In the increasingly vulnerable Australian context the 1980s brought thrusting pressures of social and environmental change, and spiralling demands from interventionist governments. Geographical teaching and research have been profoundly transformed, and in the universities and colleges autonomous geography departments are fast becoming the exceptions. The 'geography and –' title is proliferating at a rate of knots: 'geography and planning'; '– and oceanography'; '– and environmental science (or environmental studies)'; or even 'geographical sciences' (that is, planning and surveying with geography). The implications for the displacement of staff and student allegiances, and the wide potential for a national atomization of the subject, have not been widely canvassed. Visitors to the Sydney International Geographical Union (IGU) Congress in August 1988 may have caught an historic glimpse of a subject in the throes of revolutionary structural change. How ironic it would be if geographical studies, demonstrably critical elements in the protean 'Australian experience', became buried under the weight of quick pragmatic decisions on every campus.

There is still time to secure the future by learning from the past. We need to improve our contacts with the paying public, but regrettably, even the best of our standard geographical literature does not attract the wide local readership commanded by many of our highly talented historians. Before the estrangement I have elected to ascribe to the excesses of logical positivism, we used to recognize them as our synthesizing cousins, and our reluctance to follow the historians into the public arena has been costly. But my comments have not made it clear that the problem is one of orientation, and that it is not new. For example, the biologist Francis Ratcliffe's *Flying Fox and Drifting Sand* (1938) is considered a classic of its type in Australian literature, and certainly it consigns Holmes's communications on soil erosion to near-oblivion. In the modern era, too, Australia's geographers were slow to seize the reins of the 'environmentalism' band-wagon, and it was left to others to supply an eager public with significant and attractive reading material (for example, Marshall, 1966, and especially Rolls, 1969, 1981).

In 1989 Australia's federal government announced its support for a decade of 'Landcare' to focus nationally co-ordinated programmes to deal with the alarming incidence of land degradation in every state. These programmes were soon accompanied by lively debates on 'sustainable development' which brought together galaxies of experts from the public service, academia, the increasingly influential voluntary conservation bodies, and farmers and graziers' associations. National and regional media appeared to underline the involvement of every Australian in a process of landscape authorship. In those senses, perhaps there was an emphatic re-insertion of geography into interpretations of our national experience. Griffith Taylor's sombre pronouncements were rediscovered and given almost reverential treatment. Regionalism, in its various guises, received a new lease in the Landcare programmes. For example, there was a widespread if occasionally tentative adoption of the river basin as a co-ordinating unit for government-sponsored research, administration and environmental education. At the same time, media coverage of the African famines, the Gulf War, devastating natural catastrophes in Bangladesh and elsewhere, disturbing predictions on the ozone layer and the greenhouse effect, ruthless invasions of Australia's traditional, well-cultivated markets by subsidized agricultural produce from the EEC and the United States, and the startling upheavals in Eastern Europe and the USSR, emphasized the critical value of a basic geographical literacy for global as well as national citizenship. For Australia's academic geographers the message should have been clear enough: there is no need to assert our subject's claims to relevance in the Australian cultural scene – except to ourselves.

And if I say that our primary mission is to serve our host society, that is not to deny our responsibilities to the subject's global constituency: on the contrary, we might make larger contributions in that field as and when we put our own house in order. The subject did not prosper in Australia when the bulk of its practitioners paid obeisance to Geography the great trans-national corporation. The better field to be tended is a young society still working towards a coming-to-terms with a difficult environment. That drawn-out engagement has been crammed with patently geographical stories awaiting sensitive narration. And that society, like any other, contains our subject's older roots: they will yield sure harvests if we care to provide a more expert husbandry.

NOTE

1 This paper builds upon several earlier contributions, notably a Presidential Address to the Institute of Australian Geographers (Powell, 1986); permission has been granted by the editors of *Australian Geographical Studies* to make further use of the author's material originally published in that journal.

REFERENCES

Butcher, W. 1984: 'Science and the imperial vision: the imperial geophysical experimental survey, 1928–1930. *Historical Records of Australian Science*, 6, 31–43.
Curriculum Development Centre 1980: *Core Curriculum for Australian Schools*. Canberra: CDC.
Forrest, J. 1988: President's address, Section E (Geography). In A. Liversidge and R. Etheridge (eds). *Report of the First Meeting of the Australasian Association for the Advancement of Science*, Sydney, AAAS, pp. 352–9.
Hill, D. 1984: 'The Great Barrief Reef Committee, 1922–1982: the first thirty years'. *Historical Records of Australian Science*, 6, 1–18.
Lawson, H. 1924: 'The old bark school'. In H. Lawson, *Humorous Verses*, Sydney: Angus & Robertson, pp. 1–3.
McKay, J. and Powell, J. 1985: 'Geography in Australian society. Part One: The heritage'. *Journal of Geography*, 84, 98–104.
MacLeod, R. 1982: 'On visiting the "moving metropolis": reflections on the architecture of imperial science'. *Historical Records of Australian Science*, 5, 1–16.
Marshall, A. 1980: 'Griffith Taylor's correlative science'. *Australian Geographical Studies*, 18, 184–93.
Marshall, A.J. (ed.) 1966: *The Great Extermination: a guide to Anglo-Australian cupidity, wickedness and waste*. London: Heinemann.
Meinig, D. 1962: *On the Margins of the Good Earth: the South Australian wheat frontier, 1869–1884*. Chicago: Association of American Geographers.
Powell, J.M. 1979: 'Thomas Griffith Taylor, 1880–1963'. *Geographers. Biobibliographical Studies*, 3, 141–53.
—— 1980a: '1888 and all that'. *Australia 1888 Bulletin*, 5, 5–32.
—— 1980b: 'The haunting of Saloman's house: geography and the limits of science'. *Australian Geographer*, 14, 327–41.
—— 1981a: 'Thomas Livingstone Mitchell, 1792–1855'. *Geographers. Biobibliographical Studies*, 5, 83–7.
—— 1982: 'Archibald Grenfell Price, 1892–1977'. *Geographers. Biobibliographical Studies*, 6, 87–92.
—— 1983a: 'George Woodroofe Goyder, 1826–1898'. *Geographers. Biobibliographical Studies*, 7, 47–50.

—— 1983b: 'James Macdonald Holmes, 1896–1966'. *Geographers. Biobibliographical Studies*, 7, 51–5.

—— 1986: 'Approaching a dig tree: reflections on our endangered expedition'. *Australian Geographical Studies*, 24, 3–26.

—— 1988a: 'Protracted reconciliation: society and the environment'. In R. MacLeod (ed.), *The Commonwealth of Science: ANZAAS and the scientific enterprise in Australasia, 1888–1988*, Melbourne: Oxford University Press, pp. 249–71.

—— 1988b: *An Historical Geography of Modern Australia: the restive fringe*. Cambridge: Cambridge University Press.

Ratcliffe, F. 1938: *Flying Fox and Drifting Sand: the adventures of a biologist in Australia*. London: Chatto & Windus.

Rolls, E. 1969: *They All Ran Wild: the story of pests on the land in Australia*. Sydney: Angus & Robertson.

—— 1981: *A Million Wild Acres: 200 years of man and an Australian forest*. Melbourne: Thomas Nelson.

Schedvin, C.B. 1987: *Shaping Science and Industry: a history of Australia's Council for Scientific and Industrial Research*. Sydney: Allen & Unwin.

Spate, O.H.K. 1966: 'The compass of geography', Inaugural Lecture, 1953. In O.H.K. Spate, *Let Me Enjoy: essays, partly geographical*, Canberra: Australian National University Press.

Wadham, S.M. and Wood, G.L. 1939: *Land Utilisation in Australia*. Melbourne: Melbourne University Press.

Worster, D. 1979: *Dust Bowl: the Southern Plains in the 1930s*. New York: Oxford University Press.

18

Geography and National Identity in Australia

Oskar Spate

A nation of trees, drab green and desolate grey
In the field uniform of modern wars
Darkens her hills, those endless, outstretched paws
Of Sphinx demolished or stone lion worn away.

They call her a young country, but they lie:
She is the last of lands, the emptiest,
A woman beyond her change of life, a breast
Still tender but within the womb is dry.

And her five cities, like five teeming sores
Each drains her, a vast parasite robber-state
Where second-hand Europeans pullulate
Timidly on the edge of alien shores.

These terrible images in a poem by my friend Alec Hope seem to me to sum up much of the geographical nature of Australia. It does not seem much of a foundation for building a national identity.

In 1952, when I had been just 366 days in the continent, I set down some first impressions. I had lectured on its emptiness, but nothing in the books had prepared me for the immediacy of it. Two years later, standing on the Razorback south of Picton and looking to the Sydney plain, Joe Jennings – on his first day – remarked that it was astonishing that what we saw had still not been settled. I replied that it was closely settled, in Australian terms.

The emptiness hit me almost as a physical blow when I first left the ghastly sprawl of Sydney's suburbs, suburbs which, unlike their European counterparts, had never had a pre-existent nucleus to give some individuality, had never been anything else but dormitories with gardens attached. Forty years on, the contrast between bush and city is not quite so sharp, but it is still there.

In the hearts of the state capitals one senses a history full of vitality and drama, even if domestic drama; but suburbia seemed to have no history and very little geography. As Robert Graves remarked, 'The suburbs of itself is Hell.' Again, hardly good soil for the growth of a national identity.

And yet, despite all the soul-searching of the intelligentsia in quest of it, there was then and is now such an identity, authentic and unmistakable. But it seems to me that, except in limited circles in the civil and military services, this is not an identification with the State, the Commonwealth of Australia. I think very few have died for the Commonwealth or the individual states, but many for the idea of Australia – a culture and a continent. The loyalty is to a way of life which has grown up in a peculiar geographical environment, an environment which has resulted in 60 per cent or more of the population living near or close to five rival capital cities which are almost little states. Again, a Pentapolis is not very conducive to an overriding sense of nationality.

The development of such a sense is obviously a historical process, but here again the record is odd. There is no heroic monarch like Henry V or Elizabeth around whom a tradition might be built, no statesman who incarnates the nation like Washington or Lincoln. At moments we have our heroes, sportsmen and -women or race-horses, but their glory is ephemeral.

In the early 1950s I was in a country café where a girl, who by no possibility could have had a lover or even a father at Gallipoli, was listening raptly to a lyric which went 'My loved one went so far away / And now he sleeps by Suvla Bay.' This illustrates the paradox that for a country which until 1942 had not to meet any significant foreign threat, the catalytic crystal which made the sense of national identity really tangible was an unsuccessful military adventure a world away from Australian shores. But for crystallization to take place the solution must be already saturated, and well before the Commonwealth proclaimed 'a continent for a nation and a nation for a continent' the mix was there.

I think the embryo took shape, to change the metaphor, in the great days of the Sydney *Bulletin* which, though produced by quite sophisticated city literary entrepreneurs, yet fostered the articulation of the inner feelings of unsophisticated dwellers in the bush. Many of these were migratory workers, shearers and so on, and these projected the image of the dinkum Aussie as a lean bronzed laconic bushman. Being migratory and hence without votes, they had little loyalty to the separate states but a sometimes passionate loyalty to the bush culture. True, more than half the population was already in the five cities; nevertheless, the tradition stands in no need of either apology or boosting.

This tradition had its masterpieces, the short stories of Henry Lawson, the bush balladry, Tom Collins's *Such is Life*. It is dead now, or rather it has suffered a ghastly resurrection in commercial parody. Any news story from a country town is likely to be introduced on radio by a cheap imitation of a bush ballad. Already in 1952 I had wondered how long it would be before conducted tours would include a reserve where one might see real live bushmen brewing tea in a real billycan. With the current cult of outback safaris this has come about.

Geographic isolation protected this more than incipient nationalism. By the 1950s this isolation was breaking down as airways reduced the travel time from Europe from weeks to hours. This shrinking of the world produced a sometimes acute anxiety lest Australianism in the arts should be diluted by the waves of overseas influences. Yet Australians have always desperately wanted to be in the swing of world movements, to show themselves universal while yet keeping their local colour and flavour. Hence this was the time of the sometimes pathetic search for the Great Australian Masterpiece, which involved a good deal of blowing up sparrows into eagles.

Some poets sought to preserve, or to create, a local genius of the place, completely at one with the environment, by jettisoning all European myths and symbols in favour of those drawn from Aboriginal dreaming, a deliberate turning away from their own times and culture. But which is more native American, Longfellow's *Hiawatha* or Masters's *Spoon River Anthology*, Benét's *John Brown's Body*? I do not think one builds a national identity by such artificial means.

So far I have been speaking mainly in terms of history, though a history firmly rooted in the peculiarities of the Australian environment. What have Australian geographers and geography contributed to the national sense?

When I came to Australia in 1951 the number of full-time fully recognized university geographers was precisely six, including myself as the second Professor of Geography in the entire country. The one fully fledged department, at Sydney, seemed to devote itself mainly to elaborating a system of regions, to serve as filing boxes for data which would never be used for their ostensible purpose of planning development. We were too few on the ground to make much impact on national thinking, though we did our best.

Those were great days; we few professionals saw ourselves as pioneers, a happy band of brothers. There was a father figure in the background, Griffith Taylor, and another nearer home, Macdonald Holmes at Sydney, though few of us were prepared to concede him this position. But the six full-time staff of 1951 numbered over 140, lecturers and above, by 1971 and there were some 20 professors. We had far outgrown the infantile stage.

Australian geographers have contributed immeasurably to the fuller understanding of their environment, physical and human. But I do not think that by and large they have occupied themselves directly with the question of national identity; they have, rather, contributed to the analysis of the economic infrastructures which are so essential a part of the life of a nation. Their most direct input into our question has probably been in studies of population geography and migrant communities. And in Papua New Guinea and the South Pacific islands they have certainly contributed to the development of national identities. Altogether they have developed academic geography – in the good sense of academic – to a pitch which seems almost unbelievable when I recall our primitive beginnings 40 years ago.

And if geographers have not contributed greatly or explicitly to fashioning a national identity, the indirect effect of greater understanding of our environment is surely of great significance in such a task. For despite the concentration of people in the cities and the slick modernity, the looming background is ever present in consciousness: the vast spaces so scantily peopled, the savage alternations of drought flood and fire. Amongst the foremost problems facing the Australian nation today is the management of its own continent in the face of pollution, deforestation, soil exhaustion and deterioration, and geographers have been to the fore in tackling these. This is surely nation-building, or at least repairing.

External geography, our position in regard to our Asian neighbours, was a very potent factor even before the bombs fell on Darwin in February 1942. In the 1950s heightened recognition of this quite often took the absurd form of labelling our continent as 'a part of Asia', proclaiming ourselves, in a parody of Sukarno's New Emerging Forces, a New Submerging Force. This obviously left out all geographical factors except mere propinquity, and one could proceed to show that on these lines everywhere on the globe is a part of somewhere else. It was a dangerous blanket slogan, a question-begging cliché burking serious thought on the very real and vital problem of learning to live with Asia, as indeed did its no less absurd antithesis, White Australia. It said much less than nothing as to our national identity.

By a tragic irony, the people who fit least easily into a national identity are the original Australians, who were too much at home in the land they loved so deeply to cope readily with the vastly different Australia that whites have made. Back in 1951 the Aborigines were hardly visible in the great cities, though they were there, as fringe-dwellers, and were certainly not counted as part of the nation – literally so, for until 1967 they were not even counted in the Census. They are highly visible now, though the difficulty of counting them in remains equally intractable. Their passionately held cultural values, so different from ours, must be respected, but yet are a stumbling-block on the road to 'one Australia'.

Internally again there has been a vast change in the ethnic origins of the rest of the Australian population; and here at least geographers have done much to elucidate the issues. No longer do people of 'Anglo-Saxon' or Celtic stock form over 90 per cent of the population. Since the first wave of displaced persons at the end of the Second World War the immigration net has been cast widely; alongside the old-established Italians and Greeks there are now significant numbers of Asians, especially Vietnamese and Lebanese and small groups from Latin America.

These are seen by some as a threat to national unity and, indeed, if badly handled they could become a divisive factor. But this is not necessarily so, and in many respects they have greatly enriched the national life, giving it a badly needed colour and diversity – and a vast improvement in cuisine. Opponents of multiculturalism call for a change in the balance of the migrant intake to emphasize conformity with Australian values. These values are not defined, and if conformity to them means a return to the petty parochialism and philistinism of the 1950s, the only reply I can give is in the words of an indubitably Anglo-Saxon song: 'Oh, no John, no John, no'.

With all its limitations of parochialism and philistinism, there was a freshness and frankness about Australian life in the 1950s which led me, and scores of thousands of others, to decide very quickly that this was the place where we would stay. Despite the heavy overlay of phoney modernity and bureaucracy, the reckless cult of bigness, the relentless pressure for uniformization – an ugly word for an ugly thing – in today's world of giant agencies for everything, freshness and frankness are still there. There *is* an Australian nation with a character all its own; history and geography – the two are inseparable in the life of a nation – have moulded it. I dislike many things in it, but I am proud to be a small part of it.

POSTSCRIPT, 1993

I do not think I need to change much, though there have been some significant changes of recent years. Perhaps a major factor, an external one, in fostering a sense of national identity is our ever-increasing involvement with South-east Asia. Paradoxically, as these ties strengthen (and those with Britain weaken), we mercifully hear much less of the old cliché that we are 'a part of Asia', reducing geography to mere propinquity, on which basis everybody could be shown as a part of somewhere else. This fallacy has all but disappeared and been replaced by a far more realistic appraisal.

The great postwar immigrations from Europe and latterly from South-east Asia, especially Vietnam, have led to a strong emphasis on 'multiculturalism'. This, though distrusted and even feared by a minority, is more and more seen as adding colour and vitality to national life, and not only through ethnic cultural festivals. It is strikingly reflected in cuisine, which in the old days was uninspired, or worse. In country towns one could get a good plain 'roast and two veg' dinner in the hotel, but for the rest there were only dim cafés and not-so-good Chinese restaurants. Now the Canberra phone book has many pages of advertisements for 'ethnic' restaurants from a score or more countries, ranging from Malaysia via Pakistan and Turkey to Mexico. Whether this contributes to a single national identity may be questionable, but it is undeniably welcome.

Another factor, on which geography has an obvious bearing but which imposes some strain on national solidarity, is the increasingly loud and sometimes strident Aboriginal voice. Gone are the days when I could truthfully write that the Koori, as they now call themselves, had been unceremoniously swept under the carpet of history. A recent decision of the High Court of Australia in the Mabo case (named for a man who raised a land claim) repudiated the notion that when Captain Cook arrived in 1770 and proclaimed British sovereignty over eastern Australia the country was *terra nullius*, not owned by anybody. This opened the way for claims by Aborigines for title to land, an opportunity which they have taken up with great enthusiasm, staking out demands, sometimes by very few people for inordinately large areas, or alternatively seeking cash compensation for loss of land. One such demand, by the few thousand Torres Strait Islanders, is for $A5 billion: three times the national debt and five times the federal expenditure in the current budget. This would obviously have vast geographical consequences and is clearly counterproductive: the backlash has already begun.

A new factor, which is in the short run also divisive but in the long term, in my personal view, likely to contribute greatly to the sense of national identity, is a quite astonishing emergence of republican sentiment, affecting both Anglo-Celtic and immigrant communities. A few years ago there was hardly a whisper, if even that, of republicanism; now it is prominent in public debate and taken very seriously indeed, being voiced by an increasing number of conservatives. Recent events in Britain have obviously not helped the monarchist cause. The transition would raise far more complex problems than are realized by many of its proponents, but there seems to be a general expectation that the last links with the British Crown will be severed by the year 2001, the centenary of the proclamation of the Commonwealth of Australia. Then indeed will the slogan of those who agitated for its creation from six separate states be realized: 'A continent for a nation, and a nation for a continent'. And what could be more geographical than that?

19

South Australia: Discoverers, Makers and Interpreters

Murray McCaskill

The identity of a discipline of knowledge is formed over time by the personalities, the writing, the teaching and the actions of its practitioners, together with the collective myths that have grown up around them. In any geographical area where there have been few practitioners, and this has been true until the mid-twentieth century for most of the lands colonized by European settlers, the impress of their personalities and preferences on the historical profile of their discipline will be strong. Apparently trivial personal circumstances can have a powerful role in moulding the identity of a discipline within a small nation and particularly so in any one of its component regions.

In nineteenth-century Australia the intellectual life as well as the commercial life of each colony proceeded in a high degree of detachment from those of their neighbours. Influences from Britain and, to a lesser degree, North America, were transmitted more directly from overseas by early training and through publication to the single university in each capital city rather than between the colonies (and after 1901 the States) of Australia.

Australian geography is indeed more than the sum of its regional parts, but the regionalism of the States is still recognizable. This chapter explores the origins of geography in South Australia, the second Australian state after New South Wales in which geography gained an initially precarious foothold in the nation's fledgling university system.

In Australia and New Zealand – and perhaps in other newly settled lands – a useful schema for examining geography's identity comprises the following sequence:

1 the **discoverers and recorders**; these were the explorers who were often the hero figures of official and popular history; their journals and reports provide the foundation for the written heritage of a regional or national geography;

2 the **makers** of geography on the ground. They include the Aboriginal inhabitants whose impacts on the landscapes were often underestimated, the pre-settlement theoreticians who set the frameworks and the ground rules for the settlement process, and the colonial civil servants and surveyors who laid out the design of the Europeanized landscapes;

3 the **interpreters**, the scholars and scientists who applied the universal concepts, techniques and theories of their discipline to their local environments, and sometimes to the interpretation of the work of the 'discoverers' and the 'makers' who preceded them;

4 the **consumers**, who comprised the audiences, whether general public or students, for the lectures, the readership for the literature, or the employers of those who received instruction in the discipline. The consumers, actual or potential, were often important influences in the decisions to establish geography in educational curricula of schools and universities.

All four groups contributed to the 'profile' or 'identity' of geography at the national and regional levels. Until about 1950 each group had a few figures who, in today's more crowded scene appear to have been commanding in their influence. The New World experience suggests the fallacy of attempting to derive the history of a discipline of knowledge simply from the work of the third group above – the full-time scholars and scientists. In Australia the written heritage of the explorers and the practice and theory of the 'makers' and 'interpreters' were well blended in the development of geography at both national and regional level.

DISCOVERERS AND RECORDERS

In the search for foundation documents for geography's written heritage, South Australians could not pass over the contributions of three men: Matthew Flinders, the navigator who charted the coastlines, and the two overland explorers, Charles Sturt and Edward John Eyre. Not only did the actions of these men have the epic quality that ensured their place as Australian heroes; they were the first Europeans to report on significant parts of the land that became South Australia – Flinders the gulfs which became the heartland of settlement; Sturt the lower River Murray, and later the north-eastern desert; Eyre the sandy wastelands and salt lake beds north of the Flinders Ranges, and the scrub-covered peninsula which bears his name. Each journal had a powerful effect in conditioning the attitudes of decision-makers and the community in general to the nature and potential of areas and places (Flinders, 1814; Sturt, 1833, 1847; Eyre, 1845).

For geography as a discipline, their journals are important for what they said and how they said it. They contain clear descriptive accounts of the land, evaluations of its potential for white settlement, speculations and deductions on the features observed, and observations, often quite sympathetic, about the native peoples whose lives were about to be disrupted.

As Carter (1987) has recently pointed out, the explorers' journals are a combination of a record of the route traversed, the trivial daily events of food, health and weather, and descriptions of the terrain, vegetation, wildlife and inhabitants. But they also contain deduction, reflective speculation and big ideas. The 'pure' geography can be sifted out, but these journals are splendid examples of the unconscious interweaving of the scientific and humanistic styles, of objective observations and subjective impressions.

The second volume of Edward Eyre's *Journals* is a noteworthy contribution to the human geography, demography and customs of South Australia's Aboriginal peoples. While in his mid-twenties Eyre spent three years as Resident Magistrate and Protector of Aborigines at a strategic point where the overland stock-droving route from New South Wales crossed the River Murray. Eyre's insight into the Aboriginal view of the European invasion of their domain, his understanding of the native economy and land tenure and his proposals for a humane and sympathetic public policy towards the native peoples, was rare amongst European officials and observers of the time. He wrote with both a keen geographical awareness and an outraged sense of natural justice:

> Without laying claim to this country by right of conquest, without even pleading the mockery of cession, or the cheatery of sale, we have un-hesitatingly entered upon, occupied, and disposed of its lands, spreading forth a new population over its surface, and driving before us the original inhabitants. (Eyre, 1845, vol. II, pp. 158–9)

> the localities selected by Europeans, as best adapted for the purposes of cultivation, or of grazing, are those that would usually be equally valued ... by the natives themselves, ... this would especially be the case in those parts of the country where water was scarce, as the European always locates himself close to this grand necessary of life. (Ibid., vol. I, p. 168)

THE MAKERS

The making over of Aboriginal South Australia into a land of European settlement owed something to a London-based group of theorists, lobbyists and speculators. It owed most, however, to the surveyors on the ground, commencing with William Light, who was world renowned for the layout of Adelaide, but the task was carried on for much of the later nineteenth century under the supervision of the colony's celebrated Surveyor-General, G.W. Goyder.

South Australia was indeed a theory before it became a place. The theory was a complex one concerning the relationship between the price of land, the cost of labour and the return on capital. A spatial component to the theory is attributed to the British political philosopher, Jeremy Bentham. He argued for 'an entirely new principle' entitled the 'vicinity-maximising-or-dispersion-preventing principle' (Pike, 1957, p. 57). Many colonial administrators attempted to implement this concept in the mid-nineteenth century with varying degrees of success.

Throughout the nineteenth century the notion was maintained in South Australia that agricultural settlement and the sale of Crown land should progress outwards from the heartland in an orderly progressive sequence; that new sales should be made at the frontier edge following detailed survey of allotments, towns and roads and should not be allowed to leapfrog ahead of the settled zone.

George Woodroofe Goyder has been termed 'a practical geographer' by Michael Williams (1978). A small man of immense energy and authority, he deserved his epithet 'Little Energy'. He joined the new Lands and Survey Department in 1851, was appointed Surveyor-General in 1861 at the age of 34 and held that post until 1893. A fast moving succession of 34 Ministers of Crown Lands usually acted on his judgement on a wide range of land policy, survey and settlement issues.

Like the eighteenth-century preacher John Wesley, he spent much of his life on horseback. His most famous geographical achievement was to plot, in the words of his official instructions, 'the line of demarcation between the portion of the country where the rainfall has extended and that where the drought prevails'. Goyder's line defined the extent of country which had been severely affected in South Australia's first pastoral drought and where pastoralists were to be given relief by reduced rent payments on their leases. It was the basis of Australia's first official policy of drought relief (South Australian Parliamentary Papers, 1865–6).

In the minds of South Australians the line came to mark the division between that country considered safe and that considered unsafe for wheat farming on account of its unreliable rainfall. It is a respectable if common task to survey a line on the ground but it is a considerable feat to etch on the minds of people and governments a line that endured for a hundred years. As a master planner at a time when the name would have been anathema, Goyder was involved both in strategic planning of the general course of settlement and planning at the micro level of the selection and layout of town sites and the demarcation of reserves for water supply, stone quarries and cemeteries. He instructed his surveyors to lay out new township sites on a rigid model with central squares and encircling parklands in imitation of Light's design of Adelaide.

The shortage of usable timber in South Australia convinced him that reserves for forestry should be allocated from Crown lands or resumed from private lands. He wrote several reports on forestry, promoting the virtues of the quick-growing *pinus radiata* against the scepticism of local botanists and arguing that the humid south-east of the colony was a better place for timber growing than the dry fringes of the agricultural zone which were favoured by South Australia's first professional forester.

Goyder's reports on land settlement and forest policy have a strong practical application but they are infused with a sound theoretical grounding and a comparative knowledge of literature elsewhere. He was acquainted with the works of Alexander von Humboldt, George Perkins Marsh and other contemporary literature from the United States, illustrating the interplay of theory, reflection and practice in a New World frontier society.

CONSUMERS – A GEOGRAPHICAL SOCIETY

A discipline needs a sympathetic audience; it needs friends if its scholars are to be encouraged, influence if it is to be established at institutions of learning. Societies for the promotion of the natural sciences sprouted on the frontiers of nineteenth-century white settlement as well as in the European heartlands. An Adelaide Philosophical Society was founded in 1856 which became the Royal Society of South Australia in 1880. Geography could have been accommodated in the colonial Royal Societies, but the 1880s saw a lusty surge for separate identity, sometimes led by the very leaders of the local Royal Societies.

In April 1883 some 700 people gathered in the Sydney Town Hall to found the Geographical Society of Australasia, which was soon granted permission to use the title Royal. The founders in Sydney intended the Society to embrace the whole of Australasia, including New Zealand. Branches were formed in Victoria in 1883 and in Queensland and South Australia on the same day, 10 July 1885. Two inter-provincial geographical conferences were held but a national body never emerged. The New South Wales Branch did not survive and was eventually resurrected in 1927 as the Geographical Society of New South Wales. The Victorian Branch merged with the local Historical Society, but the Queensland and South Australian Branches survived, if at times precariously. The South Australian Branch has produced a journal continuously since 1885.

The founder members were businessmen, teachers, lawyers, journalists and senior public servants – the cream of the Adelaide establishment. The lectures and published articles strongly emphasized the themes of exploration, the customs, livelihood and ceremonies of the Aborigines, recollections of early colonial life and accounts of physical features. At the inaugural presidential lecture in the Adelaide Town Hall in October 1885 Sir Samuel Davenport, a leading pastoralist, gave a marathon survey of world exploration since the days of Columbus. The printed version amounts to 23,000 words and must have taken three hours to deliver, although relieved with an interval for an organ recital.

Despite its active programme of publication the Geographical Society did little to promote the cause of geography in secondary schools until after the Second World War and it is hard to trace any direct influence of the Society on establishing geography at university level. Geography nudged its way into the University of Adelaide in circumstances which remain obscure. In 1904 the University Librarian, Mr R. J. M. Clucas of the Registrar's Department, gave a course of lectures on 'Commercial Geography and Technology', a compulsory subject for the Advanced Commercial Certificate, later the Diploma of Commerce. The course title change to 'Economic Geography' in 1918, but it was not until Clucas died in 1930 that a one-year course, simply labelled 'Geography', was admitted to the Arts degree and the new Bachelor of Economics degree. For almost 50 years geography had a shadowy half-life, tolerated on the fringes, as a subject for commercial students and primary-school teachers but not yet an acceptable subject for full academic recognition in the parlously funded university. Other universities in Australia and New Zealand can recall similar prehistories for their geography departments but nowhere else was the twilight so prolonged as in South Australia.

INTERPRETERS

The translation of geography from the shadows to the sunlight owed much to the quiet preparation and example of the State's first two geographical scholars: Dr Charles Fenner and Dr Archibald (later Sir) Grenfell Price.

Charles Fenner was a distinguished graduate in geology from the University of Melbourne who had published on the geology and geomorphology of Victoria. From 1916 to 1939 he was Superintendent of Technical Education in South Australia and Director of Education until 1946. On his many tours of inspection of schools throughout the State he kept his eyes open and remained active in research if, understandably, not prolific. For the Geographical Society *Proceedings* he produced some delightful articles on minor historical themes. His most important papers, however, he directed to the *Transactions* of the Royal Society of South Australia in two lengthy articles published in 1927 and 1929.

The first, 'Adelaide, South Australia: a Study in Human Geography' (Fenner, 1927), is four-fifths physical geography, a detailed account of regional physiography and drainage and its apparent relationship to the spread of the urbanized area. The remainder deals with 'the effects of geographic controls' on the distribution of population. It shows the conceptual influence of W.M. Davis and Ellen Semple but quotes Vidal de la Blache, Jean Bruhnes and Vaughan Cornish who were 'possiblists' rather than 'environmentalists'. The final sentence views the city 'as an "island" of population mutually co-operating with the great productive country areas on which it is dependent, and for which it constitutes a Market, a Gateway and a Garden'.

The second paper (Fenner, 1929) is entitled 'A Geographical Enquiry into the Growth, Distribution and Movement of Population in South Australia 1836–1927'. It was the first detailed analysis of regional population change of any area in Australia and the first attempt to map intercensal population increase and decrease on a close geographical mesh. Fenner also devised a so-called 'prosperity graph' which charted the historical record of net inwards and outwards migration for the State – a barometer of generalized well-being and adversity. Whereas Griffith Taylor at the same time was addressing the issue of population and resources with a broad brush on a continental canvas, Charles Fenner was at work with scalpel and fine brush on South Australia. His son, the distinguished virologist Professor Frank Fenner, told me that he recalls as a schoolboy the maps and diagrams being prepared on the kitchen table at night after the family meals. The two papers form a neglected classic of Australian geography.

In the 1930s Charles Fenner became part-time lecturer in geography at the University following the death of Clucas and departmental records list 60 to 130 students each year. By 1939 Fenner had prevailed upon the University Council to approve a second-year course taught by two part-time lecturers, Mr C.A. Martin and Mrs Ann Marshall. As Director of Education he had to withdraw from most lecturing but remained nominally as Lecturer-in-Charge until the arrival of Graham Lawton as full-time Reader-in-Charge in 1951. By such low-cost improvisations did geography at the University of Adelaide survive into the 1950s.

Sir Archibald Grenfell Price was something of an Australian Renaissance man who was at ease with several fields of knowledge, geography being but one of them. Except for Griffith Taylor, his biographical details are rather better recorded than those of any other pioneer Australian geographer (Marshall, 1969; Kerr, 1983). Born in Adelaide in 1892, he was sent by his family in 1911 to read history at Oxford, England. Following his honours degree and while studying for a British Diploma of Education he first came into contact with geography as a discipline. In 1917 Price was back in Adelaide teaching at the prestigious St Peter's Collegiate School where he was asked to teach geography to the sixth Modern Form.

With his customary zeal for new experiences Price became a geographer by practice and declaration, but even late in life he was at pains to insist that he was not a professional geographer. With the brashness that only youth can summon he remedied the lack of a suitable text by writing his own *Causal Geography of the World* (1918). It ran through eight editions in nine years. While teaching at St Peter's, Price completed his first work of scholarship, *The Foundation and Settlement of South Australia* (1924), the first comprehensive account of South Australian history to be written from primary source materials. Price examined the early colonists' collective misconceptions of their physical environment but also placed emphasis on the formative influences of the action of individuals in landscape change. The work marked the start of his lifelong interest in geographical history and historical geography.

From 1925 until 1957 Price was Master of St Mark's College, an Anglican residential college for university students. He never held a full-time university position and, given the conditions of Australian universities of the 1920s and 1930s, had he held one he would doubtless have published less and travelled less. A trip to Java, Burma and Ceylon in 1929 whetted his interest in the tropics. Rockefeller grants in 1932–3 took him to tropical Australia, the West Indies and Costa Rica and from this trip emerged probably his most important book, *White Settlers in the Tropics* (1939). Published by the American Geographical Society, it brought Price a well deserved international regard, the second Australian geographer after Griffith Taylor to be so recognized.

Australia's first full-time academic geographer was Griffith Taylor, Associate Professor at the University of Sydney from 1921 to 1928. His vigorous opposition to proposals for the closer settlement of arid Australia earned him much criticism from politicians and newspapers and may have encouraged his decision to accept posts in North America, first at the University of Chicago and later at Toronto. Twenty years later Taylor revisited Australia in what became a triumphant progress of a geographical prodigal son. His visits to universities stirred some administrations on behalf of geography and in South Australia he urged that Grenfell Price should be appointed to teach a third year course as an interim measure before full establishment of the discipline.

Thus in 1949, at the age of 57, Grenfell Price was appointed a part-time lecturer at the University of Adelaide and with three part-time lecturers, each costing £150, the University got a full three-year sequence for what was then the price of one full-time assistant lecturer at a British university. Price's infectious enthusiasm and his genial persuasion of the university authorities led to the development of an honours school in 1951 when the Australian-born Graham Lawton, then of the University of Washington, Seattle, was appointed as full-time Reader-in-Charge and in 1959 as Professor. As in many parts of the world the growth of university student numbers during the 1950s proved a boon to the growth of geography in South Australia. Within a decade the University of Adelaide Department of Geography grew to a major teaching and research department with links to several State government departments. In 1957, for example, some 220 of its students recorded the use of all land and buildings in the metropolitan area in a series of maps which formed the basis of Adelaide's first metropolitan development plan published in 1962.

CONCLUSION

Charles Fenner and Grenfell Price were worthy academic founders for geography in South Australia. Fenner, a geologist by training and an educational administrator by practice, found time to do pioneer local research and develop some part-time university teaching. Price, an historian by training and a geographer by inclination, found time to write many books of local, national and international scope and to lead a busy life that took him briefly into national politics and for many decades into library policy administration.

Together, Fenner and Grenfell Price helped to establish academic geography by maintaining its visibility as a scholarly enterprise through the decades of financial penury and academic scepticism and indifference. Geography in South Australia received its eventual imprimatur in 1951 by the establishment of an independent university department with full-time staff. So successfully did they build on the Fenner and Price foundations that the plans for the University of Adelaide's daughter institution at Bedford Park provided for geography from the outset. Thus, when Flinders University opened in 1966 it was the first Australian university to commence teaching with a chair in geography. As in so many other places the educational explosion of the post-Second World War decades was favourable to the growth of geography as an academic discipline in South Australia By 1975 the two university departments of geography accounted for 20 full time posts with another ten or so in the Colleges of Advanced Education and some 3000 secondary school students were offering geography as one of their subjects for the final year public examinations. There was also an active Geography Teachers' Association and a Royal Geographical Society of renewed vigour and publishing activity.

In 1875 the Society's centennial was celebrated by a joint meeting with members of its Queensland counterpart at Haddon Corner in the remote north-eastern desert where South Australia abuts Queensland. This sunset ceremony gathered up the four components of geography's 'identity' suggested early in this chapter: the desert landscape, first described for Europeans by the nineteenth-century 'hero' figure, Charles Sturt; the state boundary lines demarcated by government surveyors; the scientists, scholars and teachers, whose business it is to practise and promote the discipline of geography; and the interested laypeople whose naive curiosity about other places provided the foundation rationale for our discipline.

REFERENCES

Note: Information on the development of geography within the University of Adelaide is recorded in the Adelaide University Graduates' Union *Monthly Newsletter and Gazette*, July 1970, 3–7.

Carter, Paul 1987: *The Road to Botany Bay*. London: Faber.

Eyre, Edward 1845: *Journals of Expeditions of Discovery into Central Australia*, 2 vols. London: Boone.

Fenner, Charles 1927: 'Adelaide, South Australia: a study in human geography'. *Trans. and Proc. Royal Soc. Sth. Aust.*, 51, 193–256.

—— 1929: 'A geographical enquiry into the growth, distribution and movement of population in South Australia, 1836–1927'. *Trans. and Proc. Royal Soc. of Sth Aust.*, 53, 79–145.

Flinders, Matthew 1814: *A Voyage to Terra Australis*, 2 vols and atlas. London: Nicol.

Kerr, Colin 1983: *Archie: the biography of Sir Archibald Grenfell Price*. Melbourne: Macmillan.

Marshall, Ann 1969: 'Sir Grenfell Price: an appreciation'. In F. Gale and G. Lawton (eds), *Settlement and Encounter: geographical studies presented to Sir Grenfell Price*, Melbourne: Oxford University Press.

Peake-Jones, Kenneth 1985: *The Branch without a Tree: the centenary history of the Royal Geographical Society of Australasia (South Australian Branch Inc.)*. Adelaide: Royal Geographical Society of Australasia, South Australian Branch.

Pike, D. 1957: *Paradise of Dissent: South Australia 1829–1857*. Melbourne: Melbourne University Press.

Price, A. Grenfell 1918: *Causal Geography of the World*. Adelaide: Rigby. (*Note*: a bibliography of Grenfell Price's publications up to 1967 appears in F. Gale and G.W. Lawton (eds), *Settlement and Encounter*, Melbourne: Oxford University Press.

—— 1924: *The Foundation and Settlement of South Australia, 1829–1845*. Adelaide: Preece.

—— 1939: *White Settlers in the Tropics*. New York: American Geographical Society, Special Publication no. 23.

South Australian Parliamentary Papers 1865–6: no. 78, 'Surveyor-General's report on demarcation of northern rainfall'; and no. 82, 'report of Surveyor-General on northern rains'.

Sturt, Charles 1833: *Two Expeditions into the Interior of Southern Australia During the Years 1828, 1829, 1830 and 1831*. London: Smith, Elder.

——1847: *Narrative of an Expedition into Central Australia ... during the years 1844, 5 and 6; together with a Notice of the Province of South Australia, in 1847.* London: Boone.

Williams, Michael 1978: 'George Woodrofe Goyder: a practical geographer'. *Proc. Roy. Geog. Soc. of A'asia, S.A. Branch,* 79, 1–21.

20

Maori Identity and Maori Geomentality

Hong-key Yoon

Na Rangi tua na (Papa) Tu-a-nuku e takoto nei (You and I are both from the sky-father and earth-mother).

(Brougham and Reed, 1963, p. 70)

You have said that the land is in pain, and that it is like a bird pierced and bleeding, and whose wings are quivering. You say also that you are feeling pain in consequence ...

(From 'Mr Firth's visit to the King Party', *Daily Southern Cross* (Auckland, New Zealand) 9 June 1869)

Whatu ngarongaro he tangata, toi tu he whenua (People disappear, land remains).

(Kohere, 1951, p. 38)

The aim of this chapter is to portray briefly the identity of the Maori people in New Zealand and to illuminate the traditional Maori geomentality through the appreciation of various Maori folklore materials including myths, legends and proverbial sayings. In Maori culture, oral tradition has always been the main method of transmitting history and cultural values; therefore they have developed oratory, folklore, genealogy and oral history to a remarkable degree. The present study has evolved and draws material from several parts of my previous work, *Maori Mind, Maori Land* (Yoon, 1986) and 'On Geomentality' (Yoon, 1991).

MAORI IDENTITY

The Maori people are the original inhabitants of New Zealand. They belong to Polynesian stock who, according to Maori tradition, migrated from Hawaiki, their original home. Hawaiki may be somewhere in Polynesia; perhaps it was

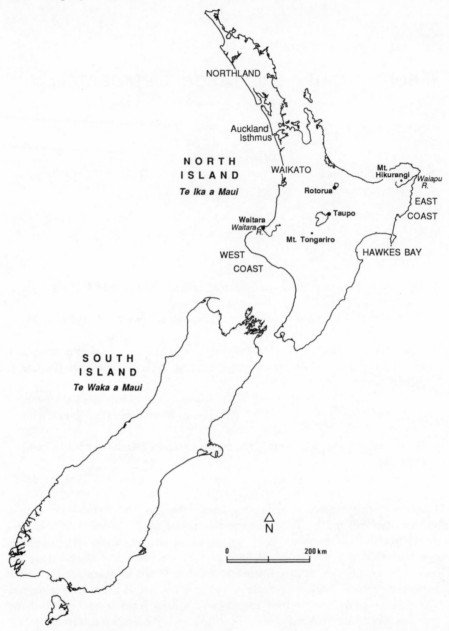

NORTHLAND

Auckland
Isthmus

**NORTH
ISLAND**
Te Ika a Maui

WAIKATO

Mt.
Hikurangi   *Waiapu
R.*

Rotorua

EAST
COAST

Taupo

Waitara
*Waitara
R.*

Mt. Tongariro

WEST
COAST

HAWKES BAY

**SOUTH
ISLAND**
Te Waka a Maui

N

0   200 km

Figure 20.1   The Maori population centres

Rarotonga or the Society Islands, although nobody knows for sure the actual location. However, Hawaiki has always been and will ever remain as the original home of the Maori people and a sacred place in the Maori mind. Since their migration to New Zealand, most Maori people have lived in the

North Island of which the most densely populated areas were Northland, Auckland isthmus, Waikato, Rotorua and parts of the east coast and west coast; the South Island supported fewer and may never have had more than 5 per cent of the Maori people (Lewthwaite, 1950, pp. 49, 51–2) (see figure 20.1).

After contact with the Europeans the Maori people experienced a population disaster due to newly introduced epidemics such as influenza and venereal diseases, internal warfare with newly acquired muskets, and increasing psychological demoralization (Lewthwaite, 1950, pp. 39–43). At one stage the Maori mortality rate was so high that their survival was thought to be in danger, but now they are a vigorous race who have ensured not only their survival but a healthy future (Pool, 1991, p. 231). Maori people, in contrast to Europeans, are known as the *Tangata whenua*, 'people of the land', as against *Pakeha* (Europeans).

Currently, most Maoris in New Zealand have both Polynesian and European heritage, and their *whakapapa* (genealogy) can be traced back to both their Maori and *Pakeha* ancestors (Yoon, 1986, p. 25). Many New Zealanders who identify themselves as Maoris are more European-looking than Polynesian: to some outsiders, such as foreign visitors, their appearance may be no different from that of Europeans.

In the summer of 1976 I attended several Maori gatherings held on the east coast of the North Island and noticed that some European-looking elders gave speeches in fluent Maori. I first thought that they were Europeans who had learned Maori. I found later that they had Maori heritage and could trace their Maori ancestors. Their Maori heritage, in comparison to the European one, was small indeed and they had European features, but they identified themselves as Maori.

Some New Zealanders now identify themselves as Maoris if they have some Maori heritage even though they may have a European appearance; while others, even if they know their Maori heritage, identify themselves as *Pakeha*, or Europeans (Yoon, 1986, p. 25). In most cases the physical appearance may be the main criterion for being Maori. Increasingly, however, more European-looking New Zealanders who have Maori heritage identify themselves as Maori. Therefore, it seems that being Maori requires Maori culture (Maori mind) more than Maori features (Yoon, 1986, p. 25). An important criterion of being Maori is, in fact, the state of mind held by New Zealanders who share Maori heritage.

Since contact with Europeans, Maori people have accepted and incorporated much European culture. The process of the Maori people's adopting and adapting to European culture was not always a happy and harmonious one. Some aspects of the newly introduced Western culture were more readily acceptable than others. For instance, Maori people were willing to embrace Christianity as their faith but they have resisted the European way of dealing with their dead by insisting on their own traditional ways of mourning, funerals and interment. Sometimes the European system was imposed on them without considering their wishes; at other times the Maori people chose to adopt the European way of living. For instance, the British legal system, including the land transaction procedure, was imposed on the Maori people, but the Maori people freely adopted Western-style foodstuffs, clothing and housing.

Some people may be tempted to see modern Maori culture as a 'conquered' culture and the Maori people as simply having adopted and adapted to Western culture. Such views do not reflect the reality of modern Maori culture. A conquered culture is suppressed, submissive and demoralized due to sweeping cultural changes imposed on the conquered by the conqueror. The Maori people are not a conquered people but the people of the Treaty. They negotiated their future with the British and became partners with them through the Treaty of Waitangi in 1840, which guaranteed the undisturbed possession of their properties, presumably including their cultural tradition (Treaty of Waitangi, Article II). The fact that the Maori people did not subscribe to Western culture outright but adopted and adapted to it selectively is evidence that the Maori culture was not entirely conquered. The degree and pattern of an indigenous culture's adaptation to a 'foreign' one is some evidence of the native people's resistance to the 'foreign' culture and of the prospects for survival of the unconquered indigenous culture. Many aspects of Maori culture have changed significantly, but the essence of the culture is surviving well. Perhaps Maori culture is no more conquered by Western culture than is Japanese culture, and modern Maori culture is not much more Westernized than that of modern Japan. However, one major difference between the two cultures is that most young Maori people now use English as the common means of communication and are not fluent in their mother tongue, while the Japanese have always spoken their own mother tongue.

The Maori people had intense and emotional relationships with the environment. They considered the system of the natural environment as a kind of family with its components, such as the sea and the forest, being family members, and the relationships between them being seen in terms of kinship relationships. Traditionally, the Maori people had distinct attitudes towards the natural environment, which can be termed Maori 'geomentality'. This Maori geomentality is the cornerstone of Maori culture which still survives. I shall explain first the concept of geomentality first and then that of Maori geomentality.

ON THE CONCEPT OF GEOMENTALITY

If a traditional French gardener is asked to make a beautiful garden, he might create a magnificent French Renaissance garden, as at Versailles. If, on the other hand, a Japanese gardener is given the same task, he might create a beautiful Japanese hill garden. The creation of these two very different gardens from the same piece of land is the result of the difference between the two gardeners' views on what constitutes a beautiful landscape. These views are two different geomentalities of garden-making, that is, the mentalities of two gardeners regarding land (Yoon, 1991, p. 387). The geomentality of the landscape-makers responsible for a certain cultural landscape can, therefore, be conjectured from the characteristics of any particular landscape pattern.

Geomentality is defined as 'an established manner (mentality) or taken-for-granted way of understanding the environment which conditions the relationships between humanity and nature. It is the mentality, held by a group of people or an individual, regarding the geographical environment' (Yoon, 1986, p. 38; 1991, p. 388). To put it simply, geomentality is a people's mind-set regarding the geographical environment.

The study of geomentality provides an avenue for a meaningful understanding of cultural landscape patterns by analysing these patterns more profoundly than can be done by conventional methods. As I have discussed elsewhere (Yoon, 1991, pp. 389–90), geomentality provides new insight into the conventional way of explaining, for instance, the urban New Zealand landscape pattern of expensive houses on elevated sites with extensive sea views, and cheaper houses on flat land without such views. A conventional explanation has been that rich people are able to afford expensive houses with good views while the poor are not. Hence a residential pattern of socioeconomic differentiation develops. This conventional socioeconomic explanation is not wrong but is arguably too superficial. Wealth certainly provides the means for some to achieve their goal of owning expensive houses with good views but this fails to explain why such houses are built on sites with extensive sea views rather than on sites with no such views. Here, one realizes that a more profound reason behind the formation of such an urban landscape pattern is not wealth itself but the *Pakeha* (European) New Zealanders' geomentality which includes the value attached to locations with extensive sea views. In this way, the study of geomentality can add a new dimension to the conventional socioeconomic explanation of cultural landscape patterns. A geomentalitist argues, therefore, that determining geomentality is the ultimate key to explaining in a profound way patterns of cultural landscapes (Yoon, 1991, p. 392).

The concept of geomentality, which is a lasting and established state of mind regarding land, has advantages over conventional concepts such as world view, environmental perception and mental map in explaining cultural geographic phenomena.

MAORI MYTHOLOGY AND MAORI GEOMENTALITY

Traditional Maori geomentality is characterized by viewing the environment as either humanity's parents (sky and earth) or siblings (tree, ocean and so on). According to Maori geomentality, humanity is seen as one of many brothers of the environmental family. Humanity is also seen as the conqueror of the immediate environment (brothers), and nature is seen as an object to be exploited and overcome, so long as people use and manage it properly. The root of this traditional Maori geomentality can ultimately be traced back to the Creation myth of *Rangi* and *Papa*. Generally in traditional societies, people tend to accept the content of their myths and legends as truth, in a similar manner to that in which Genesis is accepted as truth by Christians who interpret the Bible literally. Traditional Maori society was no exception.

There is convincing evidence that the Maori myths and legends were accepted as truth by the Maori people in traditional times and their belief is clearly reflected in their behaviour. In the following abridgement of the original 1855 English translation (Grey, 1974, pp. 1–11), the plot of the Maori evolution of the world is presented.

AN ABRIDGEMENT OF THE MYTH OF *RANGI* AND *PAPA*

In the beginning, all things originated from *Rangi* and *Papa*, or Heaven and Earth, upon which darkness then rested. There was darkness from the first division of time unto the tenth, to the hundredth, to the thousandth. These divisions of times were considered as beings, and were each termed a *Po*. There was as yet no world with light, and *Rangi* and *Papa* were both enclaved together in the darkness and they begot children who huddled between them in the dark. The names of the six children were:

> *Tane-mahuta* (forest, abridged as Tane);
> *Tu-matauenga* (people, abridged as Tu);
> *Rongo-matane* (cultivated foodstuffs, abridged as Rongo);
> *Haumia-tikitiki* (wild foodstuff, abridged as Haumia);
> *Tangaroa* (ocean and fish); and
> *Tawhiri-matea* (wind and storms).

The children were worn out by the continued darkness and consulted amongst themselves, saying: 'Let us now determine what we should do with *Rangi* and *Papa*, whether it would be better to slay them or to rend them apart.' *Tu-Matauenga*, the fiercest one said, 'Let us slay them.' Then *Tane-mahuta* (forest: the eldest child) said, 'It is better to rend them apart, and the heaven stand far above us, and the earth lie under our feet. Let the sky become as a stranger to us, but the earth remain close to us as our nursing mother.'

Five brothers consented to the separation of their parents, but one of them, *Tawhiri-matea* (wind) would not agree to it and grieved greatly at the thought of his parents being torn apart.

In turn, the five brothers attempted to separate their parents, in the order of *Rongo-matane, Tangaroa, Haumia-tikitiki, Tu-matauenga* and *Tane-mahuta*. One by one all failed until it came to the turn of *Tane-mahuta*. With his hands and arms he first tried to separate the Heaven and the Earth, but he failed. Then, he planted his head firmly on his mother the Earth and raised up his feet against his father, the Sky. Disregarding his parents' shrieks and cries, *Tane-mahuta* steadily pushed the Sky upward with his feet and thus separated them.

Tawhiri-matea (wind and storms) desired to wage war against his brothers who plotted to separate the parents. So he followed his father, the Sky, and there he quickly bred hurricanes and storms to wage a war of revenge against his brothers on the Earth.

Tawhiri-matea first attacked *Tane* (forest) with hurricanes, thunderstorms and squalls which devastated the forest and its inhabitants. In his wrath he next attacked *Tangaroa* and swooped down upon the seas and lashed the ocean. *Tangaroa* was the god of the sea and its inhabitants, who were terrified of the large waves and fled to the ocean. *Tangaroa* had by then already begotten *Punga* who in turn had two children, *Ika-tere* (fish) and *Tu-te-wehiwehi* (reptiles). The children disputed together as to what they should do to escape the storms: *Tu-te-wehiwehi* and his party wanted to flee inland, while *Ika-tere* and his party wanted to go to the sea. Without delay, these two races of beings separated. The fish fled in confusion to the sea, the reptiles sought safety in the forests and scrubs. Consequently, *Tangaroa* (sea) lost some of his children to the realm of *Tane-mahuta* (forest). Ever since then, *Tangaroa* has waged war against his brother *Tane* who has fought back. Hence *Tane* supplies people with canoes, spears and fish-hooks made from his trees and with nets woven from his fibrous plants, that they may destroy the offspring of *Tangaroa*, the fish and the reptiles. *Tangaroa*, in return, swallows up the offspring of *Tane* such as insects and and other animals that inhabit forests, overwhelms canoes with the surges of his sea and floods.

Tawhiri-matea next rushed to attack his brothers *Rongo-ma-tane* and *Haumia-tikitiki* (represented by sweet potato and fern roots) but *Papa* gently held the two children and hid them in a place of safety. So well concealed underground were these children that *Tawhiri-matea* sought for them in vain.

Having vanquished all his brothers but one, *Tawhiri-matea* last attacked *Tu-matauenga* (humanity) to complete his victory. *Tawhiri-matea* exerted all his force against him, but could neither shake nor prevail against him who stood firm and unshaken upon the breast of his mother Earth: he was the only one who had shown himself brave and fierce in war and defied the power of the wind. So *Tawhiri* recognized that he could not defeat *Tu-matauenga* and abated the storms.

After resisting the wind's fierce attack successfully, *Tu-matauenga* decided to take revenge on his cowardly brothers because they had not assisted him or fought bravely when *Tawhiri-matea* had attacked them. He first sought some means of injuring *Tane-mahuta* (forest and its inhabitants), so he began to collect the leaves of trees; he cut down the trees and hunted the birds. He then revenged himself on *Tangaroa* by fishing in the waters. After that he punished *Rongo-matane* and *Haumia-tikitiki* by uprooting edible tubers. Thus *Tu-matauenga* devoured all his four brothers on the Earth, and consumed all of them, in revenge for their having deserted him and leaving him to fight alone against the forces of the wind.

However, *Tu-matauenga* (people) could not vanquish *Tawhiri-ma-tea* (wind) by eating him as food, like the other four brothers, so the force of the wind remains an enemy of man. Still with a rage equal to that of people, the wind continues to attack in storms and hurricanes, endeavouring to destroy him by sea and land.

The explosion of the wrathful fury of *Tawhiri-ma-tea* against his brothers was the the cause of the disappearance of a great part of the dry land; during that contest a great part of mother Earth was submerged and only a small portion (island) projected above the sea.

The Sky remains separated from his spouse the Earth. Yet their mutual love continues – the soft warm sighs of her bosom still rise up to him, and men call these mists; and the vast Heaven, as he mourns the separation through the nights, drops frequent tears upon her bosom, and men term them dew-drops.

AN ANALYSIS OF GEOMENTALITY IN THE MYTH AND IN OTHER MAORI TRADITIONS

In Maori geomentality the environment was personified and was recognized as having a family structure somewhat similar to that of human beings. *Tu* (representing people) is a child of the parents *Rangi* and *Papa* and a brother to his siblings, such as forests, winds, sea and foodstuffs. Thus, people are part of the family of the environment. In Maori mythology all environmental factors such as the earth, sky, forest, bird, fish, sweet potato and wind are personified and have the human characteristics of expressing love, anger, sorrow, pride and forming a family (*whanau*). In Maori geomentality each environmental element was considered to be a relative of people. This Maori geomentality was clearly expressed in traditional Maori behaviour in the way the Maori treated environmental elements such as the sea and forest. A veteran ethnologist, Elsdon Best, commented that Maori people treated the trees and birds as their siblings as they entered the forest. Best (1977, p. 6) wrote:

> Inasmuch as man, birds and trees are descended from a common source, it is not surprising that, when the Maori entered a forest, he felt himself to be among his own kin, albeit somewhat distant relatives. This atti-tude was productive of a singular train of thought and also curious superstitions and quaint ceremonial.

Another example occurred when Maori travelling parties met; they introduced themselves to each other by saying that they were the descendants of *Rangi*, the Sky-father, and *Papa*, the Earth-mother:

> Thus, should a travelling party meet a number of strangers, or should a people be attacked by persons they did not recognize, their principal man would call out the inquiry '*Na wai taua?*' ('From whom are we?' – 'sprung' or 'descended' understood). This query was not spoken simply, but was intoned. The reply would be delivered in a like manner: 'We are from Rangi above and Papa beneath.' Then would follow some explanation as to who the speaker was. (Best, 1974, p. 14)

Maori geomentality is also very effectively expressed in proverbs, because they are the distilled and concise expression of people's ideas and values. We can cite two popular traditional Maori proverbs expressing personal and family relationships with the environment:

Ko Papa Tuanuku te matua o te tangata (Earth is humanity's parents).
(Kohere, 1951, p. 16)

Na Rangi tua na (Papa) Tu-a-nuku e takoto nei; ko ahau tenei, ko mea a mea (You and I are both from the Sky-father and Earth-mother; this is me, so and so, son of such-and-such). (Brougham and Reed, 1963, p. 70).

The ideas reflected in these two proverbs are directly from the myth of *Rangi* and *Papa* and support the view that the Maori people actually treated the earth as their beloved mother, as recorded in the following case. Traditionally, when a Maori has long been away from his native land and returns to it after his parents have died, he may cry out with great emotion, 'I greet my only surviving mother in the world, the land' (Firth, 1972, p. 368; Best, 1941, vol. 1, p. 397)

Because the environment was personified and treated as a member of their own family, Maori people have personal relationships with the environment and maintain a strong emotional bond with the land. They have tremendous affection for their own local environment.

Secondly, people were considered as conquerors of the earth, and the dominion of nature was justified. This idea was reflected in Maori mythology, for instance in *Tu*'s fight against the wind and his punishment of his brothers *Tane*, *Tangaroa*, *Haumia* and *Rongo*. This idea of the Maori people's conquest is also clearly reflected in Maui's Myth. Sir George Grey's version of the myth includes a series of different themes, such as Maui's family background, his quest for fire, his fishing which raised the North Island, his taming of the Sun, and the origin of the dog. An abridged version of 'Maui's taming of the sun', which reflects the idea of people's conquest of the environment, follows (Grey, 1974, pp. 25–8):

Maui thought that after the rising of the sun, it sank down too soon which made the days appear too short to him. So he decided to make him move more slowly. He had his brothers prepare the ropes needed for catching the sun in. Then Maui with his brothers travelled eastward and finally they reached the place out of which the sun rose. Then they set to work and the brothers of Maui made the loops of the noose and lay in wait on one side of the place and Maui himself lay in wait upon the other side.

Maui held in his hand his enchanted weapon, the jaw-bone of his ancestress and said to his brothers to hide themselves and not to be exposed to the sun in order not to frighten him away.

At last the sun rose up out of his place and got caught in the snare, then they pulled tight the ropes, and the sun began to struggle and the snare jerked backward and forwards as he struggled. Then Maui rushed forth with his enchanted weapon and struck the sun fiercely with many blows and the sun screamed aloud. They held the sun for a long time while beating him up soundly, and then they let him go. Since then, weak from wounds the sun crept slowly and feebly along its course.

In Maori thought, humanity is considered to have enormous power, which can alter the environment by fixing the speed of the Sun as well as the conditions of the Earth. People (the descendants of *Tu-matauenga*) are, especially in the terrestrial environment, the chief or *Rangatira* of all environmental elements, because all the environmental elements on the earth were conquered by *Tu-matauenga*. According to the law of conquest in Maori culture, people have the right to use the environment, and hence the Maori proverb said:

> *Te toto o te tangata, he kai*
> *Te oranga o te tangata, he whenua.*
> (Food supplies the blood of people;
> Land supplies food (material welfare) for people.)
>   (Kohere, 1951, p. 16; Brougham and Reed,
>   1963, p. 63)

This proverb is a statement on the economic value of land. In Maori tradition, *whenua* (literally, land) can include the sea and the coastal environment which produce food. We can interpret this proverb as the Maori people's acknowledgement and expectation that the environment will support their material well-being – their very life.

Thirdly, in Maori geomentality people are advised to utilize the earth in appropriate ways. Although people are considered as the centre of the terrestrial ecosystem and have acquired the right to use the environment as conquerors and the *rangatira*, the chief, of the environment, humanity is still an integral part of nature by being only one of many siblings. When people use the environment, they should treat it properly and use it in a proper way by observing all the necessary protocol; otherwise the environment may react. This idea is reflected in the Maori's traditional resource management. For instance, when they gather seafood such as *pawa* (New Zealand abalone), they should only gather as many as they need and not collect an excessive amount. After gathering foodstuff, they should leave the environment as they found it, by replacing stones which they disturbed while gathering food. The traditional Maori idea that nature will respond favourably to people only when they use it properly is perhaps reflected best in the legend of Rata. An abridged version from Grey's collection, 'Adventures of Rata, the enchanted tree: revenge for his father's murder' (Grey, 1974, pp. 84–8) follows:

> Rata's father, Wahieroa was killed by a chief named Matuku-takotako, and his father's bones were still in the hands of the enemy. In order to penetrate the enemy and to recover the bones of his father, Rata wanted to make a good war canoe. So he went into the forest and felled a tall and straight tree and cut off the top end to fashion the trunk into a canoe. All the children of *Tane*, the insects and birds which inhabit there and the spirits of the forest were very angry at this, and as soon as Rata had returned to the village at evening, they all came and took the tree and raised it up again replacing each little chip and shaving in its proper place. Early the next morning when Rata came back to work, he was amazed to find that the tree he felled yesterday stood up again.

However, he felled it to the ground once more and he cut off its fine branching top again and began to hollow out the hold of the canoe. When it became too dark to work, he returned to his village.

As soon as he was gone, the inumerable multitudes of insects, birds, and spirits came back and raised up the tree once more and made it stand sound as ever in its former place in the forest.

The next morning Rata returned once more to work at his canoe. When he reached the place, he was again amazed to see the tree standing up in the forest, untouched, just as he had at first found it. But he hewed away at it again and toppled the tree. As soon as he felled the tree, Rata went off as if going home and then turned back and hid himself in the undergrowth where he could peep out and see what took place. He soon heard the innumerable multitude of the children of *Tane* approaching the spot, singing their incantations. When they arrived at the place, Rata ran out and shouted: 'Ha, ha, it is you, is it, then, who have been exercising your magical arts upon my tree?' Then the children of *Tane* all cried aloud in reply: 'Who gave you authority to fell the forest god to the ground? You had no right to do so.' When Rata heard them say this he was ashamed of himself for what he had done (without going through the necessary protocols). Then the offspring of *Tane* (insects, birds and spirits) again all called out aloud to him: 'Return, O Rata, to thy village, we will make a canoe for you.' Rata, without delay, obeyed their orders, and as soon as he had gone they all fell to work and they no sooner began to adze out a canoe than it was completed. Then Rata and his tribe lost no time in hauling it from the forest to the water and named the canoe, '*Riwaru*'.

An important moral behind the above legend is that when humans do not have appropriate (harmonious) relationships with nature, they will be adversely affected by nature; but when people are in harmony with nature by obeying the rules of the natural environment, they will benefit from nature. The legend can also be seen to reinforce the idea of human dominion over nature and the sense of people's right to use the environment, so long as they use and manage it properly.

Fourthly, in Maori geomentality the natural environment, in different degrees, is considered to consist of living organisms and separate functioning systems. This idea is partly based on the fact that all environmental factors are personified. The universal personification of the natural environment in the Maori culture was noticed by Elsdon Best (1974, p. 44). The first piece of evidence that the environment is considered to be a living organism is from the myth of *Rangi* and *Papa*, in which the Earth is considered to be a living mother. *Papa-tuanuku*, the Earth-mother, treats her children gently and lovingly: when the sweet potatoes and fern roots are fleeing she gently protects them by hiding them in her bosom. The second piece of evidence is clearly reflected in Maui's fishing up of the North Island. In the legend, the North Island is considered a fish. This may reflect the common Polynesian perception of the land as small and valuable islands in the vast sea, comparable with fish in the water. An abridgement of part of Grey's version of 'The Legend of Maui' (Grey, 1974, pp. 30–4) is as follows:

> Maui (*Maui-tikitiki-o-Taranga*) went fishing with his brothers. Soon after reaching the open sea, his brothers caught a canoe full of fish. When his brothers said that they should return home, Maui told them that he also wanted to have a chance to throw his hook into the sea, and asked for some bait. However, his brothers refused it and then he struck his nose violently with his own fist so it would bleed. He smeared his hook with his own blood for bait, then he cast it into the sea. As Maui's hook sank down to the bottom of the sea, it caught in the sill of the doorway of the house of that old fellow Tonga-nui. As he hauled in his line, there ascended on his hook the house and then there came up foam and bubbles from the earth, as of an island emerging from the water. Seeing this, his brothers opened their mouths and cried aloud.
>
> Maui was chanting forth his incantations while hauling in his line in order to make the heavy weights of his caught fish (land) light. When he had finished his incantations, there floated up, hanging to his line, the fish of Maui, a portion of the earth (the North Island of New Zealand) on which their canoe lay aground.
>
> Maui then left his brothers with their canoe saying, 'After I am gone, be courageous and patient; do not eat the food until I return, and do not let our fish be cut up, but rather leave it until I have carried an offering to the god, and the necessary rites be completed in order. We shall thus all be purified. I will then return, and we can cut up this fish in safety, and it shall be fairly portioned out to this one, and to that one, and to that other; and on my arrival you shall each have your due share of it ...'
>
> However, as Maui left his brothers, they began to eat food and to cut up the fish (land). The gods were angry at his brothers cutting up of the fish without having made a fitting sacrifice. Then Maui's caught fish began to toss his head about from side to side, and to lash his tail, and fins upon his back, and his lower jaw. That is the reason that the North Island (*Te Ika a Maui*: Maui's Fish) is now so rough and uneven with mountains, plains, valleys and cliffs. If the brothers of Maui had not acted so deceitfully, the huge fish (the island) would have lain flat and smooth, and would have remained as a model for the rest of the earth. This, which has just been recounted, is the second evil which took place after the separation of Heaven from Earth.

In conjunction with the Maori accounting for the unevenness of the earth's surface and the existence of less-hospitable landforms, it is interesting to note a similar Christian view during the Middle Ages of the contemporary state of the earth being the result of deterioration since the Fall of Man (Glacken, 1967, pp. 162–3, 379). This view is based on Genesis 3: 17–18, where the Lord said to Adam:

You listened to your wife and ate the fruit which I told you not to eat. Because of what you have done, the ground will be under a curse. You will have to work hard all your life to make it produce enough food for you. It will produce weeds and thorns, and you will have to eat wild plants.

With new knowledge of gravity, tides and other physical phenomena, when the European scientists in the seventeenth century attempted to theorize about the origin of the earth they postulated a deteriorating condition of the earth at its crucial stages, such as at the Creation, before and after Adam's Fall, at the Flood and after it subsided (Glacken, 1967, p. 379). Some scholars were especially interested in comparing and contrasting the antediluvian and the postdiluvian world to explain the undesirable effect on nature of the sin of man in forming the present earth surface (ibid., p. 407). A famous clergyman, Thomas Burnett (1635–1715), asserted that the face of the antediluvian world had been smooth, regular and uniform; it was in fact a paradise. But because of the sin of man, the postdiluvian world became a physical ruin and a cursed wasteland with inhospitable mountains, seas and deserts which were neither useful nor beautiful; they were doleful reminders that sinful mankind deserved no better (ibid., pp. 407–8). This Christian view of the deterioration of the earth paralleled the Maori myth that explained the uneven North Island of New Zealand. Because of the sins of Maui's brothers, the fish (the North Island) became an uneven and less hospitable land.

As seen in the myths, the earth as a whole as well as elements of the natural environment such as trees, insects, birds and forests, as well as fish, reptiles and oceans, are considered to be living organisms and separate functioning systems.

Fifthly, the Maori people consider the present environment to be the result of an evolutionary process – unlike the Judeo-Christian mentality, which regards the environment as an outright creation of God. The Maori concept of evolution is somewhat comparable to the Chinese concept of Yin and Yang.

According to Maori tradition, the environment (universe) was not created by an almighty god according to his divine design, but gradually evolved through time. The main driving force behind the evolution of the environment is the interaction of relationships, through fighting, loving, revenging and struggles for dominance between various personified environmental elements. Each element of the environment was basically egocentric and struggled for self-benefits (Yoon, 1986, p. 32). In the Creation myth, the Heaven and the Earth express love for each other and behave in egocentric ways by keeping their children in the cramped space between them, so that they may enjoy their perpetual embrace. However, when the mother Earth saw her fleeing children, sweet potato and fern root, she embraced them and hid them with love. The children, in turn, separated the parents for their own selfish gains and waged war among themselves for their own selfish ends. Exacting revenge in order to achieve justice among the siblings within the family of the environment was an important driving force behind its evolution. Various environmental elements in the Maori myth fought each other for revenge and struggled for

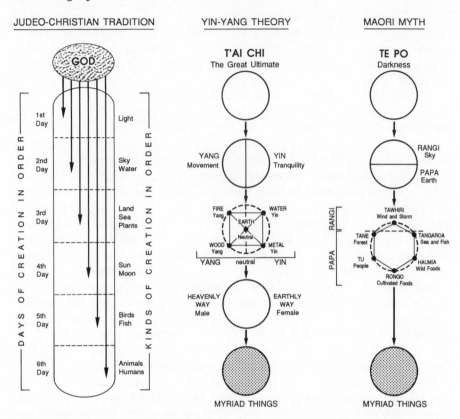

Figure 20.2   Maori and other creation myths

(*Source*: the diagram of the Yin-Yang theory is adapted from *Tai-chi Tu-shuo* [An Explanation of the Diagram of the Great Ultimate] by the Sung Dynasty Neo-Confucian scholar Chou Tun-I; both it and the Maori diagram are redrawn with minor changes from Yoon, 1986.)

the domination of others. In a way, the Maori concept of evolution, especially the diverse environmental interaction through each member's struggle for selfish gains and dominance within the family of the environment (ecosystem), shares a somewhat similar way of thinking with the Darwinian concept of evolution, especially that of the struggle for existence (dominance).

According to the Chinese tradition of Yin–Yang and the Five Elements theory, the present world is the result of a process of evolution somewhat similar to the Maori tradition. However, the environment in the Chinese tradition is not personified as in the Maori myth. The Chinese Yin–Yang theory, which was fully systematized by the Sung Dynasty Neo-Confucianists, was based on ancient Taoist tradition and is highly metaphysical and abstract. The essential outline of Neo-Confucian metaphysics and world view was best presented in the famous diagram of Chou Tun-I (1017–73) in his brief discourse, *Tai-chi Tu-shup* (An Explanation of the Diagram of the Great

Ultimate) (Chan, 1973, pp. 460–5). Figure 20.2 compares an adapted version of Chou Tun-I's diagram with a diagram illustrating the Maori concept of evolution.

Both the Maori and the Chinese tradition assume that at the beginning of the world, the two elements (dichotomy) emerged from the one undivided and unknown substance. From these two, the present world has evolved through a series of evolutionary stages. A brief and highly tentative comparison of the Maori and the Chinese tradition is made here from my earlier work (Yoon, 1986, p. 36).

Both traditions attempt to explain the evolution of the world through a process of regenerative-transformation. The term regenerative-transformation process is adopted to imply that parents do not produce their kind (as with human beings), instead, the following generations emerge in a series of dramatically transformed species (as more evolved sub-systems of the parents' generation). For example, in the Maori myth Punga the son of *Tangaroa* (ocean) begets *Ika-tere* (fish) and *Tu-tewhiwehi* (reptiles).

Each regeneration (development into the next generation) indicates another stage in the evolutionary process. The regenerative transformation process in Maori myth is rather explicit and illustrative, whereas the Chinese theory is rather implicit and mysterious. Where the Yin and Yang forces mysteriously transform into the five elements (metal, earth, water, wood and fire), *Rangi* and *Papa* beget, clearly analogous to humanity six principal environmental elements (forest, people, oceans, cultivated foods, wild foods and winds).

The well-known Judeo-Christian view markedly contrasts with both the Chinese and the Maori traditions: the Christian belief is that the contemporary world is not a result of a gradual evolution through regenerative transformation from the original substance, but an outright creation by God in fully developed form according to his own divine plan (design). According to Genesis 1: 1–31, the environment was created one by one in the following order by the God the Creator:

First day, light (separating light from the dark)
Second day, sky and water
Third day, land, sea and plants
Fourth day, the Sun and the Moon
Fifth day, living creatures: birds and fish
Sixth day, animals (domestic and wild) and humans.

God made each one of them separately, not letting one day's creation evolve into the next stage of creation as it does in Maori myth and the Yin-Yang theory. A comparison of the three traditions is expressed in figure 20.2 in order to elucidate the Maori tradition from different perspectives.

CONCLUSION: MAORI GEOMENTALITY IN ACTION

Traditional Maori geomentality views the elements of the natural environment as persons having family relationships within the ecosystem. Each member of the ecosystem struggles for its own power, and humanity is recognized as the leader (*rangatira*) of the terrestrial environment, who reserves the right to use it so long as they use it properly. An important aspect of Maori geomentality is that the present environment is the result of an evolutionary process, not that of creation by God according to his divine plan.

The Maori people have maintained very personal relationships with their immediate surroundings. In my opinion, it is mainly because the Maori viewed the elements of the environment as persons who are related to them. They have much affection for their home territory (the home environment) and they maintain very strong emotional bonds with it. They identify themselves in terms of the natural environment of their home territory. In a Maori gathering (*hui*), Maori orators still often introduce themselves by naming the important natural features of their home territory, such as mountains, rivers, lakes or harbours (Yoon, 1986, p. 48). For instance, if the orator is from the east coast of the North Island, instead of using his own name he might introduce himself by reciting the following tribal saying (motto-maxim):

> *Ko Hikurangi te maunga,*
> *Ko Waiapu te wai,*
> *Ko Ngatiporou te iwi.*
> (Hikurangi is the mountain,
> Waiapu is the river (literally, the water),
> Ngatiporou is the tribe.)
> (Kohere, 1951, p. 50)

Some other Maori tribes include the name of the chief in addition to the above standard formula of naming one's mountain, water and tribe. A well-known example is:

> *Ko Tongariro te maunga,*
> *Ko Taupo te moana,*
> *Ko Tuwharetoa te iwi,*
> *Ko Te Heuheu te tangata.*
> (Tongariro is the mountain,
> Taupo is the lake (literally, sea),
> Tuwharetoa is the tribe
> Te Heuheu is the chief.)
> (Karetu, 1977, p. 42)

The usage and the content of these tribal sayings reflect the Maori people's personal relationships with the natural environment of their tribal territory, and how much they are attached to and identify with it. Within Maori people's geomentality, the landmarks (important natural features) of a people's territory are the beloved yet somewhat distantly related persons who provide them with their spiritual identity and material support. Therefore, the Maori people were most reluctant to depart from their places (land: local environment) and strongly resisted any attempt to alienate their land (landscape) from them, both pre- and post-European settlement in New Zealand. Hence, the Maori proverb says,

He whenua, he wahine, ka mate te tangata (A man will fight to the death for his land and his woman). (Best, 1941, vol. 1, p. 297)

Indeed, the sale of tribal land was sometimes referred to as the death of the land; this was recorded as early as 1860 at the sale of the Wiaitara block (Martin, 1861, p. 39). The Maori mind has always been on the Maori land (Yoon, 1986, p. 125). In Maori geomentality, the land (natural environment of the tribal territory) was thought to be a more durable and outlasting relative than human relatives who soon pass away; as a Maori proverb says, 'People disappear, land remains' (*Whatu ngarongaro he tangta, toi tu he whenua*) (Kohere, 1951, p. 38; Best, 1941, vol. 1, p. 400). That was why a Maori coming home from a distant place after the death of his mother could cry out, 'I greet my only living mother, the land.'

ACKNOWLEDGEMENTS

I am grateful to my friend Bill Boyd of the University of New England, Northern Rivers, Australia, for critical reading and comments on this manuscript; and to Jan Kelly of the University of Auckland for her professional cartographic help with the drawings of the map and the diagrams.

REFERENCES

Best, Elsdon 1941 (1924): *The Maori*, 2 vols. Wellington: Polynesian Society.
—— 1974: *Maori as He Was: a brief account of Maori life as it was in pre-European days.* Wellington: Government Printer.
—— 1977 (1942): *Forest Lore of the Maori.* Wellington: E.C. Keating, Government Printer.
Brougham, Aileen E. and Reed, A. W. 1963: *Maori Proverbs.* Wellington: A.H. and A.W. Reed.
Chan, Wing-Tsit 1973: *A Source Book in Chinese Philosophy.* Princeton, N.J.: Princeton University Press.
Facsimiles of the Declaration of Independence and the Treaty of Waitangi 1976. Wellington: A.R. Shearer, Government Printer.
Firth, Raymond 1972: *Economics of the New Zealand Maori.* Wellington: A.R. Shearer, Government Printer.
Genesis, The Book of. In *Good News Bible*, Wellington: Bible Society in New Zealand, 1976.

Glacken, Clarence 1967: *Traces on the Rhodian Shore*. Berkeley, Cal.: University of California Press.

Grey, Sir George 1974 (1906): *Polynesian Mythology and Ancient Traditional History of the New Zeanders*. Christchurch: Whitcombe & Tombs.

Karetu, Sam 1977: 'Language and protocol of the Marae'. In M. King (ed.), *Te Ao Hurihuri*, Hicks Smith/Methuen New Zealand, pp. 31–45.

Kohere, Reweti T. 1951: *He Konae Aronui, Maori Proverbs and Sayings*. Wellington: A.H. & A.W. Reed.

Lewthwaite, Gordon 1950: 'The population of Aotearoa: its number and distribution'. *New Zealand Geographer*, 4(1), 35–52.

Martin, Sir William 1861: *The Taranaki Question*. London: W.H. Dalton.

'Mr Firth's visit to the King Party'. *Daily Southern Cross*, 9 June 1869. Auckland.

Pool, Ian 1991: *Te Iwi Maori*. Auckland: Auckland University Press.

Yoon, Hong-key 1986: *Maori Mind, Maori Land*. Berne: Peter Lang.

—— 1991: 'On geomentality'. *GeoJournal*, 24(4) 387–92.

21

The Continuing Creation of Identities in the Pacific Islands: Blood, Behaviour, Boundaries and Belief

Ron Crocombe

The range of identities relevant to Pacific peoples, as to their cultures and politics, keeps getting *wider* (both geographically and in the number of persons connected), *deeper* (in terms of number of layers), *more differentiated* between groups and individuals within groups, and *more complex*. One concern here is with the effects of population movement on identity, though these cannot be isolated from media images, investment or technology. But mobility is so high in this region that it has much more influence on identity than in most of the world.

Perhaps the most pervasive trend is for growth in external sources of identity among the peoples of Polynesia, Micronesia and outer Melanesia (that is, Fiji, New Caledonia and Irian Jaya). This is very much less so in the case of inner Melanesia (Papua New Guinea, Solomon Islands and Vanuatu).

To focus on changes in identity, particularly those mediated by the movement of people, let us divide the region into three broad categories in relation to the intensity of such influences (see figure 21.1):

1 *Highest – those with free access to a metropolitan country:*
   (a) all islands north of the equator, and American Samoa, to USA;
   (b) Cook Islands, Niue, Tokelau and Norfolk Island to both New Zealand and Australia;
   (c) New Caledonia, French Polynesia, Wallis and Futuna to France (and any other French territory in the world);
   (d) Pitcairn to UK and in practice to New Zealand;
   (e) Rapanui (Easter Island) to Chile; and
   (f) Irian Jaya to other parts of Indonesia.

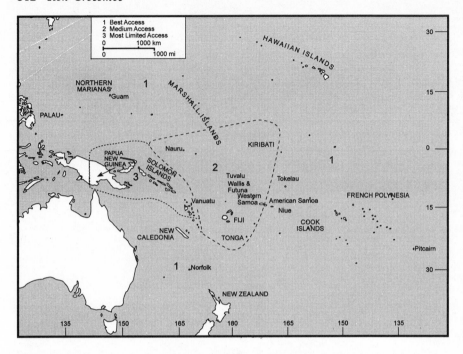

Figure 21.1   Relative accessibility in the Pacific Islands

2 *Medium – those with significant access to one or more metropolitan countries:*
  (a) Western Samoa has special access to New Zealand, and through that can obtain later access to Australia, and via American Samoa many have achieved US residence or citizenship;
  (b) Tonga has significant access to New Zealand, Australia and USA through legitimate and other processes that have spread Tongans very widely;
  (c) Tuvalu and Kiribati have limited special access to New Zealand, and extensive labour mobility elsewhere;
  (d) Fiji's Indian population have resettled in large numbers in Canada, USA, Australia and New Zealand as well as in India and Europe; Fijians have not moved nearly so much, but the numbers in Australia, New Zealand and USA are substantial, and migration of all other categories of Fiji citizens (European, part-European, Chinese particularly) has been very high;
  (e) Nauru was offered free access to Australia, but declined. But Nauruans have relatively high incomes, and are highly mobile people.

3 *Very limited external access.* This concerns mainly the central Melanesian states of Papua New Guinea, Solomon Islands and Vanuatu. Kiribati is between this category and the previous one at present, though likely to fit increasingly into category 2. The central Melanesian states have the

numbers (that is, there are many more central Melanesians than all other Pacific islanders put together) and fortunately they have room to move and expand (that is, population densities are low: only about 8 persons per square kilometre as against 75 in tropical Polynesia and 130 in Micronesia).

To some extent the above categories are reciprocal: the countries with the highest proportions of their people abroad, generally also have the highest proportion of others coming in – whether as permanent residents, business people, short-term workers or tourists. Thus influences of all kinds are maximized in the places with high rates of interaction and minimized in central Melanesia, with the exception of a small elite.

BASIC IDENTITIES

All human identities are to some extent arbitrary and changing, and are adapted relative to context. A living inventory of ingredients provides sources from which identities are in a continuous process of construction and adaptation. In usage, particular facets are emphasized or minimized to derive maximum benefit from particular situations. Some of the foci which are vital to most Pacific people in the first and second categories above, and to a much lesser extent in the third, include:

1 a multi-island, multinational kinship network;
2 a series of localities with differing kinds and intensity of identification;
3 one or more nations (legal and residential identities);
4 a nesting set of languages/dialects, and of related cultural patterns;
5 an international religious denomination;
6 an internationally linked profession or occupation;
7 over 100 international voluntary organizations;
8 consequences for identity of education on the Pacific Rim;
9 regional identities – ethnic, geographical, political.

Perhaps the most notable feature is that every category now has a strong international aspect. But the extent and nature of that differs markedly between Polynesia and Micronesia on the one hand, and Melanesia (particularly central Melanesia) on the other.

None of the categories is distinct, and each influences the other. For example, one cannot divorce one's meaningful localities and physical boundaries from one's genetic boundaries and kinship network, one's social boundaries of language(s) and culture(s) or the political boundaries of nation(s)

1. A multi-island, multinational, kinship network

The people of each country have a distinct pattern of movement. For most Polynesians from the tropical South Pacific (but not nearly so much those for Hawaiians or Maoris of Aotearoa/New Zealand), this network spreads across many islands within their country, often two or more countries within tropical Polynesia (particularly Western Samoa/American Samoa, Cook Islands/ French Polynesia, Tokelau/Samoa, Tonga/Samoa/Fiji, Tuvalu/Samoa/Fiji, French Polynesia/Wallis and Futuna/New Caledonia), and several countries beyond. For Samoans and Tongans it is New Zealand/Australia/USA. For Cook Islanders it is New Zealand/Australia; for French Polynesia, Wallis and Futuna, and New Caledonia it is France; for Rapanui (Easter Island) it is Chile and much of South America and Western Europe.

Many *people* move along the networks (in from the metropole to the islands for funerals, weddings, community celebrations and other short visits; out usually for much longer periods, whether for work, study, medical treatment or otherwise). There is much use made of *telecommunications* (by 1989 all islands, states and territories used international satellite communication; and usually reverse charges are paid by the relative in the metropole, except in the case of students calling home). *Cash remittances* are vital, for example 8 times more money – US$39.2 million – came into Western Samoa in 1990 in remittances from Samoans abroad, than from exports – about 42 per cent of it from New Zealand, 33 per cent from the USA (but only 1.9 per cent of that from Hawaii, 10 per cent from American Samoa, 11 per cent from Australia). There is also a lot of *other material goods* such as clothes, building materials, second-hand vehicles and selected foods.

Ideas, customs and values travel with the migrants when they return home or when they phone, write, send videos and photographs or gifts of money and material goods. They revise concepts of what constitutes a good life back home and how it should be lived. The most obvious impact is on consumption – both regular consumption and that for special occasions. This includes clothing, accessories, cosmetics and food patterns (for example, Cook Islanders on the flight from Auckland to Rarotonga stock up with Kentucky Fried Chicken because it is so popular at home; and even though wedding cakes can easily be made locally, most brides and their families insist on one flown in by jet from abroad). Some of those who have lived abroad for a long time introduce a more economically orientated pattern of time allocation and a more restricted pattern of financial distribution to meet customary obligations.

Some who have not adjusted to life abroad return home with the antisocial skills they have acquired. Overseas experience has been a major factor in changing the identity and enhancing the antisocial activities of delinquent youth in several Polynesian societies. They are accused of contributing to the increase in crime, narcotic drugs, solvent sniffing, sex deviation and gang formation. Many more have returned with radically changed ideas about gender relations, inherited rank, religious beliefs and other values.

Most islanders with long-term free access to a Pacific Rim country have taken it; but there is a constant flow back and forth. In addition, most men in their thirties from Fiji have worked abroad, mostly on seasonal migrant-worker schemes between the New Zealand and Fiji governments. Many Tongans participated in similar schemes. Many who went for a short assignment stayed for the rest of their lives. The cumulative experience of those who returned to the islands probably modified their vision of themselves and the outside world – though precisely how has not been studied.

A special category derives from inter-island movement last century and early this century. Many were effectively cut off from their homelands. In recent years, however, as a result of better communications and higher mobility, many connections have been re-established and identities reinforced. For some Solomon Islanders and ni-Vanuatu this involves contacts with descendants of relatives who went as indentured workers last century to northern Australia, Fiji and Samoa – even a little with Tonga. For i-Kiribati it involves inter-action with Fiji, Solomon Islands and Vanuatu, where many of their people have long been resettled due to overpopulation at home. There are many long-standing contacts, from both before and after European contact, linking Fiji, Tonga and Samoa, and also between the Cook Islands and French Polynesia.

Others forced to migrate in more recent times include Papuans escaping the Mt Lamington eruption, the Lopevi people in Vanuatu resettling at Maat after their island exploded, the Phoenix Islanders migrating to the Solomons due to prolonged drought, the Banabans moving to Fiji due to phosphate mining, and the Bikinians having to make way for American nuclear weapons testing. All have had to integrate with 'strangers' and set up wider kin networks.

But while the various kinship networks are getting wider, they are also getting shallower, that is, a reducing percentage of one's total transactions seem to be with kin. Kin remain important for many people for life crises, for ceremonies, for capital accumulation for close kin or for special community projects, and for symbolic and psychological support.

The functional core of the kinship network is made up of indigenous blood ties. But the non-indigenous genetic ties are of growing importance, both because of increased rates of out-marriage and because more people are actively recognizing their earlier non-indigenous ancestry. While Europeans held political and other power, it was not uncommon for islanders with European ancestry to emphasize that identity. Likewise, with independence and the rise of indigenous power, many such persons equally naturally emphasized that side. But it is not a simple relationship, and there is much room for adapting identity to context for particular purposes.

With the rise in the status of China in the region in the 1970s and 1980s, Chinese heritage was more actively recognized. The media models of Black Americans as well as the coming of Black American evangelists correlates with a recognition of the very extensive African blood that was introduced to the region by persons of African descent from the whaling ships. Another difference with those of African ancestry is that it cannot be traced to existing relatives in the source country, whereas European and Chinese ancestry often can be and is.

This is a continuing process. The rate of out-marriage appears to be increasing among the elite and among the mobile of any status level. Almost all Cook Islanders have close relatives by marriage who are European and/or other Polynesian, and in a few cases Asian. I once booked into a guest-house in Samoa and an Australian tourist asked me to be best man at his marriage to our waitress. Their first contact was on his arrival five days earlier. The

Samoan lady who ran the place told me that she only had two or three waitresses at a time, but that five had married tourists in the last three months and had moved abroad with their new husbands. More new networks are established.

The concept of what constitutes 'indigenous' continues to change and to be conditioned by demography as well as by power. Thus for most purposes, in places where the indigenous community is in a minority and lacking in power, any indigenous ancestry now constitutes an indigenous person. The Fiji constitution adopted at independence in 1970 accepted as Fijian any person descended in the male line from any Melanesian, Polynesian or Micronesian. Fijians were then a minority and felt the need for this support. Following the 1987 military coup in which indigenous Fijians took control from a newly elected government which was dominated by Indians, a new constitution was promulgated in 1990. 'Full' Fijians were now firmly in control, and the new constitution accepts as Fijian only those of specifically Fijian ancestry and that in the male line.

Solomon Islands legally accepts Melanesians from Papua New Guinea and Vanuatu as equals for citizenship purposes, but not those from other parts of Melanesia. Those three central Melanesian countries also distinguish legally, and in some rights and privileges both constitutionally and otherwise, between persons of full Melanesian ancestry and those of part Melanesian and other ancestry. As these are the three Pidgin-speaking countries, that language is becoming a stamp of common identity and legitimacy for those countries.

Migrants usually leave a situation in which they had the confidence of being on their home ground, with a comprehensive kinship network and membership of the majority, to a situation where they are members of a minority. However, many Niueans, Tokelauans, Samoans, Cook Islanders and Chamorros from Guam and the Northern Marianas now have more relatives in the countries where they have settled than they have back home. And they have higher standards of income, consumption, personal mobility and sometimes personal freedom. Thus the overseas group often becomes the reference group for some values and types of behaviour. Those at home constitute the reference for a diminishing range of behaviour – particularly that relating to land and traditional custom.

Those abroad are often of lower *relative* income, education, prestige and status within their new society than they were at home. But they are probably accorded higher esteem, and certainly more rights and privileges, when abroad in metropolitan countries than if they migrated to other islands states (and in some respects even within their own country). The restrictions on Tuvaluans and i-Kiribati in Nauru, on i-Kiribati in Solomon Islands and Vanuatu, and on Solomon Islanders in Fiji are much more serious than in most metropolitan countries.

Intermarriage is highest among migrants. More than is generally acknowledged, persons of multiple ancestry occupy gatekeeper and mediator roles in the region. Many are persons who have also spent long periods away from home. They are also important models.

2. A series of localities with differing kinds and intensity of identification

The intensity of identity usually varies with the proximity to that land in which the individual feels (not always accurately) that he/she has the strongest rights and can derive the most potential benefits, whether material or psychological. At wider levels of decreasing intensity, it includes particular plots of land, a village (usually of birth or parent's birth), a district, an island and a nation. Each of these remains significant, but in most of the eastern and northern Pacific the particular island retains a high level of prominence.

Particularly in situations of high mobility, and in situations (such as Eastern Polynesia) where a previous unilateral descent system has been replaced by a bilateral one which becomes quadrilateral in the next generation and so on, there will be several such localities. But each individual is usually clear about the order of priority in which they belong, even though that priority will change according to circumstances.

The important issue for this discussion is what localities people who identify themselves as Pacific islanders identify with beyond the nation of origin. Many islanders own homes in New Zealand, Australia or USA (very few indeed in other islands countries), and many island babies are named after the streets, towns or cities where relatives live or have lived abroad. At least 10 per cent of Tongan landowners are citizens of other countries, where they live, but they retain their land rights in Tonga.

For Maori and Hawaiians the nation they identify with is effectively dominated by non-Polynesians. But when island Polynesians come into those countries most do not identify particularly closely with the Maori and Hawaiian populations. This is for two main reasons: the Maoris and Hawaiians tend to be concentrated in the less prestigious categories; secondly, they tend to assert that they, as the indigenous settlers, should receive various privileges over immigrants. That gives the Polynesian and Micronesian immigrants more in common with other immigrants in those places. In both cases there has been a significant move by immigrant Polynesians to press further afield – from New Zealand to Australia and from Hawaii to mainland USA.

The total national population of the Cook Islands is only 18,000, but more than half that number travel between the Cook Islands and abroad (mainly to Aotearoa/New Zealand) every year. American Samoans and Chamorros likewise travel very frequently to the USA

It is not necessary to have been to a place to identify with it. I once picked up a man on the road in Fiji. In the course of chat I asked him: 'Where do you come from?' and he replied: 'I come from Rotuma.' 'When were you last in Rotuma?' 'I've never been to Rotuma.' This is true for many islanders, who still consider home to be the place their ancestors came from. Until the 1950s many third and fourth generation New Zealanders still referred to the United Kingdom as 'Home', though they had no intention of living there. They were still happy to send soldiers to defend the UK.

3. One or more nations

For most Pacific Island nations, most people identify with more than one nation. And the more elite, the more multiple the identification. This is not always a willing identification – for example, most Melanesians of Irian Jaya would probably prefer not to be Indonesians, which by citizenship they are. And the majority of Kanake of New Caledonia would prefer not to be French – at least, not exclusively so. Tahitians have consistently voted to stay French (and at the recent election candidates from one of the two independence parties all lost, while the other doubled its roles, but only to 10 per cent of the total vote), but they still want more local autonomy. Wallis and Futuna seem committed at present to retaining French nationality and local autonomy. Easter Islanders are also Chileans and likely to remain so for the foreseeable future. Pitcairn Islanders show no sign of wanting to change from being British, nor do their Norfolk Island cousins want to lose their Australian citizenship, even though they want to retain their local autonomy.

All islanders north of the equator have American identification of one kind or another; either as citizens, as nationals or through the Associated State relationship they have an automatic right of entry, work and perpetual residence. All Cook Islanders, Niueans and Tokelauans have automatic New Zealand citizenship and automatic right to enter, work and live in Australia as well.

Cook Islanders, Niueans and Tokelauans have for practical purposes triple nationality (their own plus New Zealand and Australia) but most constitutionally independent Pacific countries require that a person hold one nationality only. Western Samoa has relaxed this requirement (and many Samoans, including many in high places, had an American or New Zealand passport under the bed years before it was lawful to do so). Many Tongans and Fijians also hold dual nationality and it is becoming more widely accepted. It is interesting that the banning of multiple citizenship was to preclude Europeans and other non-indigenous persons doing so in the islands, but in practice it is mainly islanders who have done this. Multiple national identity extends to voting in two countries for some Samoans and Cook Islanders in New Zealand and Australia.

4. A nesting set of languages/dialects and related cultures

The average Pacific islands language has only 5000 speakers. Depending on how one defines dialects/communalects, the number of persons in a group identifiable by some aspect of its language would probably be less than 1000. But the key issue here is what *external* languages and dialects are spoken and what meaning they have for those who speak them.

The idea of a nesting set implies that one's 'root' language/culture comes first, and national or international ones later. But that is changing fast. If it has not happened already, then very soon for most persons of Samoan, Cook Islands, Niuean, Tokelauan and Chamorro ancestry at least, it is the other way round, with English as the first language (though not all those to whom this applies would admit it). For a growing proportion of Tahitians, French and/or English is the most-used language.

English has become the key language for the region and is taught in schools in all countries. Tahitian and New Caledonian children, for whom the main international language is French, are all taught English as well. More than 300 Tahitian children go to New Zealand every year to practice English. Others go to the USA and elsewhere.

The main external language/culture models tend to be those of working-class people of European origin in the USA, New Zealand and Australia. Maori in New Zealand and American Blacks (but not the more numerous Hispanics) in the USA tend to be secondary models. The external models tend to reflect back to the home society through travel in each direction, and through telecommunications and home videos which are becoming a widespread means of communication between islanders abroad and those back home – particularly in relation to life crisis events such as marriages, hair-cutting ceremonies, babies and death.

5. An international religious denomination

Pre-Christian religion survives in very few pockets of Western Melanesia, and then always in interaction with Christianity and/or Islam. Nevertheless, some elements of the diverse pre-Christian religions have been incorporated into Christianity in the islands as they were in Europe. In each case the recipient culture made the final product into a unique amalgam with its own stamp of authenticity. For example, Vince Diaz, himself Chamorro, has studied the deep and intricate relationship between Catholicism and Chamorro identity.

The earlier established denominations in the South Pacific drew their primary external influences from immigrants from Britain (mainly the London Missionary Society or LMS), or from Australia/New Zealand (Anglican, Wesleyan/Methodist, Presbyterian, Seventh-day Adventist). There was also some religious training or religious visitation by islanders to those countries – even as far back as the first half of the last century.

Catholic sources were more diverse (France, Netherlands, Spain, Australia, New Zealand) and with little reverse flow of islanders. Partly as a consequence of the difference, to belong to the LMS involved an identification with Britain, but to be served by Dutch priests (as were the islands of eastern Polynesia) did not mean a corresponding identification with The Netherlands.

The second wave of Christianity has involved a strong US identity. In the early days they were from the north-east, mainstream and white (especially the American Board of Commissioners for Foreign Missions), but later from the west or south, fundamental and minority. The Church of Jesus Christ of Latter Day Saints (Mormon) accepts the USA as the Promised Land; it took a special interest in Polynesia to which it felt it had a special mission as the home of one branch of the Lost Tribe of Israel. The Mormons were relatively wealthy, encouraged islanders to become missionaries beyond the islands, and facilitated education for islanders in the USA. This is the most strongly nationally identified denomination in the Pacific.

American priests took over from Spanish in Micronesia in the late 1940s. Their impact was strong, but it was for one generation only because the USA, like other industrialized countries, is running out of priests. Some of them may be replaced by Polynesians, who are going into the priesthood in considerable numbers. This is one factor that is likely to draw Micronesia a little closer, but not much closer, to Polynesia. It is likely to be offset by an even greater number of priests from the Philippines.

Then the Apostolic/Pentecostal/Assemblies of God/Revivalist movement spread from southern USA to the islands in many forms and with great vigour. Their commitment to tithing (giving one-tenth of one's income to the church) yields considerable funds. There is much travel of American evangelists to the islands and of islanders to the USA, and the influence is reinforced by US videos, songs, styles and even accents and terminology. This is also true of the Mormons.

Perhaps the most significant difference with some of the Apostolic movement is that the origin, personnel and orientation are to Black American culture rather than White. However, despite differences of detail, that merely shows that Black culture in the USA differs only in detail from White – at least, with the export model. There is no Hispanic influence in the islands.

The churches that localized most were those that were most 'moderate' in doctrine: the LMS derivatives, Wesleyan and Presbyterian. Having localized, however, they seem to lose their flocks at the fastest rate to the new, well-financed, highly organized, more fundamentalist, US-orientated denominations, which also have the highest proportions of non-indigenous personnel.

An interesting reversal is that now there are many more islanders who come to preach Christianity to New Zealand than New Zealanders going to the islands. But there are no islanders taking any islands religion to New Zealand; it is simply a re-export of repackaged Christianity.

Two non-Christian faiths are of increasing significance. The Baha'i faith, which has spread throughout the islands since its introduction in 1954, gives its adherents a strongly international orientation. It is also orientated to 'green' values. Islam is more recent in having Pacific Islanders as members, but in some places it is the fastest-growing faith. With growing support in Tuvalu, Tonga, Kiribati, Samoa and elsewhere, it orientates its followers to the Middle East and Pakistan, where islander missionaries are now being trained. Part of the process is the adoption of Arabic names, a significant feature in cultures where names are vital symbols. In future it is likely also to be orientated to Indonesia, Malaysia and Brunei.

6. An internationally linked profession or occupation

Almost all of these are orientated to the Pacific Rim. There are Pacific Islands associations of some professionals, but much stronger are those with Rim associations. Generally, the higher the status of the profession in terms of years and level of training, the more foreign the orientation and the formal linkage, for example, among journalists, diplomats, politicians, senior officials, higher education staff, judges and lawyers. The linkage above the equator is mainly to the USA and its way of doing things professionally, and to Australasia below the equator.

Even the basic workers, if unionized, are linked into the international trade union movement through the South Pacific and Oceanic Council of Trade Unions as well as through bilateral ties. Here again, the ties are mainly with Australia and New Zealand in the South Pacific, and the USA in the North.

7. Over 100 international voluntary organizations

Most of the voluntary organizations to which islands people belong are international – over 100 of which operate in the islands. They include youth organizations such as Scouts and Guides; women's movements such as the Young Women's Christian Association (YWCA) and the Pan Pacific and South East Asia Women's Association (PPSEAWA); health, nutrition and welfare non-governmental organizations such as the Red Cross, the International Planned Parenthood Federation and the St Vincent de Paul Society; service organizations such as Rotary and Lions; political and ideological movements such as the Nuclear Free and Independent Pacific and the South Pacific Association of Progressive Parties; cultural and media associations such as the Pacific Arts Association and the Pacific Islands News Association, and a host of others. The great majority are linked primarily to White industrialized countries on the Pacific Rim. A member of such an association has at least some contact with, orientation towards and often identity with the country from which the activities originate, are financed or promoted.

8. Consequences for identity of education on the Pacific rim

Education outside the nation is much more extensive than for other parts of the world. The diverse experiences of the educational elite modify islands identities, but some sources and models are changing. Thus almost all Cook Islanders undertaking tertiary education do so in New Zealand, the USA, Australia, Fiji, Papua New Guinea, Samoa and a few in East and South-east Asia.

Few islanders go beyond the Pacific Rim; those who do, go mainly to Europe. The largest categories are the Tahitians and New Caledonians to France, and Commonwealth citizens to Britain. The main exception is that many Fiji Indians study in India, which remains a powerful if uncertain model for them, but this influence does not permeate beyond that community.

Most English-speaking South Pacific people who completed high school from the Second World War until the 1980s were taught, at least in part, by New Zealanders. The only exception was in Papua New Guinea, and they were taught by Australians. Anyone north of the equator was taught by Americans.

The USA undertook the world's most intensive educational airlift ever, and flew all the brightest and best from Micronesia and American Samoa to the USA for higher education. And in preparation for that experience, the US Peace Corps and other US personnel were posted to every primary and secondary school, determining as well the curriculum and training. Even since self-government this pattern is still strong. Almost all political and professional leaders in islands north of the equator and in American Samoa have been educated in the USA.

New Zealand had a much less intensive programme of education abroad, but the first heads of government of the Cook Islands, Fiji, Kiribati, Solomon Islands, Tuvalu and Vanuatu were all educated in New Zealand. The first Prime Minister of Western Samoa was not educated there, though his wife and his successor were. Many Papua New Guineans went to Australia for secondary and tertiary education, but the proportion was not nearly as high.

The New Zealand educational influence (both of islanders going to New Zealand and New Zealand teachers going to the islands) waned in the 1980s and will not resume its former importance. It remains significant, however: 1233 Pacific Islanders studied at the University of Auckland in 1991, and many more in the country's other six universities and twenty or so community colleges and polytechnics. Most of them are likely to remain abroad. But the US and Australian influences expanded and are likely to continue to do so.

An East Asian influence has become apparent, and there are now islands students in the Philippines, Singapore, Malaysia, Thailand, Taiwan, China, Japan and Korea. One also sees expansion of islands students throughout Europe and even a beginning in Latin America. The main area of expansion for the next generation, however, is likely to be North-east Asia.

9.   Identity and Pacific regionalism

There was very little co-ordinated contact between the islands until after the Second World War. Then a complex system of regional organizations was set up by the withdrawing colonial powers; they initiated almost all such organizations and continue to be the main sources of finance, though the rising powers of North-east Asia are likely to take over that role, and the influence that goes with it, before the turn of the century. Some of the consequences of regionalism that are relevant to this topic are the effects that islanders living abroad, and their experiences, have on their home countries.

Regional education. The first relevant experience comes from living together in regional educational institutions. This began with the then Central Medical School (later to become the Fiji School of Medicine), set up in Suva by the Rockefeller Foundation for all the Pacific islands. Doctors from all English-speaking South Pacific countries, and later from some North Pacific islands also, studied there. While they formed a network of community and common interest, it was only after air transport became widespread and international conferences were financed from abroad that their common background became a functional network.

Then followed the South Pacific Commission's one-year courses in community development for Pacific women, the Derrick Technical Institute (financed by the United Kingdom for students from all UK territories), then the University of the South Pacific with its main campus in Suva and its agriculture campus in Apia. Next came the South Pacific Regional Telecommunications College (soon taken over by the government of Fiji, but still training some students from elsewhere), and a range of others. The University of Papua New Guinea took many Melanesian students, some from Polynesia and a few from Micronesia. The College of Micronesia, Xavier High School and the University of Guam helped to build a Micronesian awareness, though all were staffed by Americans and many more Micronesians went to the USA than to any of those.

In religious training, the Holy Spirit Seminary and the Pacific Regional Seminary between them trained most islander priests. The Pacific Theological College taught a small proportion of Protestant ministers from many denominations and countries. These three theological colleges are the only institutions to have significantly overcome the English-speaking/French-speaking divide. No educational institution has bridged the divide between the rest of the Pacific and the people of Irian Jaya or those of Rapanui (Easter Island). In addition to the 50 or so national theological colleges, there were some multinational ones, such as Tangintebu for Kiribati and Tuvalu.

Seventh-day Adventist ministers for the region are trained together at the Pacific Adventist College in Port Moresby. Mormons giving their two years of missionary service as church elders get a more international experience: they may be sent anywhere in the world.

There are now over 30 regional educational institutions, the main ones being based in Fiji, Hawaii, Papua New Guinea and Guam. More Pacific Polynesians, however, are educated in universities and other institutions of higher education in New Zealand and the USA than in any of these.

In addition to those who live together for several years studying for formal qualifications, much larger numbers are brought together for regional short courses by the 40 or so United Nations agencies and other international agencies, such as the Commonwealth, the Asian Development Bank, the World Bank and so on.

Pacific universities are intended to strengthen territorial identity, usually that of a recently formed nation – as with the University of Papua New Guinea, the PNG University of Technology, the National University of Samoa, and the proposed or emerging institutions in Solomon Islands, Tonga, Federated States of Micronesia, and Tonga. The American Samoa Community College and the University of Guam serve both territorial and national identity functions. Cenderawaseh University in Irian Jaya, and the *Université française du Pacifique* in Tahiti and New Caledonia strongly promote Indonesian and French identities, respectively.

Identity above the national level is recent, limited and fragile. The United Kingdom wanted to bring its scattered Pacific territories together as a single national federation, but most island leaders approached about it resisted strongly. Reluctantly, it had to accept separate approaches to independence. One of the United Kingdom's aims in setting up the University of the South Pacific was to enhance a regional identity for the former United Kingdom and New Zealand territories in the South Pacific by bringing young, potentially influential people together. That goal has been achieved to some degree but much less than was expected or feasible, because of the centralist philosophy and consequent imbalance in the nationalities represented. This left Fiji as by far the greatest beneficiary, at the expense of the smaller nations. Within Fiji, the Indian population were the greatest beneficiaries. This has been a cause of nationalistic reactions both within Fiji and beyond, but the external donor countries see the centralist model as to their benefit, and subsidize its continuity in that form.

Once the USA had withdrawn from its control of the College of Micronesia (COM), the Micronesian countries preferred to emphasize national rather than regional identity and operation. Thus the COM campus in the Marshall Islands has now been changed to the College of the Marshall Islands. The main campus is becoming *de facto* a college of the Federated States of Micronesia, and within that mainly of Pohnpei, the state where the campus is located. Chuuk state, rather than send its people to Pohnpei, tends to send them to Guam or Hawaii, to the US mainland or elsewhere, in the absence of its own state college. The Palau campus of COM is becoming *de facto* a national entity.

The primary identity of the Pacific Adventist College is religious, the regional one being secondary. However, as the only university-level institution covering all Pacific islands, its regional role has been effective. *L'Université Française du Pacifique* planned to project a French image throughout the South Pacific, but is unlikely to achieve any identity function beyond the French territories. Even there the distinct territorial/national identities could diverge more than converge. Extensive scholarships for Micronesians to US universities and the supply of US staff to Micronesian institutions was part of a conscious policy to strengthen Micronesian identity with the USA. This was intensified following the Solomon Plan in 1964, which aimed to give Micronesians a permanent and ineradicable US identity. Though not formally adopted, the main recommendations of the Plan were implemented, though whether that was in the Micronesian interest is another issue.

The greatest success in universities enhancing a Pacific regional identity will come from the extensive exchange of students between island universities in such a way that students from all countries study in the widest possible range of countries.

The rapid growth in regional education has been accompanied by growth in education outside the region. No detailed study has been made of what mix of the two and where is most desirable. National education promotes national identity, which every Pacific country rates as a priority. But with a highly internationalized world, too much concentration on the nation can leave it isolated and parochial, and not well placed to look after its interests in the wider world.

Regional education has much to commend it but that increases the regional identity; that is not where the real power relative to the islands resides either. Moreover, regional education does not necessarily promote the kinds of islands identity its founders had in mind. For example, Sir Michael Somare, former Prime Minister and now Foreign Minister for Papua New Guinea, recently pointed out that most senior academic staff at the nation's universities now come from the Indian subcontinent, Africa and other countries with which his country has no major interaction in terms of trade, aid, political, media or other relations, nor does the government see them as models to be emulated. And most academic staff at the main campus of the University of the South Pacific are ethnic Indians or others from the Indian subcontinent (as they are at several other regional educational institutions in Fiji); the next largest number is of Europeans. Pacific Islanders are few. Many more Pacific Polynesians teach in New Zealand and American universities and colleges than at the University of the South Pacific.

Pacific universities do put students from the various countries in contact with one another. This has many advantages, but the important question is what contacts and identifications will serve the island nations best in the long term? To the extent that they leave the home country, there is also a good case for a considerable proportion to study in those countries with which theirs has most interaction. In the past this has been mainly in the USA for the islands of the North Pacific, and in Australia or New Zealand for the South. But from now on it should include a strong element of study in Japan and other nations of the North Pacific Rim.

Working in regional organizations. Another experience creating cross-Pacific linkage is working for one of the 300-plus regional organizations. Only the large ones have a multinational staff, particularly institutions such as the South Pacific Commission in Noumea (New Caledonia), the Forum Secretariat in Suva (Fiji), the Forum Fisheries Agency in Honiara (Solomon Islands), the South Pacific Alliance for Family Health in Nuku'alofa (Tonga), the regional headquarters of the International Planned Parenthood Federation in Apia (Samoa), and so on. The extent of these employment opportunities is largely determined from abroad, for almost all the regional organizations are financed mainly from overseas. The growth or otherwise of such opportunities, therefore, is not a matter which can be determined within the islands. Moreover, island governments have not been prepared to contribute much to regional employment, and none has been prepared to facilitate the employment mobility for member country citizens as has been done in the Caribbean. On the one hand, these organizations familiarize staff with the region, but on the other, they also familiarize them with international pay-rates, housing standards, working conditions and perquisites They constitute models which almost none of their respective home countries can afford to copy.

The sporting regions. There is a series of networks. For example, rugby football links Fiji, Western Samoa, Tonga and New Zealand. Boxing has been the only sport that effectively cuts across the language barrier, and connected New Caledonia, Vanuatu, Fiji, Tonga, Samoa, the Cook Islands and French Polynesia. Cycling connects French Polynesia and New Caledonia with New Zealand and Australia.

Then came the South Pacific Games, linking all the South Pacific Commission region. But the big recent change is the entry of the Oceania Olympic Committee and close to 20 Oceania associations for a number of sports codes. These link all independent islands, including Australia and New Zealand, in a network of games and other contacts, including the sending of professional coaches. And of course, islanders are becoming ever more involved in sports connections involving the whole Pacific Basin and the rest of the world.

Whether one measures in terms of numbers participating, hours spent or money and other resources, it is clear that vastly more is devoted to foreign sports (and very little indeed to indigenous sports as yet – though Samoan cricket is a partial exception) than to local cultural activity. But like Christianity, rugby and some other sports have become integrated into most Pacific cultures.

The youth, women's and cultural organizations. These exist by the dozen, and everything from Pacific Scout Jamborees to Pan Pacific and South East Asia Women's Association meetings bring people together across the region. The contacts are generally brief, but the feeling of identity with the international organization is sometimes very strong. In practice the organizations are in most cases world-wide, but the Pacific linkage is with regional divisions, including Australia and New Zealand and sometimes also the North Pacific Rim.

What are the regions with which the identity is made? All this interaction builds up feelings of common interest and identity, but what are the boundaries of the regions concerned? Unlike nations, where the unit concerned is usually precise, there are many definitions of the Pacific region. Each definition is based on different boundary markers. Each includes and excludes different countries and people. The differentiating criteria include:

1 *Dependent/independent.* This roughly sets off the South Pacific Commission and its subsidiaries (which include all island states and territories) from the South Pacific Forum and its subsidiaries (which include only the independent and self-governing states). Australia and New Zealand are constitutionally members of both, just like other island states, but in practice their status is different as donors, as settler-dominated societies and as former colonial powers. The South Pacific Forum began as the Commonwealth Pacific. However, Fiji's withdrawal from the Commonwealth and the Marshall Islands and Federated States of Micronesia membership of the Forum, and with other changes coming, mean that the Commonwealth and the models that go with it are of reducing significance.

2 *Inside or outside the British/French/US system.* Irian Jaya and Easter Island and the Bonins come into none of the significant frameworks of regional organization. The 'parent' countries have discouraged contact, with the recent exception of Chile in relation to Easter Island.

3 *International language (and related symbol system).* English-speaking/French-speaking/Indonesian-speaking/Spanish-speaking constitute fairly closed blocs. And with each language goes a package of activity and identity

patterns. The South Pacific Commission has long made linkages across the English/French divide, but they are still minor in effect and mainly in the cultural area such as the South Pacific Games or the Festival of Pacific Arts. Power linkages across the language divide are few. All regional activity is undertaken in English and in what might be termed a 'Commonwealth' framework of structures and principles. National symbols of the host country are added at the opening and in other ceremonial contexts.

4   *Ethnic identity.* Polynesia/Melanesia/Micronesia are relatively recent as meaningful categories because, until recent times, there was little interaction between them. Now they have become significant categories and identity markers. They may become even more so in the coming decades, in both national and regional affairs. Ethnicity frequently does not correlate with sovereignty, despite the original meaning of 'nation' as an ethnic community. Western Samoa and American Samoa are constitutionally divided though ethnically the same people. Pukapukans are ethnically closer to Tokelauans than to fellow Cook Islands citizens. The people of Nui are ethnically Kiribati but Tuvalu citizens. All Melanesian states contain Polynesian minorities, as do some Micronesian, as well as diverse Melanesian or Micronesian cultures.

Indigenous/non-indigenous also remains an important functional division. As with any category, it has flexible boundaries. For example, persons of Kiribati parentage born and brought up in Fiji are indigenous Pacific Islanders but not indigenous Fijians. And the very large number of people of multiple islands/non-islands ancestry (vastly more than census figures disclose) are sometimes constrained and sometimes given more opportunities by the existence of such potentials.

The extent to which non-indigenous people wish to achieve, have achieved or are likely to achieve various levels of identity with the rest of the Pacific Islands, merits a study in itself. These are matters of feeling and symbolism rather than of trade (intra-regional trade is only 1 per cent of total trade) or power relations. Nevertheless, the one is likely to feed into the other.

One can see among Europeans in New Zealand and New Caledonia and, in smaller numbers throughout the region, among Indians in Fiji and in smaller numbers elsewhere in the islands, also among Chinese in many island countries, a grappling with the complex factors which allow elements of choice in evolving their new identity. Until a generation ago, the grappling was a matter of the indigenous people adjusting to external forces. Today, all are involved and reaching for a new place and new patterns for the twenty-first century.

5   *North Pacific Islands/South Pacific Islands.* Hawaii and the Micronesian entities north of the equator enjoy a Pacific identity, but there is a tendency to want to join the South Pacific, which is seen by many people as the 'real' Pacific Islands. It is likely that the North Pacific partici-

pation will increase in Pacific Islands affairs, but it will never be a major issue fur the North Pacific islands in the way that the North Pacific Rim will be. And the orientation of these islands is likely to shift from being predominantly to the USA to being predominantly to North-east Asia. Anything of the Pacific Islands will be minor in most contexts, though major for some symbolic purposes.

6 *Is an ASEAN/FORUM identity likely to emerge?* I would not be surprised if forces from further north push ASEAN (the Association of South-east Asian Nations) and the South Pacific Forum countries closer together over the coming generation. Interaction between the two increased during the 1980s and early 1990s. This is almost certain to continue. Most Malays, Indonesians and Filipinos share some common ancient cultural and genetic origins with Polynesians, Micronesians and some Melanesians. Although they separated over 3000 years ago and have been isolated one from the other since then, the commonality will be given enhanced emphasis as it becomes advantageous to do so.

7 *The Pacific Basin.* The coming significance of the Pacific Basin cannot be doubted, but it may be too large, too diverse, to be much of a focus for identity. But we now begin to see more Pacific Basin organizations forming, and statements forthcoming, such as that by the recently established Pacific Ecumenical Forum, which lists as the first of its priorities 'To establish and affirm our common identity as people of the Pacific'.

WHO ARE THE PEOPLE WITH WHOM THE IDENTITY IS FELT?

Pacific identity is largely something for the elite – it is mainly non-Pacific island sources that finance them to travel within the Pacific for education, conferences, workshops and so on, and thus they get to know one another and to share some elements of common identity. And the Pacific Islands gatekeepers and mediators are, much more than the average, of multiple genetic, cultural and educational background. Such people are valuable to the indigenous communities for their foreign knowledge, and to foreign people for their insights and knowledge of the indigenous world.

Many more Pacific Islanders have to pay their own fares. In contrast to those whose fares are paid for them, those who pay for themselves travel overwhelmingly to the Pacific Rim countries. There they meet expatriate Pacific Islanders, or go just to visit. They have been overwhelmingly to Australia, New Zealand and the USA – trips to Disneyland being very popular with people of the Eastern Pacific. But the Pacific coasts of Asia are fast becoming attractive destinations too, particularly for those from the North Pacific islands. Then there are those who go to train or to confer. If going within the islands they are usually paid for by donor countries on the Rim. Those going to Rim countries are usually paid for by the country to which they are travelling.

Again, East Asia has recently been added to the range and is the area of fastest growth of contact.

All relations which islanders pay for are overwhelmingly with the Rim. The pattern of telecommunications shows that no island country has its main telecontacts with any other islands country. For most, only a very small percentage of contacts is with other island states and territories. Likewise the pattern of mail flows: for example, for every piece of mail posted in the Cook Islands to an address in the Cook Islands there are seven pieces posted abroad, and only the tiniest fraction goes to other islands. Patterns of trade and transport show the same tendencies. Twenty years of effort to enhance intra-regional trade have had very little effect.

Concepts such as 'The Pacific Way' can express a rhetorical unity, but unless there are more specific material foci of identity, and with the boundaries of the region so variable for various purposes, such unity at a regional level is likely to be largely symbolic.

CONCLUSION: WIDENING IDENTITIES AND PROTECTIVE REACTIONS

Whether Pacific islanders have more say over their identities than people elsewhere in the world depends on whether one looks by country or by the person. Island *countries* have the least say – they are too small and open to be able to determine more than a fraction of the aspects of their own identity. But island *individuals*, on the contrary, have more say than most other individuals in the world living in such large units that they have little room to manoeuvre.

Ethnic identity in the islands has over the past hundred years come to focus on increasingly larger units above the local community, particularly on the nation or territory. The national level of identity has been intensified since independence. It is ironic that independence from colonial powers correlated with closer integration with those powers on many fronts, including in some cases identifying more strongly with aspects of those powers than ever before.

The tremendous growth in external sources of identity, and efforts to integrate more closely with world systems, has been offset by strongly asserted protective reactions focused on the selective adaptation of some elements of indigenous culture, from various periods of the past, in the quest for unique and functional identities for the future.

The protective reactions emphasize selected aspects of symbolic and ceremonial life: language in new roles, emphasis on the expressive arts, revaluation of concepts of legitimacy of membership in ethnic and other categories, a new emphasis on focal ceremonies and festivals – funerals, hair-cuttings, tattooing, accession to traditional titles, *tere* and *malaga* (travelling parties) and so on.

With so much of material life (technology, consumer goods, even sports) being standardized throughout the world, and with the island states being linked as small, relatively minor units in larger systems, it is to be expected that the aesthetic, cultural and expressive aspects of life will be given a higher profile.

During the colonial era, with clear European supremacy on a number of fronts, adjustments of identity were mainly a matter for the indigenous people. Today, however, this quest applies to Europeans and Asians in the region as well as to indigenous people. They now tend to be minorities out of power or, like Australia and New Zealand, majorities at home but within a hemisphere and a region in which their kind constitutes a small minority. Especially in New Zealand, one sees a conscious search for a new identity, incorporating elements of ancestral immigrant cultures as well as elements of those indigenous to their geographical region.

National identities are in most cases strengthening in specific areas, particularly as the symbol systems on which they depend become more clearly defined and implemented. In many cases, that definition is being refined and retuned at home by islanders who have spent many years abroad. Those islanders who have emigrated also depend to some degree on the symbol systems of the home society.

Identities below the national level are generally reducing, though still significant. However, as the Fiji case indicates, internal ethnic identities can also grow. And were it not for larger countries trying to maintain existing boundaries, many Pacific states would lose minorities who would prefer to set up their own state.

Blood (in the sense of recognized kindred) and genetics (in the sense of common identity and often common cause with those who are accepted as being from a common genetic stock) are in some cases stronger forces than they were a decade or two ago.

The strong drive for national identity, plus the pressures of world forces of universal culture, leave the quest for regional identity somewhat vague and indeterminate. It is likely to evolve further, but not very far, before larger forces in the Pacific Basin become more significant. They are likely to shape the most potent forces in the identity mix, and to provide parameters within which local and cultural factors in identity will operate.

ACKNOWLEDGEMENTS

Margaret Mackenzie, Cluny McPherson, Malama Meleisea and Michael Ogden kindly commented on the draft of this chapter. Their help is much appreciated, but remaining errors and inadequacies are mine.

22

Tradition, Culture and Imposed Change in Indonesia

Cheri Ragaz

INTRODUCTION

Indonesia is a new united state in the sense of being a political legal concept. What really existed before this was made up of many states or *negara* whose numbers ran into the hundreds if not thousands. This new united state consists of a plurality of old societies and it is in this respect that nation-building presents itself as one of the most fundamental problems. Any kind of identity that Indonesia tries to find has to take into account this two-fold reality. If it only stresses the unity, the population will fail to be 'emotionally committed', on the contrary, overemphasis on plurality will fail to motivate people to work for a common goal. The Indonesian society today reflects two contradictory elements: the traditional and the modern. Through a planned state ideology called the *Pancasila*, Indonesia is trying to achieve unity through diversity. Its motto is *Bhinneka Tunggal Ika*: variety yet one; diverse but united. This expresses the strong desire to achieve unity despite the immensely heterogeneous character of this newly formed state which achieved independence in 1945. The threat of disintegration is real, but integration is also feasible and not beyond reach. *Lain ladang lain belalang, lain lubuk lain ikannya*: different field, different locust; different pond, different fish – that is the way Indonesians usually describe their rich cultural inheritance.

Both the parts and the whole have their own structures, their own systems and their own dynamics, but are at the same time interrelated. This also reflects, for example, the Javanese attitude towards nature which sees both human beings and nature as subjects, as co-equal and interrelated in a mutually interdependent system. Nature is neither worshipped nor exploited. Each individual person is *jagad cilik* (a small world or microcosm) who represents the *jagad gede* (the big world or macrocosm) and that is why people live in harmony with one another (*rukun*) and with nature. The unity and diversity of Indonesia are manifested, sustained and guarded by *Pancasila*, the national ideology.

There are four major epochs in the history of Indonesia, which have had a formative effect on the moulding of its present national character. These are: (a) the ancient empires; (b) the experience of Western influence and colonial control; (c) the Japanese occupation and subsequent struggle for independence; and (d) the period of independence. At the same time it can be said that in viewing almost any aspect of Indonesian history in the past 400 years, it is difficult to avoid consideration of 'external' influences and the international context. These have been fundamental and constitutive.

The aim of the following interpretative exposé is an attempt to trace the development of national integration since the period of independence, focusing on development and national image after the coup of 1965.

DEVELOPMENT AND THE SOCIAL PRODUCTION OF NATURE

If there were one major experience which we could single out as being responsible for the setting of contemporary views and visions of the world, it would most probably be the development of industrial capitalism. It is therefore an extremely difficult task to assess adequately the social production of nature in one territorial unit unless it is grasped in relation to the dynamics of the capitalist world-system as a whole. During the past decade, the region of Southeast Asia has embarked on a new phase of economic, political and social development situated within the context of the New International Division of Labour, in which it is playing an important role. The kinds of policies which are emanating from the planning ministries of the countries therein, in connection with those of international financial institutions, point to the changing international structure of development dynamics with over-accumulation of capital on the one hand and over-accumulation of labour on the other. This seems to substantiate the view that underdevelopment is always underemployment.

The notion of the social production of nature should include theoretically the interaction between technological change and labour, social institutions and economic growth within their specific historical relations (the approach which historicizes nature is to be found in the writings of Lukács, Gramsci and the Frankfurt School).

On 1 June 1945, Sukarno delivered his famous speech, which was, from then on, known as 'The Birth of *Pancasila*'. He proposed that Indonesia would be neither an Islamic state nor a secular state, but a *Pancasila* state. According to Sukarno, *Pancasila* means 'five pillars' or 'five principles', arranged in the following way: (a) Indonesian nationhood, or Indonesian nationalism; (b) internationalism/humanitarianism; (c) unanimous consensus/democracy; (d) social welfare; (e) the One Lordship. He maintained that these five principles could be compressed into three (*Trisila*): the principle of socio-nationalism, of socio-democracy and of the One Lordship, which could then be further compressed into one (*Ekasila*): the principle of *gotong royong* (mutual co-operation or expected reciprocity).

The notion of *gotong royong* is the most essential part of the Republic's ideological representations, encompassing a subtle dialectic between state and community. Its function is to reproduce particular social realities and remain in a dialectical relationship with them, having at the same time an associated systematic manipulative function. Apart from being a catchword for unity, it is the key concept in relation to the ordering of the state's fundamental administrative units. It means goodwill and co-operation in the field of local government to an extent whereby, through a network of rules and sanctions, the *kampungs* (villages) identify themselves as being *gotong royong* communities. Events of significance, although constructed by using local imagery, must have links with official sub-village administrative organs in order to be recognized as *kampung*-communal.

Indonesia exists in a modern world and the country must have institutions capable of responding to its international, technological, economic and political environment. This environment can be increasingly coupled with what has been called 'the see-saw movement of capital' (Smith, 1984) which lies behind the larger uneven development process. With the government of 'The New Order' presided over by President Suharto and his generals since 1965, capital has found the agents with which it can mediate between conditions of existence and the reproduction of nature, which are 'meso-level' explanatory variables responsible for the expression of the fundamentally different impact that the capitalist world-system has on different places in space and time. In this respect, the ideology of *Pancasila*, as used by the government, reflects and vindicates activities and decisions which both incorporate a territory and which go far beyond its boundaries. Concrete attempts at systematic economic planning and development in Indonesia were only begun in 1967, some time after the abortive coup and the formation of the New Order government of Suharto who stood for political stability and economic development.

THE NEW ORDER STATE-SYSTEM

The hegemony of the New Order state consolidated in the public mind an image of the Old Order as a period of chaos, disorder and violence from which nation and civil society had to be delivered. The Sukarno model of attaining national unity and integration by symbol-wielding and revolution was rejected in favour of political stability, order and economic development. The price to be paid was the imposition of military rule, which presented itself as a truly 'national force' associated with the interests of the community at large. The doctrine of the *Dwi Fungsi* (Dual Function) was produced by the army in 1965 to justify their role as simultaneously a military force and a socio-political force addressing 'ideological, political, social, economic, cultural and religious fields'. The general image projected was that the country needed strong leadership of the kind which only an army can provide.

The New Order in Indonesia represents the culmination of a struggle for power by a new urban elite whose values are reflected in a kind of 'metropolitan superculture' with its close association to foreign countries, especially the United States and Japan, and by the public reliance of key political figures on the advice and economic patronage of members of Indonesia's Chinese minority who form much of the country's independent bourgeoisie. Patronage substantiates the political base of the regime by way of foreign and *cukong* (that associated with Chinese businessmen) capital, neither of which could directly challenge the military's political hegemony. Funds are also used to obtain the acquiescence of potential dissidents and to maintain an extensive apparatus of repression.

Societal harmony can be preserved by stressing the vertical dimension of society. This kind of arrangement believes that uniformity and conformity to a single policy are the key to harmony. Here, the emphasis is not so much on the interdependency between the individual members as on the dependency of the members on their leaders. Individual activities have to conform to the will of those in power. 'Obedience' is the key word. But if obedience is emphasized, then the legitimacy of the authority becomes the central problem. Obedience is only possible if the individual members of the society are convinced that it is right and good and proper to obey the existing authority. Individual initiatives are discouraged. It is difficult to change a policy or abandon a wrong one. An Indonesian saying goes, '*asal bapak senang*' (As long as the boss is happy, why bother him with bad news). If the problem of legitimacy is unsolved it is to be kept by force. The New Order legitimizes its monopoly of power and authority as being a necessary prerequisite to the economic development of a modernizing state which at first requires control of 'disruptive' social and political forces. The social foundations of the New Order state have been made possible largely due to the reinterpretation of broad economic and political alliances with international finance capital.

The development of the international economy together with new information and communication technologies and the powerful transnational organizations linked by subordinated decentralized networks, steer the process by which territories represent structural aspects within the framework of interdependent operations. With its unifying ideology, the ruling class of the New Order is liberated from the necessity to make concessions to other economic and cultural segments of the population and has given full play to its own ambitions and preferences, the result of which has been an amalgamation of nominal Islam (*abangan* mysticism), Western consumptionism, bureaucratic politics and technocratic elitism. The New Order elite is culturally, politically and economically distant from the mass of the population. The task facing this elite is to alleviate the country's socio-economic problems, despite its alienation from the predominantly Muslim mass of the population. The somewhat vulnerable government with its dependence on non-Muslim foreign and domestic allies has had to change its inward-orientated industrial strategy to an outward-looking strategy which seeks to integrate Indonesia into the New International Division of Labour and to make investment decisions on the basis of efficiency and comparative advantage – a reflection on the ideological level of what constitutes, at the concrete level, the logic of capital accumulation by international corporations operating at a global scale. It is not only the World Bank but international corporate interests which have urged deregulation and reform and an end to 'corrupt' practices of the bureaucracy in the last few years. What they are really doing is to make power 'placeless' with the aid of information flow technologies. This represents a major challenge to the existing structure of political and economic power in Indonesia up to the present. It threatens to alter the balance of ownership and control between different elements of the Indonesian bourgeoisie. The interests of domestic business groups of military bureaucrats and private capitalists, hitherto firmly entrenched in the import-substitution sector, whose existence has been based on government intervention involving protection, subsidy and the mediation of integration with foreign capital, is no longer guaranteed. In the wake of the slump in oil revenues, fundamental structural changes became necessary.

At another level, whilst certain dominant interests may lose their legitimization in the development process, 'community' social relations and socio-political mobilizations continue to operate according to a *local* place-orientated logic. Implicit in Gramsci's work is an awareness of ways in which differences in the forms of knowledge that predominate in various communities present barriers or resistance to their assimilation into hegemonic ideological communities. Different conditions of life over time and social relations differ in their bases of the grounds of knowledge, forming 'epistemic' communities which cannot be assimilated into 'ideological' communities without a great deal of struggle. The problem of penetration and assimilation of those 'epistemic' communities presents a challenge to any group which seeks to recruit them to its cause, whether they be religious leaders or what Althusser calls Ideological State Apparatuses. Gramsci acknowledged that people adhere to more than one group and, because of this, their thinking is episodic and incoherent, containing stratified deposits of past philosophies which leave in people what he called 'an infinity of traces without an inventory' (see Hoare and Smith, 1971, p. 234). He claimed that 'coherent' philosophies, if they are imposed from the outside on the masses of subordinate groups, have no direct influence on the people. What they have is not through ideological integration but 'as an external political force, an element of cohesive force exercised by the ruling classes and therefore an element of subordination to an external hegemony' (see Hoare and Smith, 1971, p. 265).

ISLAM IN INDONESIA

The case of Islam will now be considered. Although there was no historical or cultural basis for national consciousness before colonial rule, 90 per cent of the population were at least nominal Muslims, Islam having been introduced to the archipelago in the thirteenth century. There are modern and traditional interpretations of Islam. Javanese nominal Muslims are usually called *abangan*, as opposed to strict Muslim *santri*, and may refer to themselves as *Islam Statistik* (statistical Muslims) and hold their own practices strongly influenced by pre-Islamic Hindu-Buddhist mysticism. Indonesia's rulers have had an ambivalent attitude towards Islam. They have tried to use it as a source of legitimacy and at the same time have held it at arm's length.

Indonesia's military rulers since the mid-1970s have clearly tried to restrict the political and social role of Islam, although they called on the religion's forces during their coup in 1965–6. The restriction of Islam under the New Order reflects both the influence of technocratic and pre-Islamic ideas and their fear of institutions which they do not control. Leaders of Suharto's New Order, themselves overwhelmingly of *abangan* cultural origins, were not only concerned with reducing Islamic militancy once their regime was established, but strove to increase the autonomy of the *abangan* religious variant as a political counterbalance. The *adat* (traditional) law which formerly regulated the entire life of the community was dominated by spirits and supernatural powers, communal life being inevitably static and deeply conservative. The roots lay in the obscurity of the past, when the ancestors laid down the *adat* once and for all, or as the Minangkabau of West Sumatra say: 'It doesn't crack with the heat or rot in the rain' (Alisjahbana, 1969, p. 5).

At the same time, societies are structured by their economies, which in turn are highly interdependent at the international level. Indonesia, like other national governments, has been faced with adjusting to the dominant logic of global capitalism in the most advantageous manner. Although the ideology of *Pancasila* is used as a tool of the state to promote 'unity within diversity', the role of the state in mediating between national and international capital is determined by concrete economic and political conflicts which derive from the development of capitalism itself.

INDONESIA IN A GLOBAL CONTEXT

In terms of population, Indonesia is the fifth largest country in the world with an estimated 192.9 million population (1993). So large a population spread out over so vast and fragmented a territory, presents wide variations in ethnic type, religion and language. There are over 250 different regional languages and just as many dialects in Indonesia – the divisions tending to follow the basic ethnic divide between coastal and interior peoples of the archipelago. Since independence, the old trader's *lingua franca*, often referred to as 'market Malay' has been adopted as the national language, Bahasa Indonesia, in all parts of the country. In terms of resources, it is one of the richest in the Third World, with sizeable deposits of petroleum, natural gas and other minerals, as well as some of the richest timber stands in the world. Its position is economically and politically strategic. Yet Indonesia is among the poorer countries of the world. According to a survey conducted by the Central Bureau of Statistics in 1990, the number of people living in absolute poverty, barely subsisting in a poor and overpopulated agricultural economy, was 27.2 million. More than 60 per cent of the population live in Java and the neighbouring island of Madura (which together comprise 7 per cent of Indonesia's total area). A great part of Indonesia's urban population is concentrated in Java, together with modern transport facilities as well as some of the most productive agriculture and the majority of the country's poorest peasantry.

One of the most relevant questions we can pose is: How is Indonesia adjusting to the dominant logic of global capitalism in the most advantageous manner? The Indonesian economy has experienced a significant structural transformation during the last two decades. There are three interrelated concerns within the framework of Indonesia's development strategy. The first is the nature of sectoral change; second is the problem of regional impact; and third, the structure of economic control and ownership. In spite of these structural shifts which manifest themselves most obviously in a dramatic expansion of the industrial and service sectors at the expense of the agricultural sector, agriculture remains one of the most important forms of economic activity. Although increasing commercialization of agriculture is gradually reducing this sector's capacity to absorb new entrants to the labour market, it continues to employ more than half of the country's labour force. Cash crop developments of the outer islands have minimal effect on the unreformed agrarian heartland of Java – transmigration in particular is having disastrous ecological effects.

INTERNAL CHANGES IN INDONESIA

During the Sukarno years Indonesia was inherently unstable economically, reflecting a combination of political obstacles to effective development. While Sukarno tried to develop a badly disrupted economy, his policies were often counterproductive, overcompromising and sabotaged by a complex of powerful vested interests. A radical restructuring of the economic and political system was essential but impossible, given the chronic antagonism between nationalists, the Muslim establishment, communists and regional interests.

After the bloody *coup d'état* in late 1965 overthrowing the Sukarno regime, President Suharto faced the daunting task of restoring economic and political stability. The legacy he inherited included: (a) inflation at 650 per cent; (b) external debts in excess of US$2 billion; (c) transport and communications nearly at a standstill; (d) production capacities in major industries at extremely low levels due to shortage of imported materials and spare parts; and (e) economic and political instability. These and other factors were the outcome of a protracted and complex process of social, economic and political conflict which mirrored, first, the weaknesses of social classes and, importantly, the failure to emerge of a powerful, national bourgeoisie. This enabled the development of a state apparatus in which political powers and bureaucratic authority were appropriated and integrated by the military, party and state officials themselves. Secondly, the bid to reconstitute the Indonesian economy from a declining agricultural export economy into a state-led manufacturing economy through import-substitution industrialization was not successful; a failure attributable to the national state and the national bourgeoisie (comprising Chinese and indigenous groups, essentially merchants unable to generate the process of accumulation necessary for industrialization). Thirdly, the failure of forces of social revolution, comprising elements of labour, peasants and intellectuals under the political leadership of the PKI (Indonesian Communist Party) to wrest political power from the conservative forces, which included rural landlords, national bourgeoisie, the urban middle classes and the military.

The centralization and formalization of political power by the military over the last 25 years has meant that all other groups have been systematically excluded from any meaningful political role. Even at the periphery, access to influence is confined to patron–client networks or to government-controlled and -sponsored corporatist organizations such as GOLKAR (the state political party) and other state organizations for business, labour and so on.

From 1966 under Suharto's direction, strategies to produce some degree of economic stability were implemented, and this with remarkable speed as the inflation rate dropped to 120 per cent in 1967, 86 per cent in 1968, 10 per cent in 1969 and 9.1 per cent in 1990. (It has since risen again to 10.5 per cent.) Government expenditure was reduced, tight monetary controls introduced and a balanced budget adhered to. Backed by the West, outstanding debts were re-scheduled and credits provided. The return to a more normal economy in 1968 enabled the government to give greater attention to development objectives, and in 1969 it launched the first of a series of five-year plans. *Repelita* I (1969–74) was directed to rehabilitation of the economy. *Repelita* II (1974–9) broadly followed the objectives introduced in the previous plan, the major ones being to provide better food, clothing and housing, to improve and expand infra-structure. The third five-year plan, *Repelita* III (1979–84), was introduced as the 'Trilogy of Development': equitable growth, equitable distribution of development benefits and maintenance of political and economic stability. On-going development objectives were stressed, including raising economic growth rates in targeted geographic areas and bolstering economically weak groups. High priority was placed on the expansion of education and health facilities, especially in the less-developed regions of the country. The fourth five-year plan covering 1984–9 (*Repelita* IV) placed emphasis on the development of industries – especially labour-intensive industries – that increased foreign earn-ings through exports, with special emphasis on those industries that added value to domestic resources. The actual amount of foreign direct investment, however, is not always a reliable guide to measuring its employment impact. For example, Indonesia has the highest amount of foreign direct investment within the ASEAN region yet the impact on employment is not as great as in other countries. In Indonesia, textiles and chemicals have been the most attractive industries for multinational companies (in Malaysia, electronics and food; in Singapore, electronics and petroleum products; in Thailand, electronics and textiles; in the Philippines, chemicals and food-processing). Of all these industries, those which are most labour-intensive and therefore have the largest employment impact are the textiles and garment and the electronics industries.

Anticipating the gradual decline in energy prices, Indonesia began in 1983–4, a series of adjustments and reforms in its fiscal and monetary policies as well as its administrative reforms, programmes of incentives aimed at enhancing the competitiveness of exports, especially of non-oil and gas products. Additional reform measures known as 'deregulation packages' have been adopted and have had a profound effect on the business environment. In terms of scope, they include the following. In the manufacturing sector, productivity

and quality incentives were introduced, costs lowered and licensing procedures streamlined. For trade in general, export incentives were created, tariffs
rationalized, nontariff barriers reduced and customs procedures simplified.
Opportunities for investment have increased, procedures simplified and greater
flexibility in terms of investment introduced. In taxation, flat-rate income taxes
and new revenue-generating instruments have been introduced, self-assessment
and improved enforcement. Within the banking sector, regulatory mechanisms
have been improved, new market openings created and greater access to liquidity
ensured. Within capital markets, a new regulatory framework has been introduced creating viable equities markets, access to foreign funds and separation of
the regulatory body from the share trading agency. Within the area of monetary
management, there have been changes in currency convertibility, debt repay-
ment, balance of payments management and control of money supply. In
addition, environmental protection has been approached with the introduction
of an environmental impact assessment mechanism, stricter land-use policies and
stepped-up enforcement of environmental protection standards. *Repelita* V lays
down the direction of economic policy and development for 1989–94. It
adheres to the Trilogy of Development and continues to emphasize the diversification of the economy away from oil and gas dependency and towards export-
orientated industries. The fifth plan calls for an annual GDP growth rate of
5 per cent. Manufacturing is expected to grow by 8.5 per cent, transportation
and communications by 6.4 per cent, construction, retail and wholesale by
6 per cent, agriculture by 3.6 per cent and mining by 0.4 per cent. The plan
anticipates a 16 per cent annual rise in non-oil and gas imports. While agriculture will remain the largest single sector in the economy throughout this
period, it is expected to decline from 27.2 per cent of GDP to 21.6 per cent
by 1994. The lead growth area in *Repelita* V is expected to be the manufacturing sector, with manufacturing's share of GDP predicted to increase from
14.4 per cent to 16.9 per cent. The mining sector's slower growth, including
oil and gas, reflects the anticipated decrease in world demand for those com-
modities. For the first time in a five-year plan, private sources – domestic
and foreign – provide the majority of investment. While in *Repelita* IV, public
investment accounted for 54 per cent of total investment, in the following
five years the government expects 55 per cent of the total investment to come
from the private sector. The recent reforms to stimulate the expansion of the
banking sector and capital markets are designed to facilitate this expansion of
private investment.

The principal investors in Indonesia are: Japan, Hong Kong, Taiwan,
South Korea, USA, United Kingdom, The Netherlands, Singapore, Germany,
Australia and Luxemburg. Local investors figure also in the latest estimates.
The trend for investment is in manufacturing projects, which comprise the
majority of current investments. Other major activities include the manufacture
of food, tobacco products and textiles and are mostly carried out by private
enterprises. Cement, fertilizers, petrochemicals and basic metal products have
gained relative importance in recent years and are produced primarily only by
state-owned industrial enterprises, either alone or in joint ventures with foreign
firms. Major gains have also been registered in Indonesia's wood-products

sector, with Indonesia now serving as the world's largest supplier of plywood. Concurrent with establishing itself as a growing exporter of basic manufacturing goods, Indonesia is also moving into the exportation of more advanced products such as aircraft and aircraft components and electronic goods. It's state-owned aircraft manufacturer, Industri Pesawat Terbang Nusantara (Nusantara Aircraft Industry or IPTN), has the most extensive and developed facilities for aircraft design and production in South-east Asia. With other aircraft manufacturers such as CASA of Spain and Messerschmitt-Bolkow-Blohm (MBB) of Germany, IPTN has participated in a number of successful joint ventures. The consumer-orientated electronics industry is a sector with tremendous growth potential. At present, assembly of products exceeds manufacture, with only some parts manufactured locally and almost all required parts and components being imported.

Through joint ventures, indigenous entrepreneurs are gaining further experience and knowledge which will equip them to participate more fully in development. According to the guidelines on foreign equity, in Government Regulation no. 17/1992, 'Concerning Share Ownership in Foreign Capital Investment Companies', the Government of Indonesia stipulates that all foreign investment must be in the form of a joint venture with an Indonesian partner. The government also stipulates that at least 20 per cent in a foreign investment be owned by a local partner and that this should be raised to no less that 51 per cent within 20 years starting from the moment of commercial production. Such PMA companies with foreign capital investment arrangements can be established with share capital being entirely owned by foreign partners if they fulfil one of the following requirements: (a) the value of paid-in capital is at least US$50,000,000; (b) they are located in provinces such as Maluku, East Timor, West Nusatenggara, South-east, Central or North Sulawesi, East, Central or South Kalimantan, Bengkulu and Jambi. Within five years, starting from the moment of commercial production, at least 5 per cent of the entire value of corporate share capital has to be sold to Indonesian partners and increased to 20 per cent within 20 years. The following exception exists to the 20 per cent minimum Indonesian share ownership requirement. A joint venture which exports 100 per cent of production and is located in a bonded zone may have as little as 5 per cent Indonesian ownership. The percentage of Indonesian ownership does not have to increase over time. The Government of Indonesia plans to combine characteristics of free-trade zones and industrial estates in several bonded areas. Export-processing zones have been established, for example, in Jakarta and on Batam Island, which is located 12 miles south of Singapore. There are other facilities, some still under construction, some being planned – notably in Surabaya (East Java), Cilacap (Central Java), Ujung Padang (South Sulawesi) and Medan (North Sumatra). Several industrial estates are in full operation.

LABOUR MARKETS IN INDONESIA

To assess progress in the agricultural sector since 1967, one must keep in mind several things: most of the Indonesians living in rural areas work within an agrarian structure characterized by small, non-viable holdings; a marked disparity in land-ownership (60–70 per cent of land is owned by 10–20 per cent of rural families); a high degree of landlessness (over 50 per cent); a high degree of under- and unemployment (about 18 per cent and 35 per cent, respectively); heavy indebtedness of over 60 per cent of Javanese peasants; a combination of high rents and high interest charges to local money-lenders have meant that the renting smallholder has retained little more than 30 per cent of his harvest. Inefficient marketing, credit and infrastructural facilities have been combined with administrative corruption. This has yielded low production, extensive poverty, malnutrition and inefficiency.

Recently, the number of 'local money-lenders' has been substantially reduced since the government introduced financial assistance packages in the form of low interest rate credits designed for farmers, small and medium-size businesses. These credits are known as *Kredit Investasi Kecil* (Small Investment Credit) for farmers and small businesses and *Kredit Modal Kerja Permanent* (Permanent Capital Investment) for medium-size businesses. Another point to be considered is that non-agricultural activities have been insufficiently developed to absorb much excess rural labour. The decline of village industries because of increasing urban industrialization, increasing capital-intensity in agriculture, and a high rate of population growth have increased under- and unemployment. (Parts of Java have a population density of about 2000 per sq. km.) This latter factor has major relevance to policies of economic planning and population control.

Indonesia has traditionally emphasized agriculture in its economic development but gradually the industrial sector has also been developed with the aim not only of increasing employment possibilities but also of increasing national competitiveness. The challenge which Indonesia has to face now is how to take full advantage of the availability of its abundance of human and natural resources to sustain the current national development efforts. Manpower problems in Indonesia are related to high population growth and a young population structure. However, according to the 1990 National Economic and Social Survey, the rate of population growth has been successfully reduced in the last two decades. In 1972 the rate was 2.3 per cent, in 1990 it was 1.97 per cent and in 1995 it is expected to be approximately 1.6 per cent. Many of those who have already entered the labour market are without much work experience. Within the span of five years during *Repelita* V, the workforce has increased by 11.9 million people. Low levels of manpower utilization lead to low incomes for the workers and low average household incomes. Indonesia has one of the largest remaining pools of inexpensive and relatively educated labour in East Asia. Wages for unskilled labour are among the lowest in the world. The industrialized nations have mapped out a new development strategy for Indonesia, the main feature of which is the great stress on 'export-orientated' industrialization. This, seen as the engine of growth, seems a little incongruous with the present reality, where the manufacturing sector employs only 8 per cent of the labour force, accounts for 5 per cent of total exports and 12 per cent of non-oil exports, and is clustered around Jakarta. Nevertheless, advocates claim that with a proliferation of export-processing zones populated by foreign multinationals, and joint ventures taking advantage of cheap labour, manufactured exports can account for a quarter of total exports (roughly Malaysia's situation at present) within ten years. These foreign 'enclaves' of multinational subsidiaries operate large-scale and often capital-intensive plants (as in vehicle assembly) or large labour-intensive plants (as in electronics assembly operations). Little linkage exists with the rest of the manufacturing economy, and there is relatively little emphasis on import-substitution (the market for consumer goods is smaller than that of South Korea, for example, which has a population one-quarter the size of Indonesia's). The World Bank, for instance, sees these 'enclaves' of foreign capital as the key to Indonesia's manufacturing role. Export-processing zones appeared in Asia at about the same time as the New Order in Indonesia and in the intervening years until now, labour costs in the original 'cheap labour' countries of East Asia have escalated sharply while Indonesia's have not. The multinational corporations are now looking at Indonesia.

The point of examining the country's industrial growth within the confines of this chapter is to estimate types of changes which are occurring within the production and reproduction of society itself.

The geographical and historical dialectic between societies and their material environments in the modern world system is best approached by looking at the systematic requirements of a global capitalist economy and how these are being identified and responded to. The impact of foreign direct investment leads to distinct histories in different places – Indonesia is not South Korea. Of immediate relevance, but extremely difficult to judge, is the performance of the economy in terms of employment.

The growth of manufacturing industries in the 1970s tended to apply modern capital-intensive technology. The reason for this choice of technology over employment was obvious. Labour-absorbing activities were mostly concentrated in the informal sectors: rural farms, household small-scale industries, seasonal labour, petty traders and other activities. The number of people who had no choice but to move into the 'undefinable' informal sector must have increased rapidly since 1987. The term 'informal sector' is frequently used interchangeably with 'traditional sector'. However, informal sector is understood here as part of the traditional sector, namely the part which consists of survival activities such as the collection of paper or metal from a rubbish heap for recycling in contrast to the other parts of the traditional sector which suffers from a very low productivity but is nevertheless more than just a sector of survival.

Reliable unemployment statistics are not available. It is difficult to trace the growth of employment in the informal sector; nevertheless the population census gave some indication by using the employment figures by status: 'self-employed', 'employer', 'employee', and 'unpaid family workers'. An 'employee' could, with some minor adjustments, be considered to be a formal-sector worker. The population census defined 'employees' as 'persons who work for another person or an institution for pay or cash in hand or in kind'. Therefore, agricultural workers, although they do not have specific supervisors, are classified as employees: 'Employees in the modern sector are those working as ...

Government employees, employees of state/private companies, hotel employees and servants. ... They are more organized than the other employment status: "self-employed", "employer" and "unpaid family workers".' Thus, the latter groupings are considered as the informal workers. Using this approach, it is not surprising that almost 90 per cent of the growth of employment in the 1970s originated in the informal sector. However, the informal sector did not receive as much attention as the formal one. Almost none of the economic facilities provided by the government and the banking system, such as the tax holiday and financial credits, reached the informal sector. Since this sector was more able to absorb the unskilled labour than the modern formal sector, it became the major source of employment for the Indonesian labour force. As a result of this development, the annual growth of employment in the informal sector, at about 3.9 per cent, was much higher than the growth of employment in the formal one, at about 1 per cent. Although employment creation through the informal sector relieved the employment pressure to a certain extent, it created some serious problems. Although the bulk of unskilled employment was absorbed by the informal sector, its productivity tended to be low and, in some activities, declining rapidly. The informal sector tended also to create an army of poor, unorganized workers. This is unavoidable because of the low productivity of the workers. The Ministry of Manpower tried to set up a minimum wage system, but it did not of course cover the unorganized informal workers. Additionally, the limited inter-island mobility of the informal-sector workers added pressure to the already densely populated island of Java. The transfer of workers from the informal to the formal sectors is very difficult due to the limited employment opportunities and because of the different characteristics of the industries. With the move towards more modern capital-intensive technology, the need for labour seems to be low. On the other hand, the skill requirements for operating these industries are high compared to the average level of education of Indonesian workers. Since the 1980s, and especially after the second oil crisis, the Government of Indonesia has realized the importance of the informal sector in absorbing unemployment. Certain measures were taken in order to encourage as well as to regulate the development of the informal sector, such as financial assistance in the form of credit. In big cities such as Jakarta and Surabaya, the local governments have built markets and small kiosks where small businessmen and traders can do business.

Indonesia's recent industrial restructuring aims to enhance employment opportunities in the industrial sector and at the same time shift to outward orientation for its growth. This shift has implications for technology and manpower policy or planning. What technology policy adopted in Indonesia will be consonant with the aim of restructuring? The answer should be labour-intensive technology. It may be questioned whether this will be possible in view of the tendency in capital-goods producing countries in the developed world towards larger capacities for higher technology. Does Indonesia (and other less-developed countries with abundant labour entering the international market) have any alternative option other than to adopt the advanced and capital-intensive technology if it doesn't want to miss out on these developments?

Manufactured exports can be generated in low-wage economies only by pursuing specific policies, that is, by liberalizing the trade and investment regime or by creating, for example, separate export 'enclaves' that are specifically designed to attract foreign investment such as the export-processing zones which have been mentioned. Both of these regimes specifically designed to stimulate manufactured exports result in what has been called 'footloose' export activities, dependent on imports of capital goods and intermediate inputs to be processed or assembled. An export of this kind has very few linkages with other manufacturing sectors as well as with the rest of the economy. Indirect production and income effects are limited.

In general, the labour force will need more specialized training. There is a substantial increase in demand for specialized skills to support structural changes. At this point the role of multinational corporations becomes apparent. Some companies have started on-the-job training and in-house education programmes. Taken as a whole, the multinational corporations do not play a major role in terms of generating employment, but they can serve as a training ground for future entrepreneurs. This also means a drain of skilled workers who prefer to work there because wages and working conditions are higher. There is no indication that multinational corporations are concerned with or can even solve unemployment problems in host countries. Most of the multinational corporations have established trade unions and concluded collective labour agreements. The laws and regulations in Indonesia provide for equality of treatment between multinational and national enterprises. The characteristics of industrial relations in Indonesia are also applied in multinational corporations, in the sense that all parties should comply with the national development and their relationship is based on the remarkable state ideology of *Pancasila*. *Pancasila* does not regard the worker as simply a factor of production, but also as a human being with integrity and dignity who must be treated accordingly. It does not consider the interests of workers and management to be contradictory but rather that these interests represent the mutual interest of both partners. Any difference of opinion between workers and management should be resolved through the process of negotiation in order to achieve concensus within a spirit of brotherhood. In relation to establishing work ethics, the Department of Manpower has published (Indonesia, 1985) a remarkable little book entitled *Manual on the Implementation of Pancasila Industrial Relations*. On the first page, the former Minister of Manpower, Sudomo, states:

The New Order Government is determined to put into practice *Pancasila* and the Constitution of 1945 in every aspect of social and political life including the work situation. Therefore the relationship and value systems between members of the production process should continuously be based upon the values contained in the *Pancasila* and the Constitution of 1945. (*Pancasila* Industrial Relations was adopted first in 1974.)

CONCLUSION

On the orthodox level, we can argue that global capital is simplifying Indonesia's unification attempts in that the economic base is becoming the motor for determining cultural life and the reproduction of nature, sanctified by a powerful state ideology. On the other hand, there are 'non-commensurable' rationalities which rise to the surface and are often suppressed by force, but which remain in perpetual conflict and threaten the foundations of the nation.

REFERENCES

Alisjahbana, S. Takdir 1969: *Indonesia: social and cultural revolution*, 2nd edn. Oxford, Singapore: Oxford University Press, p. 5.
Anderson, Benedict R.O'G. 1990: *Language and Power: exploring political cultures in Indonesia*. New York and London: Cornell University Press.
Arief, S. 1983: 'Industrial restructuring, technological development and implications for manpower planning: the Indonesian case'. *South East Asian Economic Review* (Kuala Lumpur), 4(3), 143–54.
—— and Sasono, A. 1981: *Indonesia: dependency and underdevelopment*. Kuala Lumpur: META.
Dürste, H. and Fenner, M. 1987: 'Wirtschaftliches Wachstum und soziokultureller Wandel'. *Entwicklungszusammenarbeit*, 2, 5–7.
Hoare, Q. and Smith, G.N. (eds and transl.) 1971: *Selections from the Prison Notebooks of Antonio Gramsci*. London and New York: International Publishers.
Indonesia's Industrial Development, report prepared by UNIDO mission to Indonesia 1991. Vienna: UNIDO.
Indonesia – Changing Industrial Priorities, Industrial Development Review Series, 1987. Vienna: UNIDO.
Indonesia. Department of Manpower 1988: *Labour Legislation in Indonesia*. Jakarta: Department of Manpower.
Indonesia. Department of Manpower 1985: *Manual on the Implementation of Pancasila Industrial Relations*. Jakarta: Department of Manpower.
Jones, G.W. 1984: 'Links between urbanization and sectoral shifts in employment in Java'. *Bulletin of Indonesian Economic Studies* (Canberra), 20(3), 120–57.
Linnemann, H., Van Dijk, P. and Verbruggen, H. 1987: *Export-oriented Industrialization in Developing Countries* (Council for Manpower Studies, Manila). Singapore: Singapore University Press.
Mathews, R. and Brown, J.M. 1989: 'Indonesia: Survey'. *The Financial Times*, 1 Sept.
Smith, N. 1984: *Uneven Development: nature, capital and the production of space*. Oxford: Blackwell.
Vatikiotis, M. 1988: 'Indonesia: worrying about idle minds'. *Far Eastern Economic Review* (Hong Kong), 142(41), 39.
Warr, P.G. 1983: 'The Jakarta export processing zone, benefits and costs'. *Bulletin of Indonesian Economic Studies* (Canberra), 19(3), 28–49.
Wong, K.-Y. and Chu, D.K.Y. 1984: 'Export processing zones and special economic zones as generators of economic development'. *Geografiska Annaler*, 66B, 1, 1–76.

23

Geographical Identity and Patriotic Representation in Argentina

Marcelo Escolar, Silvina Quintero Palacios and
Carlos Reboratti

Every nation-state has a territory under its political sovereignty. Nations that turn into states must have a particular geographic configuration, but the same could not be said, however, about their history. On the contrary, no nation which is recognized as such by its members can have an identity without having a particular and exclusive history. Therefore, nation-states necessarily require a political geography, while nations cannot exist without a history of their own.

These basic premises are a useful point of departure to approach the subject of the formation of a geographical/symbolical national referent with which every member of a nation identifies. This process is produced both with social efficiency (the 'nationalized' society becomes a specific community) and political legitimacy (the national community justifies its exclusive sovereignty). It is, in other words, the history of the construction of a national identity (Chebel, 1986, pp. 77–8).

The juridical act of the foundation of a national state can legitimate only the appropriation of a territory, but not its past. From that legal and institutional base the traditional sources of nationality can be recovered under the form of historic nations (Guiomar, 1990; Wallerstein, 1990, p. 106; Hobsbawm, 1984, pp. 21–3). But they can also become an obstacle to the state's capacity for nationalizing the population under its domain and for consolidating its own coherence when confronting other states which appear as competitors in terms of its territory and communal 'identity'.[1] This can explain the multiple processes that a social formation can go through in its history as a sovereign nation-state (Recalde, 1982, pp. 9–37).

Within the indisputable frame of a particular territory, the state slowly builds up a national community using all-inclusive factors such as race, language or ethnicity (Wallerstein, 1990), looking for an identification between a territory and a population that very often originally belonged to different national communities (Rokkan and Urwin, 1983, pp. 1–18). In this way, the original symbiosis – almost merely a juridical convention – is dressed up with the attributes of an imagined historical identity (whether recovered or invented) which will allow the personification of the national state in its geography and its people (Chebel, 1986, pp. 56–61).

There is an abundant bibliography of conceptual and empiric work on this subject, whether considering the formation of the national identity as 'the invention of tradition' (Hobsbawm, 1984), the production of a 'national myth' (Citron, 1989), the 'imagined communities' (Anderson, 1983) or 'fictional ethnicity' (Balibar, 1990). All these works use a historical perspective to analyse the different discourses, institutions and cultural practices in the formation of the modern nation-state. But, possibly because territory is taken for granted, a geographical perspective is absent. It was by means of a process of subjective representation, recognition and cartographic design, however, that the invention of the contents of a 'natural' state territory took place and that a legitimate discourse about national sovereignty was developed.[2]

An implicit separation between geography and national community lies behind the exclusion of the geographic-ideological components of nationality. Geography, on the one hand, can be materially real, as the realm of the nation's territorial appropriation, and it can also be ideologically real as a territorial collective representation. The national community, on the other hand, also has two components – first, the material one: the number and quality of the people that belong to it; secondly, the ideological component: the historic foundations of the group identity. Both dimensions can therefore become a reference for the constitution of individual representations, whether spontaneous (when the subject considers himself a part of the whole, without questioning its foundations, a kind of 'geographic common sense'), or reflexive, when there is involved an explicit justification of identity.

Any community with its own identity and territoriality (Sack, 1986, pp. 30–43) can be considered as a 'national community'.[3] Conflicts may arise in order to legitimate the national character of a particular state nationality, some discourses pretend to 'explain' the territory or the national community, or to establish why people belong to a geographical territory (Guiomar, 1974; Bourdieu, 1980). This conflict becomes evident when referring to a national community. But when it relates to the national geographical representation it is generally less clear, and perhaps more subtle, and it only becomes apparent when boundary problems arise.

Which is the role of a territorial and anthropological discourse dealing exclusively with a particular community and territory, that is, the nation-state? Leaving aside the subject of non-state national identities and their possible means of representation (whether collective, individual, subjective or objective), the analysis can be circumscribed by two parallel and mutually conditioned dimensions: on the one side, that of the state territorial formation, and on the other, the elaboration of a legitimate territorial discourse.[4]

The nation-state is, then, largely an historical construction that results from the recovering of pre-existing nationalities or from the invention of a new national tradition. In any case, the community which is limited by its boundaries will develop an excluding and exclusive collective representation of national identity (Escolar, 1991a). In order to achieve that, various official institutional discourses developed in the armed forces, the school, the cultural industry and the family. The nationalist mobilization works around specific issues such as the homogenization of language, a history, patriotism and geography. In the latter case, the method consists of organizing a systematic and symbolic patriotic tradition. This is directed to 'naturalize' the territorial representation, using a mythological personification of the natural and potential characteristics of the Fatherland.

The patriotic representation can be considered as the collective feeling of identification by the communities of a nation-state with certain historical and abstract symbols, including a naturalized territorial realm, considered as legitimately appropriated.

In the following pages an attempt will be made to explain how in Argentina geography played an important role in the definition of the national territory as a patriotic representation, mainly through the participation of discipline in the educational system. At the same time, the process of the institutionalization of geography, and its role in establishing a legitimate territorial discourse, was closely linked with the different stages and characteristics of Argentina's nationalist movements and discourses. So, as an objective way of territorial justification, geography can be connected with the founding of the modern national identity.

NATIONAL AND TERRITORIAL IDENTITY IN ARGENTINA'S EARLY HISTORY

At the beginning of the nineteenth century, the fall of the Spanish monarchy under Napoleonic rule brought about the dismemberment of its colonial empire. From then on, and throughout the century, many opportunities arose for the formation of what today are the Latin American states. The traditional national historiographies and mythologies go back to this historical moment to discover in the different declarations of independence the original milestone of the national identity, which would later develop teleologically into the various nation-states. This interpretation established a genetic and evolutionary continuity between the presumed original identities and the modern national ones that relate to today's Latin American states.

Today, however, with regard to the territory of the Viceroyalty of the River Plate, it is hard to ascertain the existence in 1810 of a national feeling of self-identification that could have inspired the movement for independence (Halperin Donghi, 1961, pp. 108–20). This is particularly relevant to the Argentine case, if we consider that no less than half a century elapsed between the first political autonomy manifest in 1810 and the definite political organization of the national state.

During this extended period, the Confederation of the River Plate Provinces was a vague political and territorial reference It was recognized externally as a sovereign state but internally it suffered from the absence of an effective state able to centralize power and to impose its rule over a heterogeneous group of local autonomous areas with a confusing superposition of identities (Halperin Donghi, 1980, p. 10; Chiaramonte, 1989). It was in fact the result of an attempt to preserve the territorial heritage of the Bourbon administration. The Viceroyalty of the River Plate had been, however, a short-lived administrative unit that covered what is today Argentina, Paraguay, Bolivia and Uruguay, and was internally very diverse. In fact, it had originally belonged to the Peru Viceroyalty and was created for strategic purposes, nucleating heterogeneous regions – many of them arbitrarily separated from their original centres – under the influence of the port of Buenos Aires.

Once royal authority disappeared from the River Plate, some regions launched their own independence movement, thus giving birth to the new states of Paraguay and Bolivia, and later Uruguay. Other areas remained somehow related to each other, partly because of a shared economic dependency on the Buenos Aires customs. But they took ever-increasing political powers, virtually becoming autonomous states (Chiaramonte, 1983, p. 165). They had their own armies, they minted their own money, and sometimes they declared themselves independent republics. But they preserved a certain unity for diplomatic purposes: they left the management of foreign affairs in the hands of Buenos Aires as a way of confronting the increasing colonial ambitions of many European nations.

Only in 1852 did the autonomous provinces agree to create a central government and organize a National Confederation. The effective formation of the state, however, would only be possible with the material support of Buenos Aires, the richest and most dynamic of all the provinces. But the reaction of local interests against that proposal was strong: Buenos Aires split from the Confederation and for ten years it remained an independent state. In 1861 its army defeated the Confederation forces, and Buenos Aires set itself to organize the state under its political and economic domination. The liberal political elite that come to power as a result of this process led the way to the construction of a nation-state (Ozslak, 1982, pp. 51–81).

The emergence of a nation-state brought about the imposition of a national identity that subordinated the various proto-national cultural identities to that which was developed in Buenos Aires. This conflict was the origin of a long history of rivalry between the 'interior' provinces and the *porteño*[5] central power, and it played a major role in the nationalistic arguments that, when searching for the lost roots of the Argentine national identity, found them in the cultural traditions ignored or destroyed by Buenos Aires cosmopolitanism (Prieto, 1988; Quijada, 1985, pp. 24–5)

In 1862 began a period that marked the formation and consolidation of a modern national state based on the progressive appropriation of territory from the now reunited River Plate Confederation. At this stage the construction of the nation was conceived of as a political project with strong voluntaristic elements, dressed up with the customary deterministic convictions of the moment. This project was perceived as the natural pattern of development dictated by the laws of human evolution, and it was not intended to create a fundamentalist mythology aimed at the imposition of a compulsory and exclusive nationality (Botana, 1984, pp. 341–54).

The formation of a nation-state was conceived of as part of the general transformation that would allow for the inclusion of Argentina in the world process of capitalist modernization. The formation of a national community and citizenry was to be based on the cultural development that would support the political consolidation of a liberal republic, integrated with the new international order. A national identity adequate for the formation of a modern nation had to be compatible with the pattern of Western progress and culture. The ideal national citizen was a liberal bourgeois, a defender of Progress and Order. Until the turn of the century the problem of nationality was not conceived of in terms of an exclusive, autochthonous and specific cultural identity: the Argentine liberal nationalism was basically cosmopolitan (Halperin Donghi, 1976, p. 196; Solberg, 1970, pp. 26–30).

From this point of view it is possible to understand that the modernizing role given in the Argentine model to massive immigration was compatible with and complementary to the project of building a modern nation out of this portion of American land and people. The main ways to achieve this end were the education and the migration policies.

At this stage the view of the national territory was not yet furnished with a discourse directed to legitimize the existence of the territory before the state, as the natural referential framework for a transcendent and essential nationality. Although geography became a compulsory subject in the curricula of the new and massive school system, it was essentially an encyclopedic and descriptive knowledge, linked to history and not intended to explain the nation's natural territorial unity (Quintero, 1991). In other words, territory had not yet become an object of patriotic representation.

The main territorial problem was the need to settle the 'empty spaces' (Reboratti, 1987, pp. 80–96) and to create a national population able to become the sovereign people required by the newly built state. At the same time, by the end of the 1870s a long series of agreements and negotiations with the neighbouring states led to the delineation of most of the international boundaries. After that, and with the help of the armed forces, the state occupied the inner spaces (Patagonia and Chaco) which until then had been either settled by the aboriginal population or simply uninhabited.

The interest in the territorial question intensified in the 1880s when, after achieving political and military stability, the ruling classes faced the ideological and material problem of the appropriation of a land that seemed inexhaustible. During this period geography, as a school subject, offered an inventory of the wealth of the national territory.

The process of internal and external consolidation had revealed that most of the territory was practically unknown. At that point geography made its first appearance outside schools, in order to fulfil an essential task – the surveying of the land. In 1879 the government created the Military Geographical Institute, an organization dedicated to the cartographic survey of the country. In the same year the Argentine Geographical Institute (IGA) was founded, and two years later the Argentine Geographical Society (SGA) as an offshoot of the former. Created in the image of the geographical societies of France, England and Germany, those institutions were not, in fact, places for the production of geographical knowledge but for its promotion. Their membership was highly heterogeneous, and possibly no member would call himself a 'geographer', but rather a naturalist, an explorer or a traveller. There were also many politicians and government officials, who thus demonstrated the interest of the ruling class in the state's territorial patrimony. A great number of travels and explorations took place between 1880 and the end of the century, most of them sponsored by the IGA. They were mainly directed towards the less-known areas of Chaco and Patagonia, although the north-western and north-eastern parts of the country were also explored in an attempt to reappraise them. All these travels were widely publicized: the travellers' conferences, for example, had large audiences and their proceedings were published and distributed among the public. Gradually, the country was being surveyed and, at the same time, a collective *imaginaire* of the national territory was growing within society.

This kind of proto-geography did not develop within an academic framework: rather, it was politically and economically orientated and did not induce the production of professional geographers. When the task of surveying the new lands was completed, the geographical societies entered into a state of lethargy. Although they had played an undeniable role in the surveying and occupation of the territory, they had no influence over school programmes. Consequently, these geographical associations had no real significance in the process of legitimation of the discipline, the development of which took place without any local academic support. Geography gained scientific status and disciplinary autonomy only after the creation of a special institute devoted to the training of schoolteachers. The institutionalization of geography was a very long process, and it was parallel to another process: that of the rise of an exclusive national-istic discourse. Both were related to the scientific legitimation of the territory that became a new dimension of the consolidation of a national identity.

SCIENCE, TRADITION, AND A PATRIOTIC CULT: THREE GEOGRAPHIES IN THE
FORMATION OF NATIONAL IDENTITY

Since the 1880s government officials in charge of educational plans and programmes favoured the traditional and patriotic perspectives when devising a nationalistic education. Such perspectives seemed an effective remedy for the Republic to 'preserve its Institutions from the degenerations that would arise from immigration' (MJIP, 1886) because 'if it is true that science has no fatherland ... certain fields of study seem more appropiate for a particular people, its language, territory, history and institutions ... they tend to arouse a fervent eagerness for the prosperity of the fatherland, an austere respect for its laws and an enthusiastic love for its traditions' (MJIP, 1888). It was not until the beginning of the twentieth century, however, that the role of the secondary school in the politicization of the middle class became evident. Then it seemed necessary to introduce some sort of rationalization into the teaching of subjects that were linked to the formation of a homogeneous nationality (Tedesco, 1970, pp. 175–8).

In 1904 the Executive established an institution for the formation of secondary schoolteachers: the National Teachers Institute (INP), absolutely separate from the University and directly linked to the government (MJIP, 1908). The INP produced, for the first time in Argentina, 'geography teachers' formally qualified as such.

The first stage in the institutionalization of a scientific geographical discourse was thus produced about 30 years after the first steps were taken towards the construction of an Argentine nation-state. At that stage the production of a collective national discourse directed to the socialization of the population seemed urgent and necessary.

By the end of the nineteenth century, it was all too evident that the project of a liberal republic based on the exclusion of wide sectors of the national population from the formal attributions of citizenship was exhausted. It had been eroded by the social conflicts arising from the appearance of an organized workers' movement, the process of urbanization and the growth of a large middle class with aspirations towards political participation (Baily, 1967; Floria and Garcia Belsunce, 1988, pp. 93–4; Terán, 1987, pp. 11–14).

There were only two possibilities to deal with these problems: strict repression in order to sustain the regime, or the orderly political incorporation of the majorities into the aristocratic republic. This meant not only allowing these people to exercise their citizenship rights but also making them share the ideals and objectives of the republic. In relation to the second possibility, at the beginning of the new century several proposals were put forward that encouraged the socialization of the people into an exclusive perspective of the Argentine nationality. Geography was meant to play a fundamental role in this process: the three main strategies devised for the construction of a national identity saw different ways of relating people and territory within the framework of the state nationality.

The first of these strategies was directed towards retaining the liberal republican model, keeping the political command of the state and assuming the conduct and control of the process of democratization. Geography, considered as a scientific discipline, was taken as a means to explain the natural relationship between territory and national character.

The second strategy pursued the construction of a new national project based on the socialization of the benefits of the old model and on the foundation of a new political subject: the people (*volk*), politically sovereign and unified by a common nationality. Although it was an explicit political strategy, it required academic support. Geography as a humanistic discipline could relate the political community to the cultural nationality through the collective internalization of a territorial identity.

From the point of view of the third strategy, the only feasible solution was the creation of an irrational discourse that could become a nationalistic and almost religious belief for the masses. In this case geography played only a minor role: it was to produce an aesthetic and sensory approach of the patriotic landscape, loosely based on empirical and tangible evidence.

The main voice sustaining the first position was that of Joaquin V. Gonzalez, a prominent member of the progressive branch of the oligarchy in power. It was Gonzalez who, as Secretary of Education, created the above-mentioned INP. This politician and intellectual was determined to reform the liberal project by expanding the political participation of the majority of the people (Botana, 1977, p. 259). He was also very interested in the development of a discipline that was, from his point of view, the future synthesis of the natural and social sciences: geography. He was much influenced by the development of physical geography in England, and by the active intervention of the Geographical Association and the Royal Geographical Society in the educational aspects of geography (González, 1901, p. 192; 1914, pp. 7–19; Stoddart, 1986, pp. 42–8; Freeman, 1980, pp. 65–8, 96–101). For the first time in Argentina Gonzalez talked of geography as an indisputable scientific discipline that could explain the historical laws of human evolution from a naturalistic approach. 'Geography is the science I found most interesting, since I understand its precious spirit to synthesize the rest of the sciences, being itself of course a natural one *par excellence* and furthermore the most defined and characterized social sciences' (González, 1914, pp. 7–8). 'It totally embraces the environment where man and nations are formed, grow, move, collide and disperse or are dissolved ... [explaining] the historical events that change the position of races, peoples and States on the Earth' (González, 1901, p. 121).

From González's point of view, the formation of a national identity required a set of arguments legitimated by their reference to a scientific discipline. Geography could be 'the centre of a vigorous reaction of the current ideas' aimed at preventing the progressive dissolution of the conservative order and at restoring the national spirit in tradition and territory. Because geography 'is almost exclusively national, it will contribute together with History, to shape the fatherland's real personality, as it defines and distinguishes what can be called the body or the form of this soul or idea' (González, 1901, p. 122). González clearly attached a formative capacity to geography as a school subject, and his preoccupation with unifying the educational discourse was evident when he said: 'called to attain such a profound transformation of the national spirit ... the teaching of geography will lead to a modification of the whole soul and substance of teaching in Argentina' (González, 1914, pp. 21–3). A scientific conception of geography and history would support the legitimate foundations of the Argentine national state.

A different position was adopted by Ricardo Rojas, an intellectual who disliked the cosmopolitan and bourgeois cultural atmosphere of Buenos Aires (Paya and Cardenas, 1978, pp. 21–2). He was concerned with the construction of the national spirit, and he saw clearly the political nature of any historical reconstruction directed toward the preservation of the cultural foundations of nationality (Blas Guerrero, 1984, pp. 35–8). Rojas thought that 'the historical sense' which played a fundamental role in the formation of a patriotic ideology 'consists in the imaginative representation of the idea of time.' (Rojas, 1909, p. 54). In this sense, the construction of a *volk* as an efficient retrospective illusion would allowed people to identify themselves with the official *imaginaire* (Balibar, 1990, pp. 117–27). Thus, he thought it was urgent 'to react and transform the school in the home of citizenship'. Due to the peculiarity of countries like Argentina where 'once the nation is constituted, we still have to wait for the settlement of the desert and the creation of the people's soul ... [this is] ... our most urgent problem ... [and] ... to it we must subordinate our education' (Rojas, 1909, pp. 135–6).

Rojas was to play a very important role in the Argentine cultural nationalism (Glauert, 1963), and his interest in geography derived from his idea of 'political patriotism'. This notion represented a rationalized approach to patriotic representation, opposed to what he understood as a 'a liturgical sickly patriotism and an absurd xenophobia' (Rojas, 1909, p. 17). He was convinced of the voluntaristic and militant nature of nationalistic education, and thought that 'this modern conception of patriotism, politically and territorially based on the Nation, is what I call *nationalism* ... nationality must be the conscience of a collective personality ... [and] Geography could provide the conscience of the territory ... [thus] the school would contribute to define the national conscience and to systematically think about true and fertile patriotism' (Rojas, 1909, pp. 47–8).

Rojas thought that a homogeneous collective image of the fatherland was a fundamental precondition for a new national project. This project, which was still liberal in an economic sense, created the conditions for the development of a fictional ethnicity directed towards the construction of a new social consensus based on the hegemonic role of the urban middle classes. 'The citizen', Rojas said, 'must be conscious of the national territory, the national history and the national language. Potentially, this is citizenship'. And so 'Geography must be not only instructive but also educational' (Rojas, 1909, pp. 148, 68). From his point of view, geography's disciplinary field was clearly included among the 'modern humanities'. Thus his territorial conception of the nation-state was based upon reason but not necessarily upon scientific objectivity. Geography, he said 'has to make possible the experimentation of aesthetic suggestions in the emotion of landscape, and of civic suggestions in the formation of patriotism'. He counted openly on common sense as a justification of his position: 'the natural environment theory represents the influence of territory over civilization' and without any prejudice he added: 'I cannot say what kind of influence this is, but it exists' (Rojas, 1909, pp. 68–9).

The third nationalist perspective was put forward by Jose María Ramos Mejía. He shared with González and Rojas the basic diagnosis, and he openly voiced his conservative concerns. He was a prestigious physician with a strong interest in sociology and history (Soler, 1968, pp. 170–4). At the turn of the century he expressed his view: 'In our political biology, the modern masses do not yet play their true role ... there is none of the ardent passion of political sentiment, nor of the love for a flag associated with the well-being of life', so 'We should re-establish the continuity between past and present, that the sudden and healthy contact with Europe [he referred to immigration] seems to have severed, thus posing a menace to our national physiognomy. Fortunately, the environment is strong ... and it would suffice to give it a little help through national education ... to clean the mould where the formative tendencies of the national character will be shaped' (Ramos Mejia, 1899, pp. 225–34).

Nine years later Ramos Mejía saw the opportunity to put his ideas to work. In 1908 he was named President of the National Council for Education, the official institution in charge of public primary schools. From this position he implemented a Patriotic Education Plan, directed towards the construction of that non-existent *volk* ('the modern masses'), and based on the organization at school of a thorough indoctrination system for natives and foreigners. The idea of the Fatherland, transformed into the fetish of a religious cult, played the role of a social agglutinant and an ideological support for the passive consensus on the regime's continuation. The Patriotic Education Plan consisted basically in the systematic introduction of references to patriotic symbols, national heroes, dates and epic events, which were introduced not only in the humanities or social subjects but also in mathematics, biology and handicrafts.

Within this version of a 'conservative regeneration' (Halperin Donghi, 1976, p. 228), geography was considered to be among the fields 'particularly suitable to influence the formation of patriotic sentiments'. It was simply a matter of emphasizing 'the natural beauties, comparable to the best ones in the world', of describing 'the Nation's sources of wealth', of showing the different landscapes of the national territory and of taking advantage of the lessons of geography to memorize the location of various events in the patriotic calendar (the birthplaces of national heroes, the main battlefields and so on).

Once the nationalist discourse rejected a rational explanation of nationality, the geographic argumentation became inoperative. The geography of the new 'Patriotic Education' not only discarded academic justification: it did not even pretend to adopt a rational discourse. This attitude was not rooted in ignorance on the part of the authors of the project about the academic and institutional progress experienced by geography. As early as 1899 Ramos Mejía himself commented on the opinions of 'the famous German geographer Peschel'; he cited Reclus' Geography, and welcome the 'monumental work of Charles Reitter [sic], the first European sage who considered Geography not only a body of descriptive and enumerative knowledge: he tried successfully to discover the intimate correlation existing between the Earth and the people who live on it' (Ramos Mejia, 1899, pp. 202–3).

In fact, geography as a school subject was evidently linked to the nationalistic issue: therefore it could easily become a platform for the system of indoctrination. In this case, however, the discourse on national territory was not at the centre of the concept of patriotic representation: it was only an instrument used to introduce the patriotic cult of national heroes and symbols.

The relationship between the formation of a national *imaginaire* and the institutionalization of geography has been quite direct in Argentina. The singularity of this case lies in the fact that the institution in charge of the training of geographers was not linked to the university. This situation led the professional body of geographers into a long struggle to create, in higher education, a proper place for geographical science.

CORPORATE GROUP AND SCIENTIFIC LEGITIMATION

The general elections of 1916 interrupted the conservative hegemony and put an end to the hopes for a passive incorporation of the middle classes within the national project of the oligarchy. On the contrary, the passing of the Electoral Law of 1912 that made the vote universal, secret and compulsory, had brought about the incorporation of the majority of people into the citizenry and it had therefore shattered the 'conservative order'. Executive power fell into the hands of the main opposition party, the *Union Civica Radical*, mostly a political expression of the middle classes.[6]

If it is true that the conservative/nationalistic project had failed from a political point of view, that did not mean, paradoxically, the end of the project of national patriotic indoctrination. The radical government used the patriotic representation built by the conservatives, and slightly modified it by some references to the political aspects of nationality and the new historical project. In 1919 the Education Secretary said: 'it is the special role of schools to spread and intensify civil and patriotic education ... [and] to praise the spirit of the new generations, not only in the idea of glorification, but also in the legitimate feeling of Argentinity' (MJIP, 1919, p. 44).

Since 1904 and with the creation of the INP, a diffuse and informal knowledge was changing into an institutionalized discipline. Geography and geographers were produced with independence from political strategies and official policies. Geography was considered a necessary school subject, and the INP was clearly the official institution in charge of the scientific support of the geographic knowledge.

Yet geography, unlike other fields taught at the INP, did not have a specific university referent and so lacked an effective academic legitimation. Furthermore, it existed basically for strictly political reasons. The autonomy of a geographical professional field (Bourdieu, 1976) would come about only when the doctrinal purpose of the discipline was replaced by the scientific interest, which would lead to the development of an objective knowledge regarding the territory. Not until 1953, and after the application of a long epistemological and corporative strategy, did geography become a university discipline with academic status.

Between 1916 and 1930 a growing body of geography teachers pressed for a place in different institutional posts more or less related to the subject: university and secondary school positions, official institutions linked to cartography and geographic surveys, and so on. The need was felt to create a meeting-place for the various individuals and institutions related to geography. Thus in 1922 the Argentine Society for Geographic Studies (GAEA) was founded, with the explicit aim to 'promote the study of General Geography, and particularly of Argentina's Geography ... to have an influence in the expansion and orientation of teaching [by means of] publications, conferences, the creation of Geography chairs at the Universities ... and the creation of the school of Geography' (GAEA, 1922). The main purpose was to create an institution to ensure the social reproduction of the new discipline, bringing together a heterogeneous group of scientists interested in geographical matters but who did not have a common disciplinary identity, being naturalists, geologists, agronomists, ethnographers and cartographers.

Four years later the members of GAEA, by then referred to as the highest authority in the discipline, were called by the Education Secretary to re-arrange the contents of the subject's curriculum in the secondary schools. This triggered a long public debate, both in parliament and the press, where the arguments put forward referred to the theoretical bases of the discipline and not to its educational goals. The existence of both the GAEA and the INP, as specific geographical institutions, had paved the way for the recognition of geography as a scientific discipline.

The battlefield was divided into two camps. On the one side, the traditional geography 'teachers' (lawyers, science, literature and history teachers or simply 'specialist' without any formal training) who opposed the plans they considered 'too scientific' (HCDN, 1926, p. 6) and excessively orientated towards the natural aspects of national geography. 'Geography is, by its own nature, strictly descriptive, a utilitarian or applied branch ... whatever exceeds this matter is then abusive' (ibid., pp. 671–2). The argument was that, when facing the new official plans, 'a strong majority of the teachers felt ... confused ... [as] strange words put in jeopardy their knowledge and their judgment', and those new words were far from 'the crystal-clear traditional language'. Finally, the conclusion was drawn that it was impossible to put the reform into practice because of the 'scarcity of teachers that understand it' (ibid., p. 62).

On the other side, GAEA representatives (who were at the same time government representatives), returned to J.V. González's old ideas. Geography must be 'a dynamic combination of physical and social laws', and they under-lined the importance of the proposed renewal because 'the Argentine republic is paying a high price for the scarcity of research regarding its territory' (ibid., pp. 672–3). Due to the scientific basis of the reform, the new plans could offer 'knowledge for the interpretation of our political and natural boundaries ... and the natural wonders that exist in the country' (ibid., pp. 674–5). And certainly, to face 'the study of Argentine Geography, [is] a very complex matter that needs specific preparation to be made [and this is done] in our Institutes' (ibid., p. 650).

The authors of the new official plans turned to the authority of science in order to recover the 'truly ... ignored ... geographical tradition ... represented by illustrious geologists and geographers' (ibid., p. 676) such as Huxley, De Martonne, Dantin Cereceda, Huguet de Villar and Ratzel. In this way, they were setting up a corporate field for 'geographers', depending both on a well-established external scientific tradition and on the institutional status of GAEA. The specialized teachers of the IPS were, from then on, the sole disseminators of the legitimate territorial discourse. The debate in geography moved from 'what for' to 'what is' the discipline.

Although territorial nationalism had failed as an instrument for the integration of the middle classes into the conservative project, it had contributed to the formation of a strong and lively patriotic representation. The party in government (the Radical Party) and also the Socialist Party shared with the conservatives a general nationalistic discourse that had, however, different formulations. After all, some of the objectives of patriotic education had been fulfilled, mainly those related with the most obvious levels of collective representation: the patriotic symbols and the national territory.

The institutional persistence of the geographical discourse had led to the emergence of the most basic component of any corporate group: the individual who identifies himself with a particular role. In the first stage, a non-university institution controlled a field that had explicit pretensions to being a scientific discipline, although the place was geared neither to train scientists nor to produce scientific knowledge: only to produce schoolteachers. The second stage saw the formation of the GAEA, which gathered together secondary teachers, university scholars and lecturers, and members of assorted institutions such as the military academies, the cartographic institutes and the museums. This step reinforced the epistemological and material existence of geography. Finally, in the third stage, the GAEA emerged as having recognized scientific authority, able to produce objective opinions about territory, with 'absolute independence' from any patriotic indoctrination goals.

CORPORATION AND AUTHORITARIANISM: A SCIENCE FOR THE NATIONAL IDENTITY

Throughout the period of the Radical government (1916–30), the nationalist discourse developed with different ideological orientations (Navarro, 1968; Quijada, 1985, pp. 41–63; Buchrucker, 1987, pp. 27–44). Yet the nationalist ideal was adopted as the basis for organized political action only by a minority group formed by a mixture of anti-reform conservatives, anti-liberal Catholics and pro-fascist officers and intellectuals influenced by Maurras. Maritain, Maeztu and other right-wing thinkers (Zuleta Alvarez, 1973, pp. 193–225; Buchrucker, 1987, pp. 45–70). The nationalist mobilization contributed in 1930 to the first civil–military coup in Argentina's modern history. Those minorities had based their claims against the legitimacy of the democratic system on nationalism, and they appealed to a historical, non-political reason to explain their right to represent the people as the true holders of the national interest. Geography was called upon, for the first time in an institutional way, to participate in the formation of an exclusive national identity (Escolar et al., 1991).

This situation did not last long. Shortly after the coup power passed to the traditional liberal sectors of the old ruling class, and the ultra-nationalist sector was displaced. The new government (1932–8) acknowledged the republican tradition only after securing political power through the fraudulent manipulation of the popular vote (Ciria, 1964, pp. 236–7). The nationalist argument played an important role in justifying this double game. The old liberal goal of the formation of a *volk*, incorporating natives and migrants alike, was replaced by a new project based upon the rejection of all ideological positions that posed a menace to the republican order. The shadows of fascism and communism introduced a new component in the old combination

of *volk*, history and territory: a specific political organization closely related
to a transcendent collective destiny (Romero, 1956, pp. 236–7). The justification for any indiscriminate political repression was thus to hand.

Geography did not play an explicit indoctrination role during this period.
The territorial identity that it had contributed to building in the preceding
decades was to provide an obvious point of departure for the delimitation of
the nation. The role of geography in the formation of a national identity was
evident and obvious to the education authorities. In 1934 the Secretary of
Education organized an Assembly for Spanish, History, Geography and Civic
Education Teachers, called to 'promote and deepen [the knowledge necessary
to] prepare the civic conscience of the future citizen and to wake in his spirit
the feeling of nationality' (MJIP, 1934b, p. 3).

The core of the indoctrination discourse of nationalism was placed in the
teaching of civics, while geography remained as the discipline charged with
explaining the objective nature of national identity. In the face of this new
role, it was very important to search for the epistemological bases through
which geography could scientifically contribute to a nationalistic formation.
The discussion among the geography teachers participating in the Assembly
shows that the role of the subject in a nationalistic formation was beyond
any doubt. The main point of disagreement was between a historical-humanistic
approach and a natural-scientific one, as the basis for an objective teaching
about the national territory (ibid., pp. 30–59).

The political situation of the 1930s was favourable to the process of the
institutional construction of geography. Individuals and institutions whose
nationalistic concerns included the territorial dimension saw in the GAEA
the main institutional reference.[7] At the same time there was growing participation by GAEA members in government offices where the policy to
deal with the formation of a national identity was being devised. In 1933
the government created a commission charged with eliminating from school
textbooks references that could seem offensive to Brazil, a country also in
the hands of a nationalistic government (MJIP, 1933, p. 71). The following
year, the government created a Geographical National Committee to promote
research into and diffusion of geographical knowledge related to national
problems; its main positions were occupied by GAEA members (MJIP,
1934a).

The GAEA as a scientific institution and the geography teachers as a
corporate group gradually merged. This process began in 1935, when the
GAEA invited all geography teachers to join its National Geography Meetings.
These were organized to disseminate research results and to give publicly
learned opinions on matters related to the settlement, defence and knowledge
of the national territory (GAEA, 1932 and 1937). From then on, the GAEA
controlled the whole geographical field, ruling on everything from theoretical
preoccupations to corporate strategies. At the same time, the scientific prestige
given to the GAEA by the non-geographer members was slowly appropriated
by the secondary-school geography teachers. This prestige became the basis
for the scientific legitimation of the discipline as an objective discourse about
the territory of the nation-state.

SCIENTIFIC REGION AND TERRITORIAL NATIONALISM: A 'NEW' GEOGRAPHY

In 1943 a new coup took place organized by army nationalist groups with near-fascist inclinations (Waldmann, 1986, pp. 131–6) who were concerned about the perspectives of democratic normalization after ten years of conservative governments and electoral fraud (Rouguié, 1978). The new government sought legitimation by adopting a populist discourse geared to attract the internal migrants who were flowing from the rural areas to the large urban centres. They were, potentially, a key element of the labour movement and, at the same time, a large electoral base (Baily, 1967, pp. 80–1; Durruty, 1969). The labour movement found in J D. Perón the ideal political leader. Perón, an army colonel with a strongly charismatic personality, reorganized the workers' unions and, with the help of the national industrial bourgeoisie, arrived in government through free elections in 1946. He was re-elected in 1952 and remained in power until 1955, when a new military coup closed a political period characterized by a populist state nationalism (Peralta Ramos, 1978, pp. 78–90; Buchrucker, 1987, pp. 335–8).

In order to legitimate its populist base of internal migrants, Peronism developed a new discourse of collective identification. It proposed the notion of 'national integration' in opposition to 'social marginalization', and linked social and cultural diversity with diversity of geographical origins (Presidencia, 1947; Perón, 1954). This connection was established by proposing the existence of objective territorial units (regions), alternative to the traditional collective representations (provinces). Regions could maintain their specificity, but only through an organic articulation with a national 'whole' (Escolar, 1991a and 1991b). This discourse could be ideologically considered as 'federalist' (Luna, 1986, p. 321), since it supported ethnic diversity within the country; in fact, it was centralist (Pérez, 1948, p. 15), because it denied all possibilities of linking administrative divisions with subnational collective identities (Perón, 1952, p. 641). This discourse was introduced into the main socialization machine: the state system of public education. Through the geography teachers of the GAEA the regional concept was put into practice.

Secondary teachers had obtained a numerical majority in the GAEA, and this gradually gave them a leading place in the institution. In 1949 and for the first time an INP geography teacher was named President of the GAEA. He was F. Daus, who had obtained his diploma from the INP in 1922 and then became a teaching assistant to the chair of geography at the university, in the School of Philosophy.[8] Daus, an early member of the GAEA, advanced his career in the Peronist period: university professor of physical geography in 1942, Director of the National Council for Education between 1947 and 1949, co-director of the Geography Institute at the School of Philosophy in 1947, Dean of the School of Philosophy in 1949, teacher at the INP in 1946, and vice-director of the main secondary school of Buenos Aires from 1946. Daus represented the growing process of disciplinary legitimation of the geographers corporate group and was, at the same time, the main nexus with Peronism. He contributed to making geography a legitimate institutional referent for the populist nationalist discourse about territory, and made a final effort to complete the long process of disciplinary institutionalization, through the establishment of geography as an autonomous discipline at the University.

In order to acquire true academic status as an independent discipline, geography had to develop its own conceptual framework about territory without relying on other disciplines. There was a long debate regarding the epistemological and theoretical definitions of the concept of region and its applications to the determination of a national regional system. Daus corporatively defended a specifically 'geographical' approach to regionalization. Confronting this view, the physiographic perspectives proposed to solve the problem by appealing to naturalistic theories, while despising the 'regional theory' as an inconsistent pretension to objectively define 'anthropogeographic' regions (GAEA, 1949). This debate confronted naturalists and geographers within the GAEA. For those who considered themselves 'geographers' it was a problem of finding at last a specific object for the discipline, because 'Regional Geography is the field where the old Earth Science can obtain its own originality and life' (Daus, 1951, p. 9).

The region, as the new exclusive scientific discourse of geography, was rapidly incorporated into the education system. The original geographical unit used in textbooks and classrooms for pedagogical description and explanation was the province (the first administrative division of the state). The provinces were grouped in 1926 into 'natural regions' (MJIP, 1926), a strictly physiographic definition that was not translated to social and economic territorial dimensions. In 1948, and particularly since 1953, the 'geographic region' was incorporated into secondary school teaching. From then on, geography would be 'the interpretation of ... the features of the natural landscapes as a result of the mutual action of the environment and the living creatures, especially man. All this results in the existence of Geographic Regions, whose study is included for the first time in the Geography plans for the middle school.' With this new concept as a point of departure, the study of Argentine geography could produce 'the integral valuation of the spiritual and material patrimony of our fatherland' and that explains 'the contribution of this subject to the knowledge of the authentic national reality' (MEN, 1953, pp. 87–8). Those words suggest an instrumental use of geography for a new indoctrinating discourse on the foundations of national identity. The new scientific status of geography, however, changed the place of the subject in relation to other disciplines; it was no longer grouped with those directed towards 'the socio-historic and national conscience formation', but rather was placed among the subjects for 'scientific training' together with mathematics, physics, chemistry and biology. When the military coup of 1955 eliminated all the indoctrinating contents introduced by Peronism in the school plans, geography remained unscathed because it was one of those disciplines dealing with 'objective realities scientifically grounded' (MEN, 1956).

In schools, the notion of the region meant a new conceptual approach to geographical harmony within the nation-state, under the umbrella of the natural nature of the state boundaries. The concealment of the geopolitical discourse by the regional approach was achieved through the construction of a legitimate foundation for patriotic representation and the use of scientific tools, that left behind the language of indoctrination.

In 1951 the GAEA President and Dean of the School of Philosophy declared that 'to systematic geography we oppose now regional geography, which is the modern expression of the geographic methodology, very much in harmony with the spirit and preoccupations of this age. This is, then, the new stage of geographical research, the stage of true geography' (Daus, 1951).

The last stage in the legitimation process was that of the creation of a geography department in the main national university. This was done in 1953, after Daus's period as Dean of the School of Philosophy, when a unit orientated to the training of researchers and teachers was created in this School of the University of Buenos Aires.[9] Similar processes took place in the Universities of La Plata and Cuyo, reinforcing the institutionalization of geography at the university.

Geography was no longer a strange blend of patriotic formation objectives, corporate interests and naturalistic scientific positions. It became a scientific discourse about the territory, only apparently disconnected from all nationalistic doctrines. The connection made above between this seemingly neutral discourse and a particular ideological perspective – condensed into the notion of region – does not mean that all regional research had, from then on, an ideological bias (Bolsi, 1988, pp. 251–5).

Geography became an apparently objective discipline, evidently important in education: after all, who can think of a school without geography?

CONCLUSIONS

Geography was an efficient channel for the dissemination of patriotic representation through schools, to such an extent that territory became a patriotic symbol and, together with others, it was reified as a religious myth, indisputable and untouchable. Throughout time and under different governments this fact has remained unchanged (Escolar et al., 1991). The geography teachers' body dressed up this myth with scientific arguments, thereby ensuring its permanence.

The 'patriotic territory' is today part of the common sense of Argentines. Whenever its glorification seems to decline, the geographical establishment takes good care to revive it. Any insignificant boundary conflict is enough to inspire inflamed declarations from those who retain the legitimate knowledge of geography in the school (Escudé, 1988).

The territorial mythology has fostered pseudo-scientific arguments such as those of the 'geopolitical experts', for whom the territorial myth is the only foundation of discourse. This discourse has remained undisputed in spite of its evident irrationality (Reboratti, 1982 and 1983). It has also given a strong impulse to the construction of a seemingly non-political geographic discourse that hides its ideological base by proposing a 'scientific' analysis of regional harmony in the nation-state (Escolar, 1991b).

The relationship between the institutional development of geography and the national identity of Argentina poses more questions than it answers. Is it possible to say that nowadays Argentina has a collective territorial identity? Did this identity help in the formation of the nation and later of the state, or was it the other way round? Was the creation of a territorial myth an important and useful step in the development of nationality, or did it only provide a fundamentalist bias to the basic feeling of social belonging?

ACKNOWLEDGEMENTS

The authors wish to thank for their co-operation Ms Patricia Souto and all the participants in the Program of Social History of Geography at the Institute of Geography, University of Buenos Aires. Dr Hilda Sabato undertook the homeric task of editing the final version.

NOTES

1 Examples of this sort can be traced in many of the national movements of the 1848 revolutionary period in Europe (Sigmann, 1970) and also among the conflicts derived from the national formation of the Central European States after the Versailles Treaty (Anderson, 1985; Hobsbawm, 1990).

2 When talking about 'the invention of a territory', we are referring not only to the cartographic representation (Allies, 1980) but also and above all to the official qualification and characterization of its material contents.

3 Understanding 'community' as a frame for objective and subjective belonging (Weber, 1922, pp. 33–5). Socialization does not prevent the formation of different communities, including the national one (Agnew, 1987, pp. 36–41).

4 We mean legitimate in the sense of being scientific and exclusive, with objective capacity for representation (Escolar, 1989).

5 *Porteño* means an inhabitant of Buenos Aires.

6 In Argentina, the term 'radical' has a similar meaning to that used in France (Snow, 1974; Reberioux, 1975).

7 Among others during this period, the GAEA gained the membership of J.G. Beltrán, a nationalistic lawyer specializing in boundaries, author of patriotic books and high-ranking official of the Ramos Mejia administration; the former Education Secretary A. Sagarna, promoter of the 1926 reform; and even General A.P. Justo, Argentina's President between 1932 and 1938 (Comisión, 1942; GAEA, 1974).

8 There had been a political geography course in the History Department since 1899 and even a Geographical Institute since 1917, but geography was not an independent department.

9 The autonomy of Buenos Aires University was at this time suspended and its government taken over by the Peronist government.

REFERENCES

Note: the * indicates that the citation corresponds to Spanish or Portuguese versions. The year placed at the end of the reference indicates the latest edition.

Agnew. J. 1987: *Place and Politics: the geographical mediation of state and society*. Boston, Mass.: Allen & Unwin.

—— 1989: 'The devaluation of place in social sciences'. In J. Agnew and J. Duncan (eds), *The Power of Place*, Boston, Mass.: Unwin Hyman.

Allies, P. 1980: *L'Invention du territorie*. Grenoble: Presse Universitaire de Grenoble.

Anderson, B. 1983: *Imagined Communities*. London: Verso.

Anderson, M. 1985: *The Ascendancy of Europe*. New York: Longman.

Baily. S. 1967: *Labor: nationalism and politics in Argentina*. New Brunswick, N.J.: Rutgers University Press. *

Balibar, E. 1990: 'La forme nation: histoire et ideologie'. In E. Balibar and C. Wallerstein (eds), *Race, Nation, Classe: les identités ambigués*, Paris: La Decouverte.

Blas Guerrero, A. 1984: *Nacionalismo e Ideologias politicas Contemporáneas*. Madrid: Espasa Calpe.

Bolsi, A. 1988: 'Geographie an den argentinischen universitaten'. *Geographische Zeitschrift*, 76(4); 238–55.

Botana, N. 1977: *El orden conservador*. Buenos Aires: Hyspamérica, 1986.

—— 1984: *La tradición republicana*. Buenos Aires: Sudamericana.

Bourdieu, J. 1976: 'Le champ scientifique'. *Actes de la recherche en science sociales*, 2(3), 88–104.

—— 1980: 'L'identite et la representation: elements pour une reflection critique sur l'idée de region'. *Actes de la recherche en sciences sociales*, 35, 63–72.

Buchrucker, C. 1987: *Nacionalismo y peronismo: la Argentina en la crisis ideológica mundial*. Buenos Aires: Sudamericana.

Chartier, R. 1980: 'Science sociale et découpage regional'. *Actes de la recherche en science sociales*, 35, 27–43.

Chebel, M. 1986: *La Formation de l'identité politique*. Paris: PUF.

Chiaramonte, J.C. 1983: 'La cuestión regional en el proceso de gestación del Estado nacional argentino. Algunos problemas de interpretación'. In W. Ansaldi and J.L. Moreno (eds), *Estado y sociedad en el pensamiento nacional*, Buenos Aires: Cantaro.

—— 1989: 'Formas de identidad en el Rio de la Plata luego de 1810'. *Boletin del Instituto de Historia Argentina y Americana Dr Emilio Ravignani*, 1, 71–92.

Ciria, A. 1964: *Partidos y poder en la Argentina moderna*. Buenos Aires: Hyspamérica.

Citron, S. 1989: *Le mythe national: l'histoire de France en question*. Paris: Les Editions Ouvrières/EDI.

Comisión de Homenaje al Dr J.G. Beltrán 1942: *Una vida de hombre útil*. Buenos Aires: RSVP.

Daus, F. 1951: 'Discurso de apertura de la XV Semana de Geografia'. In *Actas de la XV Semana de la Geografia*, Buenos Aires.

—— 1957: *Geografia y Unidad Argentina*. Buenos Aires: El Ateneo.

Durruty, C. 1969: *Clase obrera y peronismo*. Córdoba: Ed. Pasado y Presente.

Escolar, M. 1989: 'Problemas de legitimación cientifica en la producción geográfica de la realidad social'. In *Territorios*, 2, Buenos Aires: UBA.

—— 1991a: 'Geografia francesa y politica alemana: Camille Vallaux'. In *Anales del I Encuentro de Geógrafos Latinoamericanos y V Seminario de Investigación*, Pasto, Colombia.

—— 1991b: 'A harmonia ideal de um territorio ficticio'. In *Actas de Conferencia A cuestao regional e os movimento sociais no terceiro mundo*, UGI, USP–UNESP, San Pablo.

—— Escolar, R. and Quintero Palacios, S. 1991: 'Ideologia. didáctica e corporativismpo'. In *Terra Livre*, 8, Sao Paulo: AGB.

Escudé, C. 1988: 'Contenido nacionalista de la enseñanza de la Geografia en la Republica

Argentina 1879–1986'. In *Ideas en Ciencias Sociales*, 9, Buenos Aires: Editorial de Belgrano.

Floria, C. and Garcia Belsunce, C. 1988: *Historia politica de la Argentina Conytemporánea*. Buenos Aires: Alianza.

Freeman, T. 1980: *A History of Modern British Geography*. London: Longman.

Glauert, E. 1963: 'Ricardo Rojas and the emergence of Argentine cultural nationalism'. *Hispanic American Historical Review*, XLIII, 1–13.

González, J. 1901: 'Problemas escolares'. In *Obras completas*, vol. XIII, Buenos Aires: Univeridad Nacional de La Plata.

—— 1914: 'Enseñanza de la Geografia Fisica'. In D. Jijena *La naturaleza y el Hombre*. Buenos Aires: Lajouane.

Guiomar, J. 1974: *L'Ideologie national*. Paris: Champ Libre.

—— 1990: *La Nation entre l'histoire et la raison*. Paris: La Decouverte.

Halperin Donghi, T. 1961: *Tradición politica española e ideologia revolucionaria de Mayo*. Buenos Aires: CEAL.

—— 1976: 'Para que la inmigración? Ideologia y politica inmigratoria en la Argentina 1810–1914'. In *El espejo de la Historia*, Buenos Aires: Sudamericana.

—— 1980: *Una nación para el desierto argentino*. Buenos Aires: CEAL.

Hobsbawm, E. 1984: *The Invention of Tradition*. Cambridge: Cambridge University Press.

—— 1990: *Nations and Nationalism since 1780: programme, myth and reality*. Cambridge: Cambridge University Press.

Luna, F. 1986: *Perón y su tiempo II*. Buenos Aires: Sudamericana.

Navarro Gerasi, M. 1968: *Los nacionalistas*. Buenos Aires: Jorge Alvarez.

Oszlak, O. 1982: *La formación del Estado argentino*. Buenos Aires: Editorial de Belgrano.

Paya, C. and Cardenas, E. 1978: *El primer nacionalismo argentino*. Buenos Aires: Peña Lillo.

Peralta Ramos, M. 1978: *Acumulación de capital y crisis politica en Argentina*. Siglo XXI, Mexico.

Pérez, S. 1948: *Filosofia del federalismo en el Río de la Plata*. Monterideo: Tipografía Atlántida.

Perón, J. 1952: 'Discurso pronunciado el 28 May 1952'. In *Doctrina peronista*, Buenos Aires: Presidencia de la Nación.

—— 1954: *Manual de doctrina peronista*. Buenos Aires: Partido Peronista.

Presidencia de la Nación 1947: *Plan quinquenal del gobierno del Presidente Perón 1947–52*. Buenos Aires: Primicia.

Prieto, A. 1988: *El dicurso criollista*. Buenos Aires: Sudamericana.

Quijada, M. 1985: *Manuel Gálvez: 60 años de pensamiento nacionalista*. Buenos Aires: CEAL.

Quintero Palacios, S. 1991: 'Geografia nacional y educacion pública: la participación de la Geografia en la formación de la nacionalidad argentina'. In *Memorias del III Encuentro de Geógrafos de América Latina*, Mexico.

Ramos Mejia, J. 1899: *Las multitudes argentinas*. Buenos Aires: Editorial de Belgrano.

Reberioux, M. 1975: *La republique radical? 1898–1914*. Paris: Editions Seuil.

Reboratti, C. 1982: 'Human geography in Latin America'. *Progress in Human Geography*, 6(3), 397–407.

—— 1983: 'El encanto de la oscuridad. Notas acerca de la geopolitica en la Argentina'. *Desarrollo Económico*, 23(89), 137–44.

—— 1987: *Nueva Capital, viejo mitos. La geopolitica criolla o la razón extraviada*. Buenos Aires: Planeta.

Recalde, J. 1982: *La construcción de las naciones*. Siglo XXI, Madrid.

Rojas, R. 1909: *La restauración nacionalista*. Buenos Aires: Peña Lillo, 1971.

Rokkan, S. and Urwin, D. 1983: *Economy, Territory, Identity*. London: Sage.

Romero, J. 1956: *Las ideas politicas en Argentina*. Buenos Aires: FCE.

Rouquié, A. 1978: *Pouvoir militaire et société politique en Argentine*. Paris: FNSSP.*

Sack, D. 1986: *Human Territoriality, its Theory and History*, Cambridge Studies in Historical Geography. Cambridge: Cambridge University Press.

Sigmann, J. 1970: *1848: les revolutions romantiques et democratiques de l'Europe*. Paris: Calmann-Lévy.*

Snow, P. 1974: *Radicalismo argentino*. Buenos Aires: Francisco de Aguirre.

Solberg, C. 1970: *Inmigration and Nationalism. Argentina and Chile, 1890–1914*, Institute of Latin American Studies. Austin, Tx.: University of Texas Press.

Soler, R. 1968: *El positivismo argentino*. Buenos Aires: Paidós.

Stoddart, D. 1986: *On Geography*. Oxford: Basil Blackwell.

Tedesco, J. 1970: *Educación y Sociedad en la Argentina, 1880–1900*. Buenos Aires: Solar, 1986.

Terán, O. 1987: *Positivimo y Nación en la Argentina*. Buenos Aires: Puntosur.

Waldman, P. 1986: *El peronismo, 1943–1955*. Buenos Aires: Hyspamérica

Wallerstein, I. 1990: 'La construction des peuples: racisme, nationalisme, ethnicité'. In E. Balibar and C. Wallerstein (eds), *Race, Nation, Classe: les identités ambigués*, Paris: La Découverte.

Weber, M. 1922: *Economia y Sociedad*. Mexico: FCE, 1984.

Zuleta Alvarez, E. 1973: *El nacionalismo argentino*. Buenos Aires: La Batilla.

Documents

GAEA 1922: Estatutos, *Boletín de la Sociedad Argentina de* Estudios Geográficos GAEA, no. 4, 1935.

—— 1932/1937/1974: *Anales de la Sociedad Argentina de Estudios Geograficos*, IV, V and XVI. Buenos Aires.

—— 1946/1949/1950: *Boletin de GAEA*, 20, 26 and 27. Buenos Aires.

CNE (Consejo Nacional de Educación) 1908: *Monitor de la Educación Común*, 426.

HCDN (Honorable Cámara de Diputados de la Nación) 1926: *Antecedentes relativos a la aplicación de los nuevos programas de enseñanza secundaria*. Buenos Aires: Imprenta de la Cámara de Diputados.

MJIP (Ministerio de Justicia e Instrucción Pública) 1908/1926: *Planes y programas de estudio de los Colegios Nacionales*. Buenos Aires.

—— 1886/1888/1919/1934a: *Memorias presentadas por el Ministerio de Justicia. Culto e Instrucción Pública ante el Honorable Congreso de la Nación*. Buenos Aires.

—— 1933: *Boletin de la Comisión Revisora de Textos de Historia y Geografia Argentina y Americana*, 1, Buenos Aires, 1946.

—— 1934b: *Conclusiones aprobadas en la Asamblea de Profesores de Castellano, Historia y Geografia Argentinas e Instrucción Cívica*. Buenos Aires.

MEN (Ministerio de Educación y Juticia) 1952/1956: *Planes y Programas de Estudio del Ciclo Secundario*. Buenos Aires.

Afterword:
Identity Resurgent – Geography Revived

David Hooson

In 1988, when the international conference which initiated this collection was held in Bundanoon, Australia, the theme of 'National Identity and Geography' was nowhere near as topical as it has since become. Australia was celebrating the bicentenary of its first European 'settlers', with little reference to the people who had inhabited that country for the previous 40,000 years or so. Geography was still closely associated with the growth of empires and with the Europeanization of the world. In the same vein there was eager anticipation, from both European and American states, of the next international geographical conference which was scheduled for 1992 in Washington, where the quincentenary of Columbus's voyage was clearly to form the centrepiece. The countries of the European Community were looking forward to 1992, when the triumphant consummation of a carefully prepared integration of the component 'nation-states' and the concomitant surrender of individual sovereignty would take place. Meanwhile the Soviet Union, with its empire still intact and its control of Eastern Europe still firm, seemed to be undergoing a vigorous renewal under the *glasnost* and *perestroika* of its new leader Mikhail Gorbachev, and looking forward to a glorious celebration of its seventy-fifth birthday in 1992.

When the magic year of 1992 actually rolled around, it turned out to be anticlimactic at best, disastrous at worst. Needless to say, the most astounding new fact of life was that the mighty nuclear bogey-man of the Cold War, the monolithic, invincible USSR, had simply ceased to exist by 1 January 1992, scarcely 2 years after the demolition of the Berlin Wall and the subsequent collapse of the dreaded Iron Curtain and 'Soviet Bloc'. Less dramatic, but scarcely less significant in the long run, coming hard on the heels of the euphoric liberation of East-Central Europe, was the deflation of the long-anticipated excitement over European integration, with its projected free movement of people and goods. The new nationalism of a unified Germany had already soured into a heavy, perplexing burden, while separatist movements, distrust of the much-trumpeted Maastricht Treaty, and restrictions on immigration had transformed optimism into suspicion and anxiety about the future of Europe as a whole.

In the United States of America, anxiety and introspection about 'the state of the nation' in 1992 were pervasive enough to unseat a president who had presided over the 'winning of the Cold War' and the shaping of a 'New World Order'. Further, the much-heralded commemoration of Columbus was distinctly muted, as indeed it was even in Spain, the Dominican Republic and elsewhere, following an upsurge of revisionist statements about the disastrous impact on the native peoples of the Americas which had flowed from Columbus's fateful landfall in 1492. The rise of 'multiculturalism' and even separatist tendencies in the United States, alongside its immigration history of the preceding quarter-century (which was heavily dominated by Asians and Latin Americans), has significantly altered the complacent 'Anglo-Saxon' identity of earlier times.

There seemed no doubt that, in the aftermath of the end of the Cold War and Superpower rivalries, nationalism and national identity were catching fire with redoubled energy all over the world, reaching even the last multinational empire, China, and flaring up throughout the Indian subcontinent and in other former colonies in Africa, South-east Asia and the Middle East. Nationalism has surely shown its ugliest face in South-eastern Europe, particularly in the former Yugoslavia and in the Caucasus, with appalling cruelties, 'ethnic cleansing' and atavistic hatreds.

However, shining brightly in all this gloom of 1992 was Barcelona, where the Olympic Games, with the broadest international representation in history, was staged. Its unrivalled pageantry and panache, with all the world watching, was *inter alia* a triumphant paean of praise to Catalonian nationalism, set up in a marvellously propitious context. It served to remind the world that national identity on the march could be benign and forward-looking, releasing infinite energies for economic development and social resilience. Just as in Switzerland and other multilingual countries around the world, a feeling of identity, infused with a common culture and shared meanings and associations, built up over centuries, *can* provide the requisite glue of self-esteem and the necessary lubrication for successful, non-threatening and purposeful community life and strength. Of course, this identity may be made up of myth and memories in even the most apparently 'self-evident' of national identities, such as France or England. These, in turn, can demonstrate condescension towards the rhetorical pretensions and hyperboles of nationalist intellectuals in twentieth-century Africa or Asia, let alone nineteenth-century Europe. But, however selective and even distorted they may be, these myths and memories which promote a feeling of 'belonging' to a community of kindred spirits, once implanted, cannot effectively be challenged by others when they have become 'dyed in the wool'. When people are prepared to sacrifice their lives for the assertion of their national identities, and the restoration of the 'historic' lands to which they 'belong', it is obvious that here we have something which goes far beyond a mere academic exercise.

Clearly, this is a potion as powerful as anything else in the modern world, which may have deep roots and cries out for study. Thus it really is rather astonishing that it (national identity) has not generated its own grand thinkers, of the order of Marx, Weber or Hobbes. Even Adam Smith and Alexis de Tocqueville, in their great works, did not focus on it. Almost banal in its familiarity, it did not seem to merit 'scientific' emphasis and seemed indeed to defy it. But in the last decade or so there have appeared a number of suggestive and well-argued books which have focused on the phenomenon of nationalism and national identity as such. One of the first, and still probably the best and most complete, of these studies was written by the late Hugh Seton-Watson, whose insights would have been invaluable for interpreting current mêlée in Eastern Europe and the former Soviet Union,[1] but who found the subject elusive and difficult to systematize. In the USA, John Armstrong, building on his earlier studies on Ukraine, analysed those same vague elusive roots of ethnic identity in mediaeval Christian and Islamic lands.[2] Ernest Gellner puts forward perhaps the most articulate statement of the view that nations and nationalisms are really modern phenomena, dependent on the formation of a mobile and literate industrializing society, able to generate the necessary ideology.[3] At the same time Benedict Anderson, building on his work in South-east Asia, developed his thesis of 'imagined communities',[4] which are contrived to serve psychological-emotional as well as economic and political needs for the existence and creation of a nation. Perhaps the most comprehensive recent work has been written by the classicist-sociologist and student of Seton-Watson, Anthony Smith, who presented an original interpretation of the origin of nations that traces their genealogy to pre-modern ethnic foundations.[5]

Sir Isaiah Berlin, in a celebrated and prophetic essay in the *Partisan Review* for 1978, remarked on the strange fact that none of the great prophets of the nineteenth century anticipated the force and resistance of nationalism as a phenomenon. Nowhere was this more glaring than in the case of Marx, who assumed that, along with religion, national identities would disappear in the wake of modernization, socialism and economic globalization. Even on the eve of the collapse of the Soviet Union into ethnic fragments, the Marxist historian, Eric Hobsbawm, in an interesting book, elaborates this thesis anew and concludes that the heyday of nation-building is over, being superseded by the imperatives of the global economy.[6] It is only fair to repeat that classical 'liberal' theory more or less held to this view, and further, that none of the academic 'Sovietologists' was able to discern any sign of the impending collapse of the Soviet Union until it became an accomplished fact. Naturally, this and related events have been an immense stimulus to the study of national identity, which now stands out as the prime phenomenon of the 'New World Order' as we approach the twenty-first century. An original and erudite book that appeared recently is by Liah Greenfield; it provides a critical and historical exploration of the development of nationhood in England, France, Russia, Germany and the United States.[7] However Eurocentric it may unwittingly be, it illuminates the subtle origins of nationhood in the most powerful and seasoned countries and identities of the modern world. There is ample scope, and a pressing need, for further studies, particularly on the rise of nationalism in Asia, where it has long been connected with ethnicity.

None of these works has dealt adequately, if at all, with the fundamental importance of the territorial factor, or the physical facts and perceptions of the 'homeland' which, celebrated in song and poetry, are often 'primordial' in the sentiments of the communities who inhabit them. This factor has been enormously strengthened and deepened in the minds of the people by the image and reality of the despoliation of their homelands and environments through pollution. It has been particularly powerful in Eastern Europe and the former Soviet Union in recent years in igniting and fanning the flames of nationalism and concentrating resentment against the imperial overlords based in Moscow.

The return of geography, like the return (not 'the end') of history, is therefore a major factor to be reckoned with in evaluating the phenomenon of national identity across the world in the closing years of this century and beyond. The chapters in this book, which discuss the impact of geography on national identity, both as a phenomenon in itself and in relation to the works of the practitioners of a particular discipline and way of thought in various cultures, provide a much-needed beginning to what could be an almost infinite exploration of particular cases and their common denominators. The geographical dimension is fundamental, ultimately and increasingly inescapable, and to be ignored at our peril.

NOTES

1 Hugh Seton-Watson, *Nations and States*. London: Methuen, 1977.
2 John A. Armstrong, *Nations before Nationalism*. Chapel Hill, N.C.: University of North Carolina Press, 1982.
3 Ernest Gellner, *Nations and Nationalisms*. Oxford: Basil Blackwell, 1983.
4 Benedict Anderson, *Imagined Communities*. London: Verso, 1983.
5 A.D. Smith, *The Ethnic Origins of Nations*. Oxford: Basil Blackwell, 1986.
6 E.J. Hobsbawm *Nations and Nationalism since 1780*. Cambridge: Cambridge University Press, 1990.
7 Liah Greenfield, *Nationalism*. Cambridge, Mass.: Harvard University Press, 1992.

The Contributors

Józef Babicz is a geomorphologist and Professor at the Institute of the History of Science, Polish Academy of Sciences, Warsaw, where he works on the history of earth sciences. His publications have been on the history of geographical discoveries, the history of geographical thought, and the history of cartography.

Mark Bassin is Associate Professor of Geography, University of Wisconsin–Madison. His research interests include the history of environmentalism, political geography, and Russia. His recent publications include 'Solov'ev, Turner, and the "Frontier Hypothesis"', *Journal of Modern History*, September 1993; 'Environmental determinism in *fin-de-siècle* Marxism', *Annals AAG*, April 1992; 'Inventing Siberia: visions of the Russian in the early nineteenth century', *American Historical Review*, June 1991; 'Russia between Europe and Asia: the ideological construction of geographical space', *Slavic Review*, Spring 1991.

Vincent Berdoulay is Professor of Geography and Director of the Centre for Society, Environment and Territory (CNRS) at the University of Pau (France). His books include *La Formation de l'école française de géographie 1870–1914* (1981) and *Des mots et des lieux: La dynamique du discours Géographique* (1988). He is a full member of the IGU Commission on the History of Geographical Thought.

Anne Buttimer is Professor and Head of Geography at University College, Dublin, having previously taught at universities in Canada, Belgium, France, Sweden, and the United States. She has published ten books and over a hundred articles on topics ranging from social space and urban planning to the history of ideas and environmental policy. Her latest book is *Geography and the Human Spirit* (1993). She has been Secretary of the IGU Commission on the History of Geographical Thought since 1988.

Paul Claval is Professor at the University of Paris–Sorbonne (Paris IV). His chief interests are the history of geographical thought, relations between geography and other social sciences, urban and regional geography. His many books include: *La Logique des villes* (1981), *Géographie Humaine et économique contemporaine* (1984), *La Conquête de l'Espace Nord-Américaine* (1990), and *La Géographie au Temps de la Chute des Murs* (1993).

Ron Crocombe is Professor of Pacific Studies at the University of the South Pacific, Suva, Fiji, now based mainly in Rarotonga, Cook Islands. He was born and educated in New Zealand, and has travelled widely throughout the Pacific. His wide-ranging publications include *The New South Pacific*.

MARCELO ESCOLAR is a member of the Institute of Geography, University of Buenos Aires. His research fields are political geography, the history of geographical thought and historical geography. His recent publications include *Critica do Discurso Geografico* (São Paulo, 1993), and 'A Harmonia Ideal de um Territorio Ficticio', *Boletim de Geografia Teoretica*, 22/43 (Rio Claro, 1992).

MARIA DOLORS GARCÍA-RAMON holds an MA in geography from Berkeley and a PhD from the University of Barcelona. She is now Professor of Geography at the Autonomous University of Barcelona. She has been a visiting scholar at several American universities. Her writings mainly concern rural geography, geographical thought and gender geography. Recently she has published (with J. Nogué and A. Albet) *La Práctica de la Geografía en España: Innovación Metodológica y Trayectorias Personales en la Geografía Académica* (Barcelona, 1992), and she is one of the editors of *Geography in Spain, 1970–1990* (Spanish contribution to the XXVII International Geographical Congress, Madrid, Fundación Banco de Bilbao–Vizcaya, 1992).

DAVID HOOSON, a graduate of Oxford and London, is Professor of Geography, and formerly Dean of Social Sciences, at the University of California, Berkeley. His chief publications have been on the former Soviet Union and the history of geography. He was Chairman of the IGU Commission on the History of Geographical Thought from 1980 to 1988.

LISA E. HUSMANN is a PhD student in Geography at the University of California, Berkeley. She has travelled widely in China, especially among the 'minority' peoples of Central Asia, including those in Soviet successor states.

LADIS K. D. KRISTOF is Emeritus Professor of Political Science at Portland State University, Oregon, USA. Born and brought up in Romania, he has a PhD from the University of Chicago. His field of research and publication has mainly concerned the politics, history and political geography of Eastern Europe and the former Soviet Union.

DAVID LOWENTHAL was Secretary of the American Geographical Society and Professor of Geography at University College, London. Among his books are *George Perkins Marsh, West Indian Societies, The Past is a Foreign Country* and *Heritage*. His current interests include authenticity in the arts, the politics of archaeology, landscape values, and the uses of historical myth.

MURRAY MCCASKILL, Emeritus Professor of Geography, Flinders University of South Australia, has edited *The New Zealand Geographer*, and currently edits *Australian Geographical Studies*. His publications include *Patterns on the Land* (1973, 1978) and *Atlas of South Australia* (1986) which he co-edited.

JOAN NOGUÉ-FONT is Professor of Human Geography at the University of Girona (Catalonia, Spain). He has written three books in Catalan: *A Humanistic Approach* (Girona, 1985), *The Perception of Forest* (Girona, 1986) and *The Territorial Dimension of Nationalism* (Barcelona, 1991). He is co-author of *The Practice of Geography in Spain (1940–1990)*, written in Spanish with Abel Albet and Maria Dolors García-Ramon. He has been a post-doctoral fellow in geography at the University of Wisconsin, Madison, and is currently doing research on the history of geographical thought in Spain and, especially, on the role of geography and geographers in Spanish colonialism in Morocco.

J. M. POWELL, a graduate of Liverpool and Monash Universities and now Professor of Geography at Monash University, Australia, has published widely on the history of environment management and geographical thought. His recent books include *An Historical Geography of Modern Australia* (1988, 1991), *Watering the Garden State: water, land and community in Victoria, 1834–1988* (1989), and *Plains of*

Promise, Rivers of Destiny: water management and the development of Queensland, 1820–1990 (1991). He is a full member of the IGU Commission on the History of Geographical Thought.

SILVINA QUINTERO PALACIOS is a member of the Institute of Geography, University of Buenos Aires, specializing in the history of geographical thought and geographical teaching. Recent publication: 'Geografía y Nación: Estrategias Educativas en la Representación del Territorio Argentino', *Territorio*, 6 (Buenos Aires, 1993).

CHERI RAGAZ is currently involved in research on the human response to environmental problems (adaptation, coping strategies and resilience). She is developing an agenda to deal with transboundary aspects of these issues (and others such as famine), incorporating concepts of nationalism, sovereignty, human rights, culture and negotiation. Dr Ragaz lectures at the University of Zurich, Switzerland, on these topics and on her regional work in Asia and the Pacific.

CARLOS REBORATTI is Director of the Institute of Geography, University of Buenos Aires, with research fields in rural geography, population and environment. Recent publications include 'Agrarian Frontier, Colonialization and Public Policies in Latin America', in W. Reinhard and P. Waldmann (eds), *Nord und Sud in Amerika*, vol. I (Freidburg, 1992), and 'Población, Ambiente y Recursos Naturales en América Latina', in Abep et al., *La Transición Demográfica en América Latina y el Caribe*, vol. II (Mexico, 1993).

MARIE-CLAIRE ROBIC is Director of Research at the CNRS unit 'Epistemologie et Histoire de la Géographie' in Paris. A specialist on French human geography, she is working on the manuscripts of Vidal de la Blache and Jean Brunhes and is particularly interested in the geography and iconography of national identity. She edited *Du milieu à l'environnement. Pratiques et réprésentations du rapport homme/nature depuis la Renaissance* (1992).

MECHTILD RÖSSLER is currently a specialist for environmental sciences in the UNESCO World Heritage Centre in Paris. Her main interests are political geography, the history of geography, ecology and feminist geography. Her PhD (Hamburg) in 1988 in geography was published in 1990 as 'Wissenschaft und Lebensraum', *Geographische Ostforschung in Nationalsozialismus* (Berlin), as well as two books and some forty articles. In 1989/1990 she joined the research centre of the Science Museum, La Villette, and the Maison des Sciences de l'Homme, Paris, and in 1991 the Department of Geography at the University of California, Berkeley, as a visiting scholar, before starting work at UNESCO.

GERHARD SANDNER (who was born in Namibia) is Director of the Department of Economic Geography at Hamburg University, Germany. Since 1973 he has been chief editor of *Geographische Zeitschrift*. His research areas are regional and political geography of Central America and the Caribbean (in recent years mainly political geography, maritime boundaries and conflict) and the history of German geography, concentrating on the political context in the period 1871 to 1945.

OSKAR SPATE is Emeritus Professor of Pacific History at the Australian National University, Canberra. He was Foundation Professor of Geography there and also Director of the Research School of Pacific Studies. After gaining his PhD from Cambridge, he taught in Burma and at the London School of Economics before moving to Australia. His publications include *India and Pakistan, Let Me Enjoy, Australia*, and a three-volume *History of the Pacific*.

IHOR STEBELSKY is Professor of Geography at the University of Windsor, Canada. A graduate of the University of Washington, Seattle, Dr Stebelsky has published numerous articles on agriculture, environment, food production and consumption and historical geography of the former USSR and, in particular, Ukraine. He co-authored 'Ukraine', in *The New Encyclopaedia Britannica*, and was the geography editor for the *Encyclopedia of Ukraine*.

KEIICHI TAKEUCHI, a graduate of the University of Tokyo and the University of Milan, is Professor of Social Geography at Hitotsubashi University, Tokyo. He has published on the social and economic geography of Mediterranean countries and in the history of geographical thought, in Japanese and Western languages, especially on the problems of the Mezzogiorno and on the history of modern geography in Japan. Since 1988 he has been Chairman of the Commission on the History of Geographical Thought.

JOSEPH VELIKONJA is Emeritus Professor of Geography, University of Washington, Seattle. His primary interest is social and political geography. His published work has focused on European migrations to the Americas, especially the Italians and Slovenes, their historical experiences and spatial patterns of their settlements.

HONG-KEY YOON was born and brought up in a small village in Korea. He graduated from Seoul National University with a BA in geography, and obtained a PhD degree at the University of California, Berkeley. Since 1976, he has been teaching geography at the University of Auckland, New Zealand. His major publications include *Geomantic Relationships between Culture and Nature in Korea*, and *Maori Mind, Maori Land*. He is a full member of the IGU Commission on the History of Geographical Thought.

Index

Related Titles: List of IBG Special Publications

Also published by Blackwell for the IBG